World Religion & History back to 70,000 BC Discovered by Remote Viewing

Including missing information on the life of Christ

M.G. Hocking
Professor of Materials Chemistry*
University of London

The author documents a little-known method (expert-level Remote Viewing) which has given useful information in chemistry & physics and has also been used by the CIA and others to obtain covert intelligence information.

It has been used for obtaining new information on historical figures, e.g. revealing lost details of the life of Christ. In the absence of other new sources, this is of great interest.

Expert level Remote Viewing is a controlled vivid-daydreaming state, which, Chapter 1 shows, cannot be dismissed as hallucinations. Depriving the normal 5 senses of all input allows the very faint sub-liminal 6th sense to be perceived. Chapter 7 shows how this can easily be done, moving into a lucid dreamstate without losing consciousness.

This book answers the following objections cited by atheists to religion:

<u>Atheists</u> say that there is no evidence for God, nor for heaven and hell; and that disasters kill apparently innocent people including children, so God is unjust if God exists. But all these will be shown to be incorrect.

Dedicated to the Remote Viewing Group of the CIA, without the veracity of whose activities, the events retold in this book could not have been successfully published.

---
*(relevant to Chapter 3 & some other parts of this book)

World Religion & History back to 70,000 BC Discovered by Remote Viewing

Including missing information on the life of Christ

M.G. Hocking

UDC: 2-1:2-18:94(3):539.1
(Relation between: religion, destiny of man, ancient history, & structure of matter)

Pages: 384
Illustrations in colour: 12
Illustrations black & white: 46
© M.G. Hocking 2011 (until copyright expiry date of 1-1-2050).
To quote extracts (free), please ask the author.
(copyright not claimed on pictures, except where stated in captions; see p. 373)

Front cover: Annual ceremony, Wesak Valley, Tibet
(© Painting by Heather DuPré, London)

The author:
The author has published 150 papers in physical chemistry, and became a Professor of Materials Chemistry in 2001 at Imperial College, London.

Also published (in the field of this book) by this author:

"Exploring the Subconscious using New Technology", published by www.4-D.org.uk/Books, London; (1993). (110 pages, 136 references)

"Remote Viewing of Ancient Times: A historical novel of actual events from 100,000 BC to 500 BC", Edited by M.G. Hocking, (covers other events not in the present book), published by www.4-D.org.uk/Books, London; (2011).
Free download. Complements the present book & adds more on lucid dreams.

See website below for other books in this field.

Acknowledgements: The author is very grateful for constructive help from Dr V. Vasantasree, Diana Kempster, and other colleagues.

All books can be ordered at low prices from website below:

**Published in the UK, 2011, by
www.4-D.org.uk/Books**
ISBN 978 0 9521099 1 4

Printed & bound by Bell & Bain Ltd, Glasgow, G46 7UQ, UK

DESCRIPTION OF CONTENTS

This book has been written to counteract the rise of atheism and books on atheism, for general readers who want to study the new evidence given.

Chapter 2 is the most important for counteracting atheism, on 3 fronts:
(1) Existence of God: new evidence-based proof that God probably exists.
(2) Explanation of seemingly unfair events which befall some people.
(3) Re-birth – evidence of past lives.
Note: No scientist has proven the non-existence of God.

The topics covered in this chapter show that atheism has studiously ignored important aspects of the above topics.

The Foreword & Chapter 1 should be read first and describe Remote Viewing, which is the source of important new information on Christ given in Chapters 4 & 5. Also: gives quick seamless ways to consciously enter a lucid dream-state from the waking state, for visionary experiences & insights.

Chapter 3 gives more evidence from Remote Viewing results, in science, on the structure of molecules, atoms and elementary particles, which cannot be obtained by other methods. The results are compared with what results are available from other methods.

Chapter 7: simple effortless ways to enter a lucid visionary dream-state without losing consciousness: flotation tank, and, lucid dreaming methods.

Chapters 4 and 5 include Remote Viewing of the times of Christ and they can be read alone, if the reader wishes to skip the introductory preceding chapters. But the reader must note that Chapters 1 to 3 provide the essential background evidence supporting: (a) Remote Viewing, and (b) rebirth, which are the basis for Chapters 4 & 5.
Pages 87-9 followed by page 117 onward, give a quick sense of Chapter 4.
Chapter 4 includes a discussion of reports of very ancient flying machines, which may have partial links to modern "stealth" aircraft (high voltage lift).

A noteworthy point relevant to Chapter 4 is that very many tens of millennia ago, people were more sensitive to the (now-dormant) 6th sense. This allowed them to develop in undesirable ways -- developing an evil "magic" to affect events. This led eventually to the destruction of Atlantis, one of the main centres of very ancient lost civilisation. A parallel in our time is our development of the possibility of nuclear destruction.

NOTE: References are listed at the end of each chapter.

CONTENTS

Foreword :	**Essential basic information**	6
Chapter 1:	**Remote Viewing**	19
	◆ Dimensionality of Space	20
	◆ Chakras	24
	◆ Dream State etc	31
Chapter 2:	**Theology**	
	◆ New evidence for the existence of God	46
	◆ Karma: "As a man sows, so shall he reap"	52,57
	◆ Re-birth	54
Chapter 3:	**Science**	64
	◆ **Remote Viewing** of the Structure of Matter & affirmation by later scientific discoveries	
Chapter 4:	**Remote Viewing of the life of Christ**	87
	◆ Beginnings – 100 thousand years ago	
	◆ Krishna, & Christ as the present World Teacher	
	◆ Previous World Teacher, back to 70,000 BC.	
Chapter 5:	**Christ's 3-year ministry in Palestine**	242
	◆ 0 AD to 30 AD	
	◆ Jesus – Christ	
Chapter 6:	**Conclusions**	258
	◆ Religion in the present day.	
	◆ Possible integration with other religions.	
Chapter 7:	**Vision Devices – the Flotation Tank**	260

APPENDICES 278
1. Heart Transplants transferring donor characteristics to recipients
2. Some CIA Remote Viewing personnel 291
3. General Relativity & Gravitation 292
4. Mass increase with velocity 302
5. Remote Viewing of atoms discussion (1983) 306
6. Linking String and Membrane theory to Quantum Mechanics and Special Relativity equations, avoiding Special Relativity assumptions: M.G. Hocking, J. Scientific Exploration 21 (1), 13-26 (2007)   332
7. The high-voltage research of Dr N. Tesla: predictions 354
8. Remote Viewing mechanism 361

GENERAL REFERENCES & INDEX   373 & 377

List of Main Figures

Fig. 0.1	Psychedelic Art.	8
Fig. 0.2	Psychedelic Art.	9
Fig. 1.1	Dimensionality of Space.	For p. 20
Fig. 1.2	Simple comparison of vision in 2-D and 3-D.	23
Fig. 1.3	Cottingley last photograph.	For p. 24
Fig. 1.4	Statue of Pharaoh Sesostris I showing eyebrow chakra.	For p. 24
Fig. 1.5	Chakras: 7 chakras, stylised view.	26
Fig. 1.6	Classic Caduceus.	27
Fig. 1.7	Actual view of Crown Chakra.	For p. 27
Fig. 1.8	Actual view of Eyebrow Chakra.	For p. 27
Fig. 1.9	A typical blue sky.	For p. 31
Fig. 1.10	Same sky at night showing what is really there.	For p. 31
Fig. 1.11	EEG (electroencephalograph - brainwaves).	32
Fig. 3.0	Dr A. Besant & C.W. Leadbeater.	65
Fig. 3.1	Arnoo: the smallest physical elementary particle.	67
Fig. 3.2	Magnified detail of Arnoo, showing coiling.	68
Fig. 3.3	HCl seen by remote viewing.	74
Fig. 3.4	Methane, CH_4 seen by remote viewing.	75
Fig. 3.5	Proton (H nucleus), seen by remote viewing.	84
Fig. 4.1	Hierarchy relevant to Table 1 on page 89.	For p. 89
Fig. 4.2	"Mars" as seen by remote viewing.	94
Fig. 4.3	"Surya" (Christ) as seen by remote viewing.	109
Fig. 4.4	"Mercury" as seen by remote viewing.	143
Fig. 4.5.	"Viraj" as seen by remote viewing.	166
Fig. 4.6	"Saturn" as seen by remote viewing.	172
Fig. 4.7	"Venus" as seen by remote viewing.	177
Fig. 4.8	"Neptune" as seen by remote viewing.	189
Fig. 4.9	"Alcyone" with "Herakles" as seen by remote viewing.	207
Fig. 4.10	"Brihat" (Jesus) as seen by remote viewing.	228
Fig. 4.11 & 4.12	4-D & 5-D views as seen by Remote Viewing	229
Fig. 5.1	Levitation.	248
Fig. 5.2	Levitation.	248
Fig. 5.3	A Coptic Icon and a Benedictine icon.	For p. 254
Fig. 7.1	A flotation tank.	261
Fig. A2.1	Admiral Stansfield Turner, Director of CIA 1977-1981, who founded the CIA Remote Viewing Group.	For p. 291
Fig. A8.1–3	4-D geometry, tesseract	364-5
Fig. A8.4 & Fig. A8.5	Cottingley fairies photographs.	For p. 369
Fig. A8.6	Cottingley fairies photographs	For p. 369

Underlined figures are in colour. *These are all located* between p. 64 & 65 *for cost reasons, but are labelled with the page they refer to:* "For page xx".

FOREWORD

Give us *a tall ship, and a star to steer her by*
                                              J. Masefield

Or in plainer English:
We are at this moment contained within our own "*tall ship*", and it pays to know the rules if we are taking it on a voyage!

Foreword: Part 1

This book is meant for the general reader, and although parts of this foreword relate to Christianity, it also includes important aspects of other religions and of no religion.

Unless additional ancient manuscripts like the Nag-Hammadi or Dead-Sea scrolls are found, we are now in a situation where little more information is likely to be found on the life of Christ and other ancient history by conventional means. However, there are people with a rare ability to use meditative methods to discover missing information. In science this has been demonstrated in the direct observation of quarks (elementary particles of physics), which agrees with modern scientific results. In intelligence gathering it has been successfully used by the CIA.

The present book extends this to the retrieval of missing information on the life of Christ, by expert-level Remote Viewing observations of that time period, and of world history back to 70,000 BC. Although some may dismiss this possibility, the author believes that at least it deserves consideration, especially as there is no other way to obtain such information and because it has produced accurate information in physics and chemistry investigations and also in intelligence investigations by the CIA, and others. Historical Remote Viewing amounts to a form of passive time travel (observation only – history cannot be changed!).

Last century, the mediaeval notions of heaven, purgatory and hell as real locations were dismissed by science as unreal imaginary fantasies. But in recent decades, modern particle physics has postulated 10 (spatial) dimensions, so now the notions of heaven and hell as <u>actual</u> locations in higher spatial dimensions can become compatible with modern science, as will be shown, removing the chasm between religion and science. (See Einstein quotation, p. 19).

The reader may be surprised that Remote Viewing can be performed by CIA personnel instead of (say) a holy monastic order! **Neither "holiness" nor "goodness" are required for Remote Viewing.** E.g. narcotic drugs can instantly open the centres of perception of Remote Viewing for anyone, whether a "good" or "evil" person, but a useless jumble of visions is then perceived. Or, long meditative drug-free training can confer Remote Viewing to anyone, "good" or "bad", but very few people persist long enough to succeed; there are "secret" methods (not generally available) which can greatly speed up this process.

Only very rigorous training can produce expert level Remote Viewing; uncontrolled and mindless use of psycho-active drugs by unprepared untrained persons is extremely undesirable and leads to incoherent and unreliable results. Other methods such as flotation tanks are recommended [3], but much training is still needed to filter out subjective effects (hallucinations) and leave only objective results; see Chapter 7. Ethylene seeping up at Delphi enabled the oracle to give some useful advice, but she was trained, unlike today's teenage glue-sniffers who see hallucinations from their own subconscious!

Much material in this book is influenced by writings of the authors of little-known observations made about 100 years ago, who described aspects of what is now called "string theory", and who first used the word "string" to describe structures of elementary particles. String theory was much later developed in particle physics. The observers, Besant & Leadbeater [1], published many books based on use of an ancient Indian "Siddhi" or ability [2], developed after long training in meditative methods, which is also in more recent use by the CIA and others for intelligence gathering. They describe a well-controlled use of this method for magnifying atomic size particles and they drew these structures for all to see [1]. Details are in Chapter 1 (CIA observations) & Chapter 3 (physics & chemistry observations).

Some artists can see things which are normally subliminal, e.g. Fig. 0.1 & 0.2 by Heather DuPré shows what she actually saw, not imagined. Usually (but not in her case) psychedelic drugs like LSD are needed to see such scenes. Chapter 1 explains that such scenes can become visible if a specific "chakra" is open (rarely naturally), or, is opened by taking psychedelic drugs which are able to bring subliminal perceptions into full consciousness (these are normally "drowned out" by everyday input from our normal 5 senses).

Fig. 0.1. A vision perceived by and drawn by Heather DuPré.
(© 2010 Heather DuPré)

<u>Note on Figures 0.1 & 0.2</u>: The artist says that if she closes her eyes, the vision disappears and re-appears on opening; and if she looks away it is still there when she looks back. This means it is not a hallucination. Similar-looking psychedelic art from independent un-connected people worldwide suggests that an objective reality exists (i.e. it is not just from within one individual's mind). Examples of such psychedelic art are available on the internet, with search words, "Psychedelic Review" [8], or, "Isaac Abrams".

Fig. 0.2. A vision perceived by and drawn by Heather DuPré.
(© 2010 Heather DuPré)

These are two of 20 remarkable pictures by this artist. She also painted the front cover picture on this book. The mechanism of such perceptions is discussed in <u>Appendix 8</u>, to avoid interrupting the text here.

The author (MGH) [4] has used Besant & Leadbeater's well-controlled Remote Viewing observations [1] to show how Einstein's Special Relativity equations can be derived in just <u>one</u> line from their observations, along with a very simple derivation of $E = mc^2$. Other observations from their book [1] (<u>18</u>95 and 1908) include that the proton is made up of 3 smaller sub-atomic particles which were very much later called quarks by particle physicists, and 3 quarks have been verified relatively recently as constituting the proton. Their book also reported that space has 10 dimensions, which was many decades later postulated by particle physicists.

We are normally confined to 3 dimensions, of course. For a Christian perspective, the King James Authorised Version of the Bible John (14: 2) says, "*In my Father's house* (= the Universe?) *are many mansions* (= dimensions?)*: if it were not so, I would have told you.*" The Gospel of Thomas* (113: 4) says, "*The Father's kingdom is spread out upon the earth but men do not see it*".

In John (8: 23) Christ says, "*Ye are from beneath; I am from above: ye are of this world; I am not of this world*". "Beneath" and "above" are the only words useable in that time, but "beneath" could mean in the 3rd (our) dimension, and "above" could mean in the 4th or a higher dimension (unknown at that time). He had difficulty in conveying to them what He wanted, because He had to add (above), "*ye are of this world; I am not of this world*". What did He mean by "*this world*"?

*Note: The Gospel of Thomas is not in the Christian Canon of the New Testament, having been removed and ordered destroyed [9] by a "self-appointed-censor", Athanasius in 367 AD. This gospel was written in Egypt, far from Rome, which made it much less likely to get into the Canon. See the Supplementary Foreword (below) on surviving copies of the early Christian texts.

Because of the success of Besant & Leadbeater's observations [1] mentioned above, the author (MGH) has used (in the present book) material reported in their other books, including world histories [5], the events in which, were observed by using the same "Siddhi" (expert Remote Viewing capability). But to avoid interrupting the main text, details of their chemistry & physics observations [1] are placed separately in Chapter 3. **The credibility of these details, and of the much later CIA Remote Viewing observations, forms the basis for giving their views on other non-scientific**

(historical) matters as also likely to be credible, or at least worthy of consideration in the absence of other information.

The idea that we have all lived many past lives is not uncommon and was widely believed in Judea at the time of Christ:
"Who do men say that the Son of Man is? They (the disciples) answered: Some say John the Baptist, others Elijah, others Jeremiah, or one of the prophets". (On this point see also page 54)
"The disciples put a question to Him: Why do our teachers say that Elijah must come first? He replied: 'Yes, Elijah will come and set everything right. But I tell you that Elijah has already come, and they failed to recognise him, and worked their will upon him'. Then the disciples understood that He meant John the Baptist".
[*Italicised text* is from "The New English Bible" (New Testament), Oxford University Press (1961), Matthew 16: 13-14 and 17: 10-13.]

Re-birth was an accepted belief in the early church, e.g. Origen and Philo of Alexandria, St Gregory of Nyssa, and the Gnostics. It was only declared a heresy by the Council of Constantinople in 553 AD. The resulting later church, 500 years after Christ, is surely much less likely to represent His teachings than the pre-500 AD church!

Many people have strongly pre-conceived ideas on this subject. Many modern Christians disbelieve rebirth, assume we are created at birth and then exist forever, but being thus semi-eternal (semi-infinite!) is illogical. An eternal entity has no end date and no start date! The Council of Constantinople of 553 AD involved much bitter argument among various bishops etc and the Roman Emperor Justinian was antagonistic to the theology of reincarnation given by Origen of Alexandria (185-254 AD) and other early church leaders and so he pressed for anathemas at the Council against that. This has nothing to do with Christ, being over 500 years after Him, and, there was no formal Papal approval of the anathema issued, but in practice reincarnation was "officially dropped" after this time. But Buddhism and Hinduism have always accepted reincarnation. Please see Appendix 1 for compelling actual evidence for the continuing existence of a soul.

Of course, most of us have no waking memory of any previous lives, which does not help credence in them! But under hypnosis, regressions of many normal people have revealed apparent descriptions of previous lives. Two recent programme series from the well-known TV presenter Phillip Schofield [6], featured many half-hour films in which some significant and previously unknown details described by subjects were subsequently independently verified. Similar research is reported by many others worldwide [7].

Unless mentioned otherwise, the present book uses New Testament quotations from the *The New English Bible (New Testament), Oxford University Press & Cambridge University Press (1961)*, which is known for its translation accuracy.

The New Testament gospels as published in the earlier King James Authorised Version seem more inspired and poetic, and that version of the whole bible is available as a free downloadable searchable file in "Word™" and ASCII formats: http://www.spr.ac.uk
Where other Gospels are quoted which are not in the "Christian Canon", their sources are referenced below in the text of this book.

The present New Testament is a partial collection of early Christian documents. Briefly, Bishop Athanasius of Alexandria in AD 367 circulated a list which he approved as the authorised "Christian Canon" and he also instructed monasteries to destroy all copies of any other Gospels [9]. This grievous instruction caused the loss of most such copies, but a monk (probably), in Egypt decided to bury a large jar containing early Coptic texts like the Gospel of Mary, Gospel of Thomas, etc, which was found in 1945. Unfortunately, the buried jar was found by an illiterate man who used some of it to light his fires. Some other copies survived in fragments for 2000 years in a rubbish dump in Egypt. These contain accounts of those teachings of Christ which did not conform to what Athanasius thought they should be, in other words this was a self-authorised censorship. Pope Leo I also burnt Gnostic gospels.

John 21:25 (King James Version) says: *"And there are also many other things which Jesus did, the which, if they should be written every one, I suppose that even the World itself could not contain the books that should be written."* We have been deprived of these.

A brief Biblical note on Remote Viewing: Three of the 4 Gospels give accounts, written decades after the Garden of Gethsemane events, which must have been obtained by Remote Viewing later, as the disciples were all asleep, and distant, at the time!
A separate Foreword (Part 2) below gives more information on the development of the "Christian Canon". Although the Dead Sea scrolls are also of interest, space precludes their inclusion here.

Note: The general reader is recommended to skip this next section:
Foreword: Part 2: Supplementary Foreword on the Christian Canon. ***NOTE: This*** (Part 2) ***is not essential reading***

The general reader should skip to Chapter 1. The information below is not by the author (MGH), except for a few added comments. It was obtained from sources given at the start of each of the two sections below. *It is given here for reference only*, for the convenience of those who may like some notes on the Christian Canon. *It is not essential reading for the purposes of this book*, and is not "easy reading"! Please see the original web pages given, for more details and for their references (which are omitted in the text below).

(A) Adapted From:
http://www.historyworld.net/wrldhis/PlainTextHistories.asp?historyid=aa11

From the middle of the 1st century AD, texts were written and later collected into an "approved" New Testament, which was called the Christian Canon, representing the revised Old Testament covenant revealed by Christ.

The letters (*Epistles*) written by St Paul from about 50 and 62 AD to several early Christian communities, are the earliest of these.

Next in sequence are the *Acts of the Apostles*, which describe the activities of Peter and others in Jerusalem and of Paul on his journeys. St Luke is thought to have written this account, probably between about 75 and 90 AD. He was with Paul on some of his travels, including his last journey to Rome. Most of *Acts* is therefore first-hand contemporary evidence of the described events.

The written *Gospels* are slightly later than the *Epistles* and *Acts*, but the Gospels contain oral texts from earlier, around 30 AD.

The earliest version chosen to be put in the Bible is St Mark's Gospel, probably written down between 75 and 85 AD, and it was used, along with other sources, as the foundation for St Matthew's and St Luke's Gospels, which were written a few years later. The Gospel of John is much later (perhaps around AD 100) and differs from the other three in concentrating on spiritual issues more than biography. Modern scholarship, by considering the grammar differences etc, suggests that the authors of St John's Gospel and the Revelations of "St John the Divine" were different people. Only well into the 2nd century, the four Gospels are given their names.

Establishing the canon: 2nd - 4th century AD

About the middle of the 2nd century it was evident that many different and sometimes contradictory texts of scripture were being used in the various Christian churches, each claiming to represent the truth. There are other Gospels, like the Gospel of St Mary (Magdalene), the Gospel of St Thomas etc and even a Gospel of Judas Iscariot. Which of these should be accepted or rejected as the official canon became an urgent debate among church leaders.

By the end of the 2nd century it was widely (?) agreed that four *Gospels*, the *Epistles* of Paul and the *Acts of the Apostles* are "approved" of. But not until AD 367, a list was circulated by Athanasius, Bishop of Alexandria, which finally "established" to his satisfaction the "imprimatur" of the New Testament. But it regrettably also caused the destruction of most manuscripts of other Gospels which he did not like. A few such Gospels fully or partially survived this prejudiced censorship, which included a destruction order [9].

Meanwhile texts were being continually copied and recopied on papyrus and later on parchment. Some fragments survive from the 2nd century, but the earliest complete New Testament (the *Codex Sinaiticus*, written in Greek, probably in Egypt, now in the British Library) dates from the late 4th century.

At this time, Jerome in Bethlehem was writing his Latin version of the Bible. The New Testament then evolves into the story of its many later translations.

(B) Adapted from:
http://en.wikipedia.org/wiki/Biblical_canon

The Christian Canons listed below are usually considered *closed* (i.e., books cannot be added or removed). The closure of the Canon reflects an (assumed) belief that public revelation has ended, and thus the inspired texts may be gathered into a complete and authoritative Canon. By contrast, an open Canon permits the addition of additional books through a process of continuous revelation.

These canonical books were developed through debate and agreement by the religious authorities of their respective faiths. Believers consider these canonical books to be inspired by God or to express the authoritative history of the relationship between God and people. Books such as the Jewish-Christian Gospels, excluded from the Canon, are considered non-canonical — however, many disputed books considered non-canonical or even apocryphal by some are considered Biblical apocrypha or Deutero-canonical or fully canonical by others. There are differences between the Jewish and Christian Canons, and between the Canons of different Christian denominations. The differing criteria and processes of canonization dictate what the communities regard as the inspired books.

The word "canon" is derived from the Greek noun **κανών** "kanon" meaning "reed" or "cane," or also "rule" or "measure," which itself is derived from the Hebrew word קנה "kaneh" and is often used as a standard of measurement. Thus, a Canonical text is a single "authoritative" edition for a given work.

The *Gospel of Thomas, Gospel of Mary, etc* were omitted from the Christian Bible and most copies were destroyed on instructions by the Bishop of Alexandria during the reign of Emperor Constantine around the 4th century AD, when the Roman Empire was looking to reconstitute and solidify its power. The Emperor and the existing power structure chose the Pauline sect of Christianity as the "official" religion, which include the epistles of Paul and the Gospels and books from his disciples that form the present-day *New Testament*.

Nonetheless, full dogmatic articulations of the Canons were not made until the Council of Trent of 1546 for Roman Catholicism, the Thirty-Nine Articles of 1563 for the Church of England, the Westminster Confession of Faith of 1647 for British Calvinism, and the Synod of Jerusalem of 1672 for the Greek Orthodox church.

A four gospel canon (the *Tetramorph*) was asserted by Irenaeus, *c.* AD 160. By the early 200's, Origen of Alexandria may have been using the same 27 books found in modern New Testament editions, though there were still disputes over the canonicity of Hebrews, James, II Peter, II and III John, and Revelation (see also Antilegomena). Likewise by AD 200 the Muratorian fragment shows that there existed a set of Christian writings somewhat similar to what is now the New Testament, which included four gospels and argued against objections to them. Thus, while there was a good measure of debate in the Early Church over the New Testament canon, the major writings were accepted by almost all conventional Christians by the middle of the third century.

In his Easter letter of AD 367, Athanasius, Bishop of Alexandria, gave a list of exactly the same books as would become the New Testament canon, and he used the phrase "being canonized" (*kanonizomena*) in regards to them. To prevent people from reading texts that he did not approve of, he issued a (deplorable) instruction to monasteries in Egypt to destroy all other texts [9].

The African Synod of Hippo, in 393, approved the New Testament, as it stands today, together with the Septuagint books, a decision that was repeated by Councils of Carthage in 397 and 419. These councils were under the authority of St. Augustine, who regarded the Canon as already closed. Pope Damasus I's Council of Rome in 382, if the *Decretum Gelasianum* is correctly associated with it, issued a biblical canon identical to that mentioned above, or if not the list is at least a sixth century compilation. Likewise, Damasus' commissioning of the Latin Vulgate edition of the Bible, *c.* 383, was instrumental in the fixation of the Canon in the West. In 405, Pope Innocent I sent a list of the sacred books to a Gallic bishop, Exsuperius of Toulouse. When these bishops and councils spoke on the matter, however, they were not defining something new, but instead "were ratifying what they believed had already become the mind of the Church." Thus, from the fourth century, there existed unanimity (among prejudiced people!) in the West concerning the New Testament canon (as it is today), and by the fifth century the East, with a few exceptions, had come to accept the Book of Revelation and thus had come into harmony on the matter of the canon.

Writings attributed to the apostles circulated amongst the earliest Christian communities. The Pauline epistles were circulating in collected forms by the end of the first century AD. Justin Martyr, in the early second century, mentions the "memoirs of the apostles," which Christians called "Gospels" and which were regarded as on par with the Old Testament.

The first major figure to codify the Biblical Canon was Origen of Alexandria. He was a scholar well educated in the realm of both theology and pagan philosophy. Origen decided to make his canon include all of the books in the current Catholic canon except for four books: James, 2nd Peter, and 2nd and 3rd epistles of John. He also included the Shepherd of Hermas which was later rejected. The religious scholar Bruce Metzger described Origen's efforts, saying "The process of canonization represented by Origen proceeded by way of selection, moving from many candidates for inclusion to fewer." This was the first major attempt at the compilation of certain books and letters as "authoritative" and inspired the teaching for the Catholic Church at the time.

Needless to say there are various theologians of the 2nd and 3rd centuries that wrote a great deal of works and used the letters of the apostles as foundation and justification for their own personal beliefs. However, there was still the problem of the Roman Empire, and while the persecutions of the Roman Empire were many and extreme, the persecution still occurred and possibly interfered with the initial canonization of the New Testament. This period in church history writings is known as the "Edificatory Period" and was followed by the "Apologetic", "Polemical" and "Scientific" Periods. Some of the Christian writers of this edificatory Period are: Irenaus, Hippolytus, Tertullian, Cyprian, Justin Martyr, Clement of Rome. This stagnation of official writings led to a sudden explosion of discussions after Constantine I legalized Christianity in the early 4th century.

A *Samaritan Pentateuch* exists which is another version of the *Torah*, in this case in the Samaritan alphabet. The relationship to the *Masoretic Text* and the *Septuagint* is still disputed. Scrolls among the Dead Sea scrolls are identified as proto-Samaritan Pentateuch text-type. This text is associated with the Samaritans, a people of whom the Jewish Encyclopedia states: "Their history as a distinct community begins with the taking of Samaria by the Assyrians in 722 BC"

The Samaritans accept the Torah but do not accept any other parts of the Bible, probably a position also held by the Sadducees. Moreover, they did not expand their Pentateuchal canon even by adding any Samaritan compositions.

Both texts from the Church Fathers and old Samaritan texts provide us with reasons for the limited extent of the *Samaritan Canon*. According to some of the information the Samaritans parted with the Jews (Judeans) at such an early date that only the books of Moses were considered holy; according to other sources the group intentionally rejected the Prophets and (possibly) the other Scriptures and entrenched themselves in the *Law of Moses*.

The small community of the remnants of the Samaritans in Palestine includes their version of the *Torah* in their Canon. The Samaritan community possesses a copy of the *Torah* that they believe to have been penned by Abisha, a grandson of Aaron.

Some of these ideas will be developed in more detail later, but it was necessary to give them in brief form in this foreword.

References for the Foreword:

1. A. Besant & C.W. Leadbeater, "Oc. Chemistry", 400 pp (first edition 1908, second edn. 1921, third edn. 1951 & reprinted 2001, publ. by TPS, 50 Gloucester Pl., London W1H 3HJ). (Parts were published earlier, in 1895 in journal form: Lucifer (London), Nov. **1895**). Available as a free download from: www.4-D.org.uk but for the fine detail of some diagrams the printed version is necessary. 230 diagrams.
2. Patanjali, "Yoga Aphorisms", published about 400 BC.
3. Flotation tanks: See Chapter 7 in the present book.
4. M.G. Hocking, "Linking String Theory and Membrane Theory to Quantum Mechanics and Special Relativity Equations, avoiding any Special Relativity assumptions", J. Scientific Exploration 21 (1), 13-26 (2007). **This is reprinted in this book as Appendix 6.**
5. A. Besant & C.W. Leadbeater, "Man: Whence, How and Whither", published by TPS, Adyar, India (1913). And:
A. Besant & C.W. Leadbeater, "The Lives of Alcyone", (2 volumes, 740 pp), published by TPS, Adyar, India (1924).
6. TV programme series in August 2007 (& earlier) by Phillip Schofield on ITV1, hypnotically regressed 25 people to recover verifiable memories of previous lives, and one case in particular proved (beyond reasonable doubt) that we have lived before: "Have I been here before", ITV Productions Granada (UK), Presenter: P. Schofield, Producer & Director: S. Richmond (2006), shown in 2007.
7. See reference list at end of Chapter 2.
8. Psychedelic Review No 9, p. 42 (1967), etc, is available as a free download.
9. According to scholar Elaine Pagels, Professor of Religion at Princeton University, "In AD 367, Athanasius, the zealous bishop of Alexandria, ... issued an Easter letter in which he demanded that Egyptian monks destroy all such unacceptable writings, except for those he specifically listed as 'acceptable' even 'canonical' — a list that constitutes the present New Testament".

Note: For any updates to this book, corrections, etc, please see the websites below. A contact form is on: www.4-D.org.uk/Books

M.G. Hocking, London, 2011

www.4-D.org.uk   www.4-D.org.uk/Books   m.hocking@imperial.ac.uk

CHAPTER 1

Remote Viewing of Molecular Structure and Quantum Physics:

Towards A Theory of Everything

- ◆ Dimensions of space
- ◆ The chakras, including CIA remote viewing
- ◆ The dream state & lucid dreams

Religion without Science is blind;
Science without Religion is lame.   Albert Einstein

Nature is not only stranger than we have thought;
It is stranger than we <u>can</u> think !   JBS Haldane

Dimensions of Space

In recent years attempts have been made to produce a "Theory of Everything", to unify the main branches of physics. 80 years ago Einstein tried to unify relativity and quantum theory but he could not. The reason may be that it is impossible to unify these in only 3 spatial dimensions (plus time), which is what he tried to do.

But there are now two groups of physicists:

(**1**) Traditional physicists who will only accept 3 space dimensions and one time dimension, which they usually combine together and call space-time.
This has severe problems in reconciling relativity with quantum theory.
(**2**) Elementary particle physicists, who have calculated that 10 space dimensions and one time dimension are essential to explain particle phenomena (Supergravity theory; M-brane theory).

Dimensions of Space: some hints from religion (as cited in the Foreword):

For readers with a Christian perspective, a Biblical verse is: *"In My Father's House* (surely, the Universe?) *there are many mansions* (dimensions?). *If it were not so, I would have told you ...".* (John 14: 2). A "mansion" is defined as a "dwelling place" (dictionary).

The Gospel of Thomas (113: 4) (Coptic, found in 1945 at Nag-Hammadi, Egypt) says:
"The Kingdom of the Father (Heaven?) *is spread upon the Earth but men do not see it."* So is it in a higher dimension?

Before the universe was created, if God exists He would need to decide at the outset how many dimensions it would have, and it is most unlikely that He would limit it to 3. Any designer of such a system would create some spare dimensions and 10 (?) might be a reasonable number.

E.g. at least 4 dimensions are needed to allow parallel 3-D universes (see colour Fig. 1.1, near p. 64). For an explanation of 4-D see Appendix 8, p. 361.

Consider a pack of playing cards. Two adjacent cards in the pack are less than a millimetre apart and are like two parallel universes in 2-D space, but access for a 2-D being from one to another is only possible by jumping through what a 2-D being would call a "worm hole" (= a 3rd dimension corridor).

In 3-D space, by analogy, if there are two parallel 3-D universes, a 4th spatial dimension is needed to jump from one to another (via what <u>we</u> would call a "worm hole").

In the following text, there is a hint of a possible use by Christ of a parallel 3-D world to which He had access, i.e. a parallel world separated from our 3-D world by a displacement in a 4-D direction, like a jump to an adjacent playing card in the above 2-D analogy. This could have allowed him to escape from an angry crowd:

The Gospel of Luke (4: 28-30) says, "--- *they were filled with wrath and rose up and thrust Him out of the city, and led him unto the brow of the hill whereon their city was built, that they might cast Him down headlong. But He <u>passing through the midst of them</u> went on His way".* (Bible, King James Version).

Similarly, John (8: 58-59) says, *"Jesus said unto them, Verily, verily, I say unto you, 'Before Abraham was, I am'.*
Then they took up stones to cast at Him: but Jesus hid himself, and went out of the temple, <u>going through the midst of them</u>".
(my underlining). See also John 20: 19 & 26-27 in The New English Bible accurate translation: Christ entered "*when the disciples were behind locked doors",* and *"although the doors were locked"* Christ came in to meet the disciples and was clearly physical because Thomas was able to touch him several times.

There are (rare) accounts of such parallel worlds in modern times, where people found themselves suddenly for a while in a totally different landscape. Two very different types of such experiences have been reported – discussed in Chapter 4. But as these are anecdotal (and cannot be repeated on demand) they are not "scientifically" acceptable by the normal criterion (i.e. repeatability on demand). But of course this does not mean that they are untrue.

-=-=-=-=-=-=-

Consider one-dimensional space: this would be a line, along which objects could move. But this is very restrictive and one would certainly like to have a second dimension available:

Then this 2-D space is a flat sheet in which objects could move in 2 dimensions, but one could roll it up into a tight helical cylinder, which then "looks like" a one-dimensional line. This is easily demonstrated using a tightly rolled-up A4 sheet. Objects moving in it in any direction in that sheet, would thus effectively be confined to linear travel in one dimension.

This very simply explains the statement by some particle physicists that we live in a 3-D world (obviously), but there are 7 higher dimensions which are "rolled up", so we don't see them. Un-rolled higher dimensions are discussed on page 73.

Three space dimensions is the very minimum needed for sufficient brain cell connections, for digestive tracts, and even for effective integrated circuit designs!

But four (& higher) dimensions would not allow atoms & molecules to exist because the attractive forces holding electrons around an atomic nucleus are too weak in 4-D space. Only elementary particles can exist in 4-D space:

Explanation of why atoms cannot exist in 4-D (& higher) space:

Consider the following peripheries:
2-D: Perimeter line of a circle = $2\pi r$.
3-D: Surface area of a sphere = $4\pi r^2$.
4-D: Surface volume of a hypersphere = $4/3 \pi r^3$.
(A hypersphere is the 4-D analogue of a sphere)

A proton of charge +1 at centre of a sphere exerts an attractive force **F** on an electron of charge -1 at distance **r** from it.

So force per unit area of sphere surface = $F/4\pi r^2$.
So attractive force is inversely proportional to r^2, an inverse square.

For a hypersphere (4D analogue of a 3D sphere) the equivalent effect is:
Force per unit volume (equivalent to 3-D case "surface") = $F/(4/3)\pi r^3$.

So attractive force is inversely proportional to r^3. So fall-off of **F** in 4-D space is an inverse cube, i.e. much steeper than for 3-D space.

So atoms cannot exist in 4-D (they would fly apart, as the attractive force F between nucleus and electrons falls off too steeply).
But elementary particles can exist there.

So 3-D space is the only choice. But 4th & higher dimensions are needed for other purposes – e.g. for 3-D parallel universes (just mentioned above) to exist without interacting with each other.

Note about 4-D space: 4-D is a direction in which we cannot point and cannot look! This is easily explained by the following analogy for a 2-D being in relation to 3-D space.
It is easy to explain why we can't see into the 4th dimension with our ordinary eyes, if we use this analogy of why a 2-D being cannot see into the 3rd dimension:

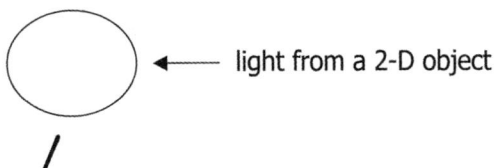

Eye of a 2-D being in the plane of this paper

Fig. 1.2.

The 2-D eye clearly cannot be swivelled down or up (meaning down into or up out of the plane of this paper), to look into the 3rd dimension!
Any light beam from an object above or below this paper in the 3rd dimension, will be angled down or up, through this plane, and so will not focus into the 2-D eye shown in the plane.

Analogously, we cannot turn our eyes in a 4-D direction!
Also if a 4-D object passes through our 3-D world, it cannot be seen, because it is not made of atoms and molecules (for the reasons just given). Any 4-D or higher dimension object is made of elementary particles (e.g. neutrinos) which are almost impossible to detect by **scientific instruments** because neutrinos etc do not interact with the ordinary matter which the detectors are made of: 60 million pass through our eyeball per second and a light-year thickness of lead would absorb only 50% of them!

An exception is that the **human mind & brain** has a link to higher dimensions, which is evident to everyone from our own dreams, and (reportedly) to a few gifted people who have what is called "second sight", a type of Remote Viewing: See Fig. 1.3, between p. 64 & 65. Why are there so many names for "nature spirits": fairy, elf, pixie, gnome, goblin, hobgoblin, brownie, salamander, undine, leprechaun, genie, sprite, dryad, hamadryad, naiad, nymph, peri, sylph, and others now archaic?

Rural people of the olden days, before the industrial revolution, may have been more sensitive than we are today. See Appendix 8.

Remote Viewing: The Chakras ("6th" sense" receptors)

◆ The CIA has specialist intelligence operatives whom they call "Remote Viewers" who have viewed, for example, the contents of locked safes in the USSR.

A few years ago a TV programme interviewed one of the CIA remote viewing group (Joseph McMoneagle) who said he was able to extend his eyebrow chakra to remotely view the details of atom bomb triggers in China.

If this is considered to be false information, there is a video DVD [10] of interviews with generals and an admiral from government intelligence agencies who describe the Remote Viewing process.

Appendix 2 shows pictures of some senior military personnel of the U.S.A. Remote Viewing programme.

A search for "remote viewing" on the Amazon.co.uk website, lists many books available [11-39 and more].

◆ 5000 years ago, the ancient Egyptians also trained young women called "Looking-girls" or "Lookers", to remotely view the borders of Egypt to warn of any invasion [54]. So Remote Viewing is not new.

Fig. 1.4 (between p. 64 & 65) shows a life-size Egyptian statue 4000 years old, showing the eyebrow chakra, from the 12th dynasty. It appears on very many Egyptian statues and famously on the Tutankhamun mask.

It is described as a "third eye" which can reportedly be extended outwards to any remote location to observe events there (Remote Viewing). Similarly, the CIA remote viewer, Ingo Swann, said he

CHAPTER 1   Dimensions of Space, Remote Viewing, Dream State

can extend a viewing ability to anywhere on the planet, while remaining in a normal conversational mode (i.e. in no trance state). (Video is available [10(3)]).

◆Besant & Leadbeater report using the eyebrow chakra to visit distant places and to magnify atoms. This is discussed in Chapter 3. Fig. 1.5 & 1.6 below show classic stylised pictures of chakras and Fig. 1.7 & 1.8 show actual remote viewing" pictures of chakras [40].

◆There are many other reports, e.g. a well-known one is Swedenborg describing details of a fire in Stockholm (hundreds of miles away) as it occurred, which was later verified (about 250 years ago).

◆About 25 years ago, before he was recruited by the CIA, Ingo Swann impressively demonstrated his remote viewing ability, here in London; he repeatedly correctly read random numbers on an electronic random number generator, which he could not see and which was provided by one of the scientists present.

◆A paper published in Nature has verified beyond doubt (to 1 part in a million) that remote viewing ESP does exist [41], by repeatedly correctly perceiving dice in a <u>closed steel box</u> shaken in laboratory conditions at Stanford Research Institute. Remote Viewing is a controlled daydreaming state and these results mean that such controlled dreams cannot be dismissed as hallucinations. Induction methods are given later in this chapter. See also Chapter 7.

◆In the New Testament, the disciples recorded events occurring while they were <u>asleep</u> in the Garden of Gethsemane. This is only possible by their Remote Viewing of it later. It was when Christ had asked them to remain awake while he went away to pray. See Matthew 26: 36-44; Mark 14: 34-40 and Luke 22: 41- 46. The same account is given in 3 of the 4 Gospels.

-=-=-=-=-=-=-=-

Traditional locations of chakras from ancient Hindu/Buddhist texts are shown below in Fig. 1.5. The geometric icons in them are not actual, but are traditionally drawn-in as meditation aids.

The <u>upper 4</u> are the important chakras:
crown chakra *(chosen for the centre of the national flag of India);*
eyebrow chakra *(so-called "third eye");* throat chakra; heart chakra:

Fig. 1.5. Classic picture of the chakras (stylised).
(See also colour Fig. 1.7 & 1.8, between p. 64 & 65)

Ancient Buddhist & Hindu texts report **Kundalini** as a life force residing below the base of the spine, but which can be made to rise up inside the spine, and vivify the 7 chakras as it does so.

It is also seen, with lost detail, in the Caduceus which is the traditional classical symbol of medicine (intertwined serpents). It is perhaps the serpent of Adam & Eve mentioned in Genesis. Fig. 1.6 below shows the Caduceus.

At the cross-over points, there are said to be chakras or centres, which are shown in the above traditional picture (Fig. 1.5) from classical Hindu/Buddhist texts.
The pictures of the 7 chakras in Fig. 1.5 are from "The Serpent Power" by Arthur Avalon, published in 1919, being a translation of traditional Indian books on the chakras. The chakras are too small in Fig. 1.5 to show their details but are available full size in recent editions, such as the Dover Edition, NY, (1974). The original 1919 edition shows the chakras as colour pictures but in Fig. 1.5 they are shown in black & white.

The chakra symbols in Fig. 1.5 are reported [40] to be absent in the actual chakras as observed by Remote Viewing methods (see colour Fig. 1.7 & 1.8, between p. 64 & 65). They are meant as aids for meditation on each chakra.

Fig. 1.6. Caduceus.

In normal people the central kundalini channel is said to be almost blocked near its lower entrance, and only the lowest or root chakra is active, manifesting as the sex drive. Also, there is a thin membrane which blocks perception of astral entities by the lower chakras, which would otherwise be very disturbing. This membrane can be disrupted by alcohol, which explains the *delerium tremens* of some inebriates, in which they see and cringe away from demonic entities in the lower "dream-world" (called "astral" due to its self-luminous quality) region. For normal people, this membrane also obstructs easy recall of dreams. Dreams are usually trivial or silly, because there is no self-consciousness in a normal uncontrolled (non-lucid) dream and so the dreamer is a passive spectator to whatever astral currents there are. Unlike a trained dreamer, an untrained dreamer is conscious in the 4-D dream state but not self-conscious.

But blockage is often not total and the tingling sensation in the spine which many people get when listening to uplifting music etc (religious experiences), is possible evidence of Kundalini rising.
Kundalini seldom rises higher than the heart centre in untrained people.

If it rises further, to the throat centre, this confers the gift of oratory. But raising Kundalini is a technique not dependent on a person's goodness -- e.g. Hitler was a good orator but entirely evil.   A useful Kundalini reference is [52].

It is dangerous to raise Kundalini if there are character flaws, and also, opening the chakras by taking drugs gives a "bad trip" if the person is not spiritually developed. Also, the ESP images appear chaotic and cannot be interpreted. Much training is needed for correct interpretation and to avoid intervention of one's imagination, which creates subjective hallucinations.

Seven chakras are independently reported worldwide, by Taoist sources in the far East, by Buddhist and Hindu sources in India and 5 by Hopi Indian sources in North America. Such independent worldwide descriptions of chakras over thousands of years is anecdotal evidence of their reality.

Leadbeater [40] gives actual pictures of the 7 chakras *(two are shown in colour Fig. 1.7 & 1.8, between p. 64 & 65).* These figures show the chakras in their fully opened state, like lotuses, but in most people the petals are almost closed and the chakras droop down.

If Kundalini is raised further, up to the eyebrow chakra, this is said to confer "second sight", a type of Remote Viewing, and according to Patanjali [42] it can be used to magnify atoms, etc.

The CIA expert-level Remote Viewers use the eyebrow chakra, and this is also what Besant & Leadbeater report using to get the atom diagrams given in their book; e.g. see Fig. 3.1 (in Chapter 3) which shows the smallest elementary particle.

It may take <u>years</u> to raise Kundalini, which involves clearing the blockages, by classic breathing exercises. Holding the breath, in these exercises creates <u>anoxia</u> (lack of oxygen) in the brain and causes CO_2 narcosis (CO_2 has a narcotic effect like taking drugs). Anoxia suppresses the normal sensory input from the 5 senses and thus allows the weak input from a <u>sixth</u> sense (ESP) to come to the fore and be perceived.

The opposite practice, of over-breathing (breathing faster than normal) is also effective for a different reason [43, 44] but as it alters the sodium to potassium ratio in the blood considerable caution is necessary. See safety notes below.

There is support for the effect of anoxia in the brain from <u>near death experiences</u> (NDEs), where some people report other-worldly visions and there is a possible link between holding-the-breath meditation exercises (called pranayama) and NDEs.

The tube which Kundalini is traditionally said to rise through, coincides with the <u>spine</u>, and this could be the <u>tunnel</u> that NDE people report <u>rising through</u>, because the soul is traditionally said to leave the body up through this tube at death, and to exit via the <u>crown chakra</u> on the head. Besant & Leadbeater confirm this by their own ESP studies [40]. Some NDE people describe the tunnel as like a ribbed hose – could be vertebrae perhaps, seen distorted.

But if an <u>NDE subject</u>, or a <u>meditator</u> using pranayama breathing exercises, or a <u>remote viewer</u>, report visions, a skeptic will say that they are just chemically generated within the brain, **but** this **ignores** their reports of later-verified events occurring elsewhere, out of sight, which clearly could not have been manufactured within the brain!

Safety Notes:
(1) Normal breathing at 4 L/minute, equates to a pCO_2 of 40 mm Hg. Hypoventilation (slow breathing/holding breath) at a rate of 2 L/minute, equates to a pCO_2 of 80 mm Hg, causing respiratory acidosis, causing potassium from the extra-cellular fluid to go into the cells. Hyperventilation (fast or over-breathing) at 8L/minute, equates to a pCO_2 of 20 mm Hg, causing respiratory alkalosis, causing potassium from the cells to go into the extra-cellular fluid. These effects alter the sodium to potassium ratio in the blood, which may have undesirable effects (medical advice needed on this point).

(2) The author would prefer not to mention an aberration of breathing exercises, in which several hundred people die each year (worldwide) from anoxia (figure from police statistics). These people have no knowledge of, nor interest in, higher states of meditation.

But in the interest of safety I feel obliged to mention that it is desperately dangerous to induce anoxia by the extreme methods they use, like putting a plastic bag over the head, or by pressing a neck artery to reduce blood flow to the brain. (Police video available). *"Don't try this at home folks!!"*

This is a side-issue, but briefly, the explanation of the above lies in over-activating the lowest chakra of Kundalini, which is the sex-drive centre.

A recent survey found about 90% of Google searches are about sex, which is a rather sad commentary on the present state of humanity!

-=-=-=-=-=-=-=-=-=-=-=-

Any danger of anoxia can be completely avoided by approaching from a different direction, which also gives much faster results:

Instead of opening the chakra by devotional meditation &c, we can greatly reduce the competing sensory input, and then even if the chakra is only slightly open, its input will be registered by the mind & brain!

What little ESP input there is from the eyebrow chakra is normally blocked out by other strong sensory input from our 5 normal senses. If all of the 5-sensory input which continuously enters our mind is itself completely blocked out, by sensory deprivation using a flotation tank, then immediate results begin to appear, or, during sleep using Tibetan dream yoga, but the latter takes many years to succeed with [see Tenzin Wangyal Rinpoche, "Tibetan Yogas of Dream & Sleep", ISBN-13: 978-1-55939-101-6, NY (1998)].

A good analogy, mentioned earlier, is that when looking up at the sky, stars are only observable at night when the daylight is absent, as shown in colour Fig. 1.9 & 1.10. Daylight is a form of "noise" (scattered sunlight) which completely masks the stars which are there all the time. Here is a blue sky, but it is an entirely false picture, because the blue optical "noise" (scattering of light) blocks out what is really there at but a very low intensity: stars: see colour Fig. 1.9 & 1.10, between p. 64 & 65. The same area at night when there is no scattered daylight, reveals stars, which of course are present all the time.

Classic texts describe about 10 methods of traditional "yogas": Raja Yoga, Hatha Yoga and Tantra Yoga, etc. Some of these paths are quicker than others. Raja yoga might take decades!

But here are 2 methods which give quick results, plus a possible 3rd method:

(1) The Flotation Tank:

Details are in Chapter 7, but a brief summary is given below:

If the chakras are partially open, a quick shortcut is to use sensory deprivation:
If inputs from the 5 senses are completely cut off, it is a very, very unusual state, and it makes the mind greatly increase its "gain" or amplification, then becoming far more sensitive to small ESP inputs which are normally drowned out by the big "signal-to-noise" ratio of everyday life (the "noise" is from the normal input from the 5 senses).

This amplification compensates for the partial blockage of the Kundalini tube, and nearly closed chakras, and thus clairvoyant vision (a form of ESP) occurs in the flotation tank.

Clairvoyant vision (remote viewing) becomes possible only if there are both beta and theta brain-waves, which a flotation tank generates by its unique property of total sensory deprivation.

Touch is the most difficult sense to escape from, but no sense of touch exists when floating in water at skin temperature (in a flotation tank), giving the very unusual state of total relaxation combined with maximum awareness (wakefulness).

If beta brainwaves disappear, we go to sleep and we have only theta waves, but both beta & theta are needed to bring the dream state into conscious awareness (e.g. lucid dreaming) [45].

It is almost impossible to sleep in the tank, so the ideal situation of both beta and theta exists and the dream state is entered with full consciousness -- a form of lucid dreaming (controlled dreaming).

An effect of the tank is also to produce balanced laterally symmetrical beta, alpha and theta brainwave amplitudes [45], the same as by using flashing light and sound. This type of brainwave pattern, with both beta (waking state) and theta (dreaming state) waves is **essential** for being self-conscious in a dream-type state. See Fig. 1.11 below.

It is in this state that Edison made his inventions and Einstein made his contributions to science.

More details of the flotation tank are given in Chapter 7.

"Stop Press": recommended additions to references list for these topics are:

(i) T. Yuschak, "Advanced Lucid Dreaming", ISBN 978-1-4303-0542-2 -- **This one is essential reading** – covers use of herbal supplements (which, unlike narcotic drugs, do not alter one's waking behaviour) to produce immediate lucid dreams and OBEs, including seamless transition from waking to lucid dreamstate consciousness ('WILD'). See page 36 for comments on suppliers. WILDs are the gateway to mystical experiences (St Teresa, etc). **See also ref. 14 on p. 241 for a free download of important further details too late to include in the present book.**

(ii) Robert Bruce, "Astral Dynamics", ISBN 978-1-57174-616-0.

CHAPTER 1   Dimensions of Space, Remote Viewing, Dream State   33

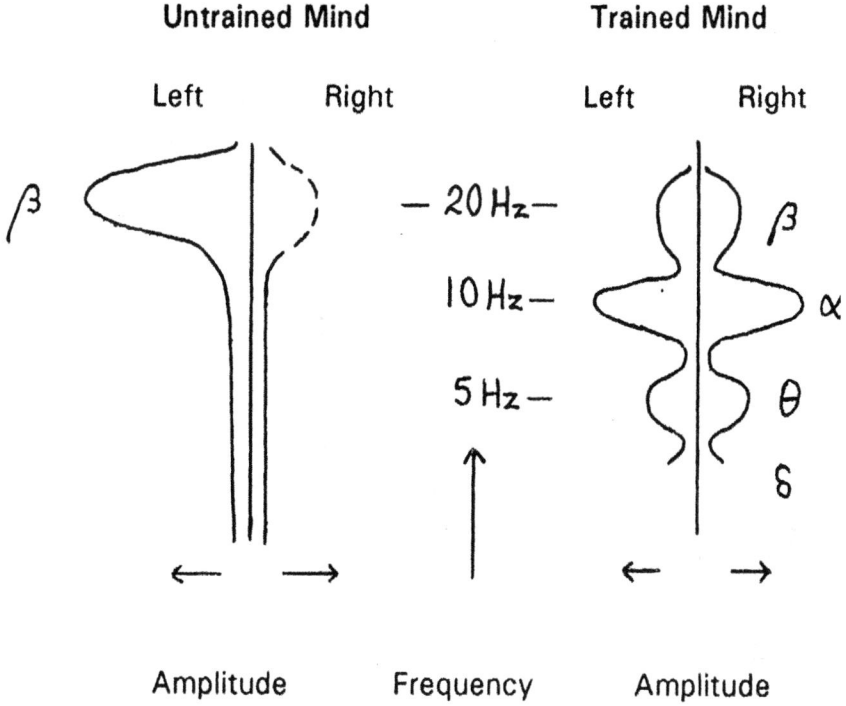

Fig. 1.11. Brainwave (EEG) graphs [see refs in 45]

Notes on Fig. 1.11:

Left brain hemisphere is logical thinking.
Right brain is intuitive artistic inventive faculty

Beta frequency is waking mind activity
Alpha is hypnagogic state (eyes closed state)
Theta frequency is dream state.

Left picture: Beta peak, shows linear logical thinking activity, typical of a man-in-the-street. Unbalanced.

Right picture: Trained meditator:
A beta peak (= awake state) links via alpha (= hypnagogic state) to theta (= dream state, where invention & intuition arise. These are brought through to the waking mind in this state).

(2) Lucid dreaming:

Dreaming ("astral" consciousness) is also a state in which we are completely unaware of our physical body and there is no input from the 5 senses, importantly, no sense of touch or pressure. There is no feeling of lying on a bed.
But unfortunately, when dreaming, although we are conscious in the dream, we are not self-conscious, so we have no volition to direct the dream with!
We are spectators only. So ordinary dreaming is useless!!

But there is a special dream state called a lucid dream, which is useful [45].
If we can muster enough self-consciousness the whole dream becomes vivid and controlled, and it is indistinguishable from waking life, and this is called a lucid dream -- we have then achieved self-consciousness in the dream state [45].

The next stage is to learn how to cut through all the illusions which characterise dreams, so that if we read a paper with some words on it and look away and then read it again, the words will not have changed, nor will the paper have transmuted into a butterfly! Only then, correct remote viewing is achieved. This is difficult and Besant & Leadbeater stress the need to achieve this for remote viewing to become objective and useful.

Per Villayat Khan of the Sufi Order gave a talk 20 years ago to the Wrekin Trust and strongly recommended being able to have continuous consciousness -- no break in consciousness on entering the dream world (an immediate lucid dream begins) and on waking from sleep one passes seamlessly into the ordinary waking state, with full recall of one's dreams. Galantamine can quickly achieve this result (see below, page 35).

The so-called "dream yogis" of Tibetan Buddhism had this ability, continuous unbroken consciousness, day & night.
An easy alternative method to achieve this effect is to use a flotation tank, but it may require several sessions.

A device called the Dreamlight (www.lucidity.com), trains the user to get dreams (not using a tank) which are indistinguishable from one's normal waking state, even though one is fast asleep!
The user can then direct the dream.

Another type of lucid dream unit monitors breathing rate, and if this rises (which it does when REM dreaming), a circuit is triggered which flashes a photoflood flash sequence; the dreamer perceives the flashes in his dream and then realises he is dreaming and a lucid dream then starts. (REM = rapid eye movements, which characterise a dreaming episode.)

Important: See [50] for a very simple lucid dreaming method.

(2a) Herbal methods for lucid dreams (LDs):

The author has found that eating a (vegetarian) dish called Lo Han ("Buddha's delight") (available in Chinese Restaurants) with soy sauce, in the evening, gives unusually vivid dreams in that night. Looking up the constituents (e.g. in Google), shows lily buds as one (galantamine), and soya sauce which contains choline.

Lily contains galantamine or galanthamine (see below), which is known to cause lucid dreams, and if either choline or Alpha GPC (see below) is present as well, the effect is enhanced. There is no effect during the day in the normal waking state (in the author's experience) and so this seems to be a morally acceptable drug in that it will not alter the taker's behaviour in any ways that may affect others nearby. And it is non-addictive.

The extract from plantation-grown, Red Spider Lily (*Lycoris radiata*) has a rich history of use in herbal treatment. For thousands of years people have used it for dream enhancement, as a headache treatment and for its positive effect on memory recall (a treatment for first-stage Alzheimers disease, refs. below).

An extract from the common snowdrop plant had long been used for improving memory in eastern European herbal lore. A synthetic version of this extract has proved, in experiments for the USA Food and Drug Administration (FDA), that the extract substance known as galantamine improves the memory of those with Alzheimers disease. These studies revealed an odd "side effect," namely that those in the study reported significantly more dreams and enhanced dream recall. Important: see footnote p.32.

NOTE: The general reader can skip the rest of this section and go to Section (3), page 41. But for suppliers, see below (on page 36).

The following paragraphs are taken, with acknowledgement, from a convenient information source, Wikipedia. The references in this section are listed separately from the others, at the end of this section.

Galantamine or **galanthamine** (trade names: **Nivalin, Razadyne, Razadyne ER, Reminyl**) is a chemical used for the treatment of mild to moderate Alzheimer's disease and various memory impairments. It is an alkaloid that is obtained synthetically or from the bulbs and flowers of the Caucasian snowdrop (Voronov's snowdrop), *Galanthus woronowii* (Amaryllidaceae) and related genera like *Narcissus* (daffodil), *Leucojum* (snowflake) and *Lycoris* including *Lycoris radiata* (Red Spider Lily). The active ingredient was isolated by Prof. Paskov 1959 (Sopharma, Bulgaria) from a species traditionally used as a popular medicine in Eastern Europe and thus the idea for developing a medicine from these species seems to be based on the local use (ethnobotany-driven drug discovery) [1, 2]. It has been used for decades in Eastern Europe especially in the symptomatic treatment of polio (poliomyelitis) and was later developed by Janssen Pharmaceutica into an Alzheimer medication. In the USA it has been sold as a dietary supplement for memory and dream support prior to being approved as a drug by the FDA. ((**Internet suppliers** include: www.smartnutrition.info/brain.html#Galantamine and www.dreamamins.com and also Vitamin Express (USA) who call it 'Galantamind' (from www.life-enhancement.com), but these are plant extracts and have no standards like BP or USP, so caution is necessary and the present author (MGH) cannot recommend or give an opinion on their use.))

The systematic IUPAC formula for galantamine is:

(4aS,6R,8aS)-5,6,9,10,11,12-hexahydro-3-methoxy-11-methyl-4aH-[1]benzofuro[3a,3,2-*ef*] [2] benzazepin-6-ol.

Galantamine in its pure form is a white powder. Galantamine is a competitive and reversible cholinesterase inhibitor. It is believed to work by enhancing cholinergic function by increasing the concentration of acetylcholine in the brain. The atomic resolution 3D structure of the complex of galantamine and its target, acetylcholinesterase, was determined by X-ray crystallography in 1999 (PDB code: 1DX6; see complex) [3]. There is no evidence that galantamine alters the course of the underlying dementing process [4]. Galantamine has also shown activity in modulating the nicotinic cholinergic receptors to increase acetylcholine release [5].

Pharmacokinetics
Absorption of galantamine is rapid and complete and shows linear pharmacokinetics. It is well absorbed with absolute oral bioavailability between 80 and 100%. It has a half-life of 7 hours. The peak effect of inhibiting acetylcholinesterase was achieved about one hour after a single oral dose of 8 mg in some healthy volunteers.
Plasma protein binding of galantamine is about 18%, which is relatively low.

Metabolism
The major route of metabolism for galantamine is through the liver, this accounts for approximately 75% of the total metabolism of galantamine. Hepatic cytochrome P450 (CYP) isoenzymes are the active enzymes for this metabolic route. *In vitro* studies have shown that CYP2D6 and CYP3A4 are involved in galantamine metabolism.
For Razadyne ER (the once-a-day formulation), CYP2D6 poor metabolizers had drug exposures that were approximately 50% higher than for extensive metabolizers. About 7% of the population has this genetic mutation, however because the drug is individually titrated to tolerability, no specific dosage adjustment is necessary for this population.

Galantamine is dispensed for mild to moderate Vascular Dementia and Alzheimer's disease [6, 7].

Available forms (for lucid dreams (LD) & for Alzheimer's)
The product is supplied in twice-a-day tablets, once-a-day extended release capsules, and in oral solution. The tablets come in 4mg, 8mg and 12mg forms. The capsules come in 8mg, 16mg, and 24mg forms. (LD dose is 4 or 8; Alzheimer's dose may be 24 mg)

Adverse events
In clinical trials, galantamine's side effect profile was very similar to that of other cholinesterase inhibitors, with gastrointestinal symptoms being the most notable and most commonly observed. In practice, some other cholinesterase inhibitors might be better tolerated; however, a careful and gradual titration over more than three months may lead to equivalent long-term tolerability [8].

Use as supplement for lucid dream & out-of-body experience
Some people who practice lucid dream (LD) or out-of-body experience (OBE) use galantamine to increase their odds of achieving an LD or OBE [9, 10, 11]. By taking a small amount of galantamine (around 4 to 8 mg) after 5 to 6 hours of deep sleep and with an induction technique such as meditation, 'MILD' or 'WILD' [4] many people report more success with galantamine [12].

There are also reports that taking galantamine without proper induction technique will not lead to LD or OBE but will result in only a vivid dream instead. It should also be noted that due to a long half life Galantamine will stay in the body for a period of up to and over 48 hours, as such it is advisable to space out the use of Galantamine over a period of three days so that the body does not build a resistance to the drug ruining its effectiveness [9].
Galantamine used with choline bitartrate or Alpha-GPC can dramatically increase one's odds of becoming lucid in the dream and increase memory consolidation during dreaming. Some people report mixing galantamine with other nootropics can enhance the degree of lucidity, but this is still controversial since some mixtures may work for some people, but lead to failure for others.

Nootropic
Like other cholinergics or acetylcholinesterase inhibitors such as Huperzine A, galantamine also has been used as nootropic or "brain enhancer" to enhance memory in brain-damaged adults [13].

Caution
The U.S. Food and Drug Administration (FDA) and international health authorities have published an alert based on data from two studies during the treatment by galantamine of mild cognitive impairment; higher mortality rates were seen in drug-treated patients [14]. On April 27, 2006, FDA approved labeling changes concerning all form of galantamine preparations (liquid, regular tablets,and extended release tablets) warning of the risk of bradycardia (and sometimes atrioventricular block, especially in predisposed persons). At the same time, the risk of syncope seems to be increased relative to placebo [15]. These side effects have not been reported in any other studies except on mild cognitive impairment. (<u>Note:</u> dosage for memory loss is more than for LD.)
Synthesis
Main article: Galanthamine total synthesis (Wikipedia).

Galantamine is produced from natural resources and a patented total synthesis process. Many other synthetic methods exist but have not been implemented on an industrial scale.

References to Galantamine (from Wikipedia, 2010):

1. Heinrich, M. and H.L. Teoh (2004) Galanthamine from snowdrop – the development of a modern drug against Alzheimer's disease from local Caucasian knowledge. Journal of Ethnopharmacology 92: 147 – 162. (doi:10.1016/j.jep.2004.02.012)
2. Scott LJ, Goa KL. Adis Review: Galantamine: a review of its use in Alzheimer's disease. Drugs 2000;60(5):1095-122 PMID 11129124: www.ncbi.nlm.nih.gov/pubmed/11129124
3. Greenblatt, HM, Kryger, G, Lewis, T, Silman, I, Sussman, JL "Structure of acetylcholinesterase complexed with (-)-galanthamine at 2.3Å resolution" FEBS Lett 1999; 463, 321-26. PMID 10606746: www.ncbi.nlm.nih.gov/pubmed/10606746
4. Ortho-McNeil Neurologics, "Razadyne ER US Product Insert", May 2006: www.razadyne.com/active/janus/en_US/assets/common/company/pi/razadyne.pdf
5. Woodruff-Pak DS, Vogel RW 3rd, Wenk GL, "Galantamine: effect on nicotinic receptor binding, acetylcholinesterase inhibition, and learning" Proc Natl Acad Sci USA. 2001 Feb 13;98(4):2089-94. PMID 11172080: www.ncbi.nlm.nih.gov/pubmed/11172080
6. Galantamine Benefits Both Alzheimer's Disease and Vascular Dementia: www.life-enhancement.com/article_template.asp?id=1919
7. Galantamine Improves Attention in Alzheimer's: www.life-enhancement.com/article_template.asp?id=1155
8. Birks J. "Cholinesterase inhibitors for Alzheimer's disease." Cochrane Database Syst Rev. 2006 Jan 25;(1):CD005593. PMID 16437532: www.ncbi.nlm.nih.gov/pubmed/16437532
9. en.wikipedia.org/wiki/Galantamine#cite_ref-Yuschak2006_8-0 en.wikipedia.org/wiki/Galantamine#cite_ref-Yuschak2006_8-1 Thomas Yuschak (2006). *Advanced Lucid Dreaming* (1st ed.). Lulu Enterprises:ISBN978-1-4303-0542-2: en.wikipedia.org/wiki/Special:BookSources/9781430305422
10. Thomas Yuschak (2007). *Pharmacological Induction of Lucid dreams*.
11. http://www.advancedld.com/f/Pharmacological_Induction_of_Lucid_Dreams.pdf
12. "Substances that enhance recall and lucidity during dreaming". *Stephen LaBerge - US Patent*. http://www.freepatentsonline.com/20040266659.html Retrieved on 2007-10-29.
13. "Galantamine LDS Profile". *Yuschak LDS Profiles*.
14. http://www.advancedld.com/galantamine.html Retrieved on 2007-10-29.
15. Galantamine Protects Neurons and Memory Following Brain Injury: www.life-enhancement.com/article_template.asp?id=1878
16. FDA alert (3/2005) on galantamine: www.fda.gov/Cder/Drug/infopage/galantamine/default.htm
17. FDA alert (4/2006) on galantamine: www.fda.gov/medwatch/safety/2006/apr06.htm

External links:
Memeron (galantamine compound website): www.memeron.com
Razadyne (manufacturer's website): www.razadyne.com
Galantamine (patient information): www.meds-help.com/galantamine/

Choline, $C_5H_{14}NO^+$

From Wikipedia, the free encyclopedia (references omitted):

Choline is an organic compound, classified as a water-soluble essential nutrient and usually grouped within the Vitamin B complex. This natural amine is found in the lipids that make up cell membranes and in the neurotransmitter acetylcholine. Adequate intakes (AI) for this micronutrient of between 425 to 550 milligrams daily, for adults, have been established by the Food and Nutrition Board of the Institute of Medicine of the National Academy of Sciences.

It is well established that supplements of methyl group transfer vitamins B6, B12, folic acid reduce the blood titer of homocysteine and so may prevent heart disease. Choline is a necessary source of methyl groups for methyl group transfer. Supplements of lecithin/choline by Central Soya scientists reduced heart disease in laboratory studies. The reduction in heart disease with lecithin supplements may however relate more to the cholesterol carrying capacity of lecithin than to the methyl group transfer role of choline.

Choline supplements are often taken as a form of 'smart drug' or nootropic, due to the role that the neurotransmitter acetylcholine plays in various cognition systems within the brain. Choline is a chemical precursor or "building block" needed to produce the neurotransmitter acetylcholine, and research suggests that memory, intelligence and mood are mediated at least in part by acetylcholine metabolism in the brain. In study on rats, a correlation was shown between choline intake during pregnancy and mental task performance of the offspring; however, the same correlation has not been shown in humans. However, this human study admits "Women in the current study consumed their usual diets. They were not eating choline-enriched diets and were not receiving choline supplementation. Therefore, our results indicate that choline concentrations in a physiologic range observed among women consuming a regular diet during pregnancy are not related to IQ in their offspring. We cannot rule out the possibility that choline supplementation could have an IQ effect."

The compound's quaternary amine renders it lipid-insoluble which might suggest it would be unable to cross the blood-brain barrier. However, despite choline's lipid insolubility, a choline transporter exists that allows transport across the blood-brain barrier. The efficacy of these supplements in enhancing cognitive abilities is a topic of continuing debate.

Acetylcholine precursors such as choline and Galantamine may help improve memory and fight the symptoms of Alzheimer's Disease (AD).

The Food and Drug Administration (FDA) requires that infant formula not made from cow's milk be supplemented with choline.

L-Alpha Glycerylphosphorylcholine (Alpha GPC, choline alfoscerate) is a natural choline compound found in the brain and in milk. It is also a parasympathomimetic acetylcholine precursor which may have potential for the treatment of Alzheimer's disease and is used as a nootropic dietary supplement to enhance memory and cognition.

Alpha GPC rapidly delivers choline to the brain across the blood-brain barrier and is a biosynthetic precursor of the acetylcholine neurotransmitter. Alpha GPC is derived from highly purified soy lecithin. (End of extract from Wikipedia, with acknowledgement.)
http://en.wikipedia.org/wiki/Galantamine
http://creativecommons.org/licenses/by-sa/3.0/
-=-=-=-=-=-=-=-=-=-=-=-=-

(3) Magnetic Field methods:

Persinger [46] developed a method using low intensity magnetic fields (1/50 of the Earth's magnetic field, but alternating at low frequency), to induce a dissociation effect (i.e. that another person is present -- but which is probably a higher aspect of one's self which is normally not perceived). It seems likely that the magnetic field stimulates/opens the Eyebrow Chakra ("third eye") and/or the Crown Chakra (see Fig. 1.5, 1.7 & 1.8 above). The effect is reported by some as being in the presence of God or an exalted being [46]. A book "The Third Man" [47] describes this unusual effect, which is also encountered by mountaineers under extreme conditions, including anoxia (see above). The method involves placing simple telephone pick-up coils (normally used to record phone messages by sticking onto a land-line phone earpiece) on the temples of the head [48] and switching audio signals on and off, using standard audio card outputs from a computer (can be standard PCI audio cards or USB sound devices which are available as standard alternatives to PCI audio cards for laptop computers).   This appears to be a very easy method (but not yet tested by the author).

Very high current methods are the other extreme, producing savant-like effects [49] (not tested by the author).

(4) Other methods:

Appolonius of Tyana (see in Chapter 4) wrapped wool around himself to obtain out-of-the-body experiences (OBEs), which possibly relates to one of the flotation tank properties of making the environment temperature equal to the skin temperature (see Chapter 7). (Note: Wool should normally be avoided.)   It is reported that wrapping oneself up completely like an Eygptian mummy also causes out-of-the-body experiences, possibly for the same reason, but the main property of flotation tanks, the removal of the sense of touch (by floating in water) is missing (or inhibited?). It may be that the custom of mummification is a badly remembered procedure from very ancient times in Egypt (see Chapter 4), where it may have been used by living people to obtain OBEs.

Three safety points:

(1) Yram [53] mentions that when she was developing out-of-the-body experiences (astral projection), a chasm opened up at her feet and she jumped violently aside to avoid falling into it, and was nearly hit by a car. In fact, there was no physical chasm and it was an illusion or astral construct generated by her subconscious!

(2) Lucid dream (astral) characters can become impossible to distinguish from actual physical ones!
This has been used as a definition of insanity, but it is an ignorant definition, as the people involved did not understand what was happening: -- an "astral" entity or illusion is just as "real" in its 4-D dream world as a physical entity is in our 3-D world!
Such entities can arise during experiences in a flotation tank, or when lucid dreaming, or by taking narcotic drugs.

(3) Safety: Only the upper 4 chakras are desirable to be activated:

A highly recommended book by monks of the RamaKrishna Order [44b] says the meditator is only "safe" after the first 3 chakras have been passed, during the rising of Kundalini.

The first 3 chakras should not be opened, as one of them gives a view of Hell -- a Dante's Inferno.

So when raising Kundalini as described in these books, it is like being in a lift (elevator) but don't open the doors until the heart Chakra is reached.

Breathing exercises are described in classic Buddhist and Sufi texts. There is also a book called "<u>Christian</u> Yoga" [51], published by Burns & Oates, official publishers to the <u>Holy See</u>, which has the Nihil Obstat and the Imprimatur from the Holy See in the Vatican, which details pranayama (<u>by that</u> Sanskrit name).

REFERENCES for Chapter 1
(most are available from www.amazon.co.uk)

10. Three DVDs of remote viewing: (1) "The Real X-Files", Channel 4 TV (UK), Presenter: J. Schnabel, Director: B. Eagles, Producer: A. Graham, "Wall-to-Wall TV Productions for Channel 4" (interviews with CIA) (1995); (2) "Strange but True", LWTV programme for ITV (UK), Presenter: M. Aspel, Producer & Director: N. Miller (interviews with Dr D. Morehouse – see also refs. 14 & 16 below) (1997); (3) Ingo Swann, Remote Viewing Conference 2006 (Intl Remote Viewing Assoc): DVD available from www.irva.org

11. Remote Viewing Secrets: A Handbook by Joseph McMoneagle (Paperback - July 2000)

12. The Seventh Sense: The Secrets of Remote Viewing as Told by a "Psychic Spy" for the U.S. Military by Lyn Buchanan (Paperback - 1 Dec 2003)

13. Limitless Mind: A Guide to Remote Viewing by Russell Targ (Paperback - 3 Mar 2004)

14. Remote Viewing: The Complete User's Manual for Coordinate Remote Viewing by David A. Morehouse (Hardcover - 8 Jan 2008)

15. Mind Trek: Exploring Consciousness, Time, and Space Through REMOTE VIEWING by Joseph McMoneagle (Paperback - Nov 1995)

16. Psychic Warrior: True Story of the CIA's Paranormal Espionage Programme by David Morehouse (Paperback - 17 April 2000)

17. Remote Viewing: What it is, Who Uses it and How to Do it by Tim Rifat (Paperback - 11 Oct 2001)

18. Memoirs of a Psychic Spy: The Remarkable Life of US Government Remote Viewer 001 by Joseph McMoneagle (Paperback - 31 July 2006)
19. Susan MacWilliam: Remote Viewing by Karen Downey, Ciaran Carson, Brian Dillon, and Slavka Sverakova (Paperback - 1 May 2009)
20. Remote Viewing: v. 5: The ESP Series by Tom Stevens (Paperback - 4 Mar 2009)
21. Remote Viewing Training Course by David Morehouse (Audio CD - Sep 2004)
22. Remote Perceptions: Out-of-body Experiences, Remote Viewing and Other Normal Abilities by Angela Thompson Smith (Paperback - Oct 1998)
23. Remote Viewing: History and Science of Psychic Warfare and Spying by Tim Rifat (Hardcover - 27 July 1999)
24. Remote Viewing and Sensing for Managers: How to Use Military Psiops for a Competitive Edge by Tim Rifat (Paperback - 17 July 2003)
25. Beyond Einstein's Horizon: Science, Remote Viewing and ESP by Ken Renshaw (Paperback - May 2009)
26. Remote Viewing: The Science and Theory of Nonphysical Perception by Courtney Brown (Hardcover - 28 Mar 2005)
27. Remote Viewing: A Theoretical Investigation of the State of the Art by Marilyn Isabelle Schmidt (Paperback - 15 May 2007)
28. Tracks in the Psychic Wilderness: An Exploration of Remote Viewing, ESP, Precognitive Dreaming and Synchronicity by Dale E. Graff (Paperback - 4 Jun 1998)
29. Remote Viewing - das Lehrbuch 1.: Technik des Hellsehens. Teil 1: Stufe 1-3 by Manfred Jelinski (Paperback - 31 Dec 2001)
30. Geheimnisse des Remote Viewing.: Auf der Spur der Matrix by Frank Köstler (Paperback - 30 Sep 2002)
31. Schritte in die Zukunft: Remote Viewing und die Gesetze der Veränderung by Manfred Jelinski (Paperback - 31 Dec 2002)
32. Remote Viewing - das Lehrbuch Teil 2: Technik des Hellsehens Teil 2: Stufe 4+5 by Manfred Jelinski (Perfect Paperback - Jan 2008)
33. Multidimensional Mind: Remote Viewing & the Evolution of Intelligence by Jean Millay (Paperback - 31 July 1999)
34. Remote Viewing: The Science and Theory of Nonphysical Perception by Courtney Brown (Paperback - 28 Mar 2005)
35. Cosmic Explorers: Scientific Remote Viewing, Extraterrestrials and a Message for Mankind by Courtenay Brown (Hardcover - Jun 1999)
36. Tracks in the Psychic Wilderness: An Exploration of Remote Viewing, ESP, Precognitive Dreaming and Synchronicity by Edgar Mitchell and Dale E. Graff (Hardcover - Oct 1998)
37. Time Traveller/Remote Viewing by Barrie Konicov (Audio Cassette - Nov 1985)
38. Remote Viewing: Invented Worlds in Recent Painting and Drawing by E. Sussman, Caroline A. Jones, and Katy Siegel (Hardcover - 28 Oct 2005)
39. J. Schnabel, "Remote Viewers".
40. C.W. Leadbeater, "The Chakras", publ by TPH, Adyar, India.
41. Targ et al, Nature $\underline{251}$, 602 (1974), et passim.
42. Patanjali, "Yoga Aphorisms", published about 400 BC.

43. "The Breathwork Experience", by K. Taylor, publ by Hanford Mead, Santa Cruz (1994)   (**Holotropic breathing**, overbreathing).
44. "Exhale", by G. Minett, publ by Floris Books, Edinburgh (2004) (**Holotropic breathing**, overbreathing).   Note:
For **Pranayama** breathing exercises, see:
44b. Swami Vivekananda, "Raja Yoga", publ Bharatiya Kala Prakashan, or others (same text) (first publ about 1880).   Note**:** does not warn against excessive use of pranayama (breath control).   And:
Monks of the RamaKrishna Order, "Meditation", publ RamaKrishna Vedanta Centre, London (1972). Recommended.
45. See refs. in book by M.G. Hocking, "Exploring the Sub-conscious using New Technology", publ CMC Ltd (1993).   Available from: www.4-D.org.uk/Books
46. M.A. Persinger, J. Neuropsychiatry Clin. Neurosci, 13: 515-524 (2001) full text free on line from:
   http://neuro.psychiatryonline.org/cgi/content/full/13/4/515
47. John Geiger, "The Third Man Factor: True Stories of Survival in Extreme Environments".
48. See website: www.shaktitechnology.com
49. Reza Jalinous, "A Guide to Magnetic Stimulation":   available from www.magstim.com   An excellent review article.   See also "Brainsight" ™
If not accessible, contact: www.4-D.org.uk/Books
50. ***STOP PRESS:***   A simpler method [Ms P. Whelan,May 2010.], needing no equipment, is to look at one's hands very frequently throughout the day and try to make them disappear, as a test of whether one is awake or dreaming. If this simple test is made habitual by being done very frequently during the day, it will automatically be continued during dreams and then one's hands will disappear and this will trigger a realisation that one is dreaming, and at that moment the dream should become lucid (self-consciousness will occur).   The dream can then be consciously directed.   Another similar daytime exercise is to continually ask if one is dreaming or not (look away and back again and see if the scene has changed -- if it has, you are dreaming!).
Other such trigger methods are in the books on lucid dreaming by S. LaBerge (Lucidity Institute).   See also some late added references on page 32. ***
51. C.H. Dechanet, "Christian Yoga", publ by Burns & Oates (1960); & others.
52. G.K. Khalsa et al, "Kundalini Rising", publ. Sounds True Inc, Boulder, CO (2009).   See also:
G.L. Paulson, "Kundalini & the Chakras", publ. Llewelyn, Woodbury Publs, MN
53. Yram, "Practical Astral Projection", publ. S. Weiser, New York (1979).
54. Joan Grant, "Winged Pharaoh", Ariel Press, Columbus, OH. **Recommended**

***A highly recommended book on lucid dreaming: page 32, footnote. See also ref. 14 on p. 241 for a free download of important details too late to include in the present book.**

-=-=-=-=-=-=-=-=-

NOTE: References are listed at the end of each chapter.
At the end of the book is another list, of general references of interest.

CHAPTER 2

Theology

- **New evidence for the existence of God**
- **Karma: "As a man sows, so shall he reap"**
- **Re-birth**

The mills of God grind slowly,
But they grind exceeding small.
                    Empiricus (160-210 AD)

Introduction

Theology:

The purpose of life is a mystery to many people. I hope to show here that it is not merely *"the pursuit of happiness"*, nor to *"live long and prosper"*.

For many hundreds of thousands of years human individuals have developed from the primitive to the advanced, but with increasingly divergent results: some have achieved much advancement while others remain quite primitive, e.g. compare philanthropists with murderers.   This difference is not explainable by the nature **vs** nurture argument [i.e. an individual's innate or

born-with character (nature) **_vs_** one's upbringing (nurture)]. There are too many exceptions to such a rule:
people born into bad family backgrounds (bad nurture) who turn out to be very advanced, compared with people born into good backgrounds who can turn out to be very wicked (bad nature).

The real purpose of life is to become an "advanced" individual (developing intellect, wisdom, will, compassion, charity, non-attachment to the things of this world, overcoming anger, etc) and to develop <u>awareness</u> of higher regions of consciousness and to develop the ability to enter those states. This will lead to "happiness", but mere "pursuit of happiness" is not the way to describe the process and is quite obscure and misleading.
At the present time, the emphasis is on education –- developing the mind, an achievement which is then incorporated into the soul, thus becoming a <u>permanent</u> part of that person's future "nature". Unbalanced development often occurs: e.g. if intellect develops without wisdom and compassion, or if malice, anger, jealousy, cruelty etc remain, then an unpleasant delay in progress occurs.

Two questions arise**:** how is "advancement" achieved and what are the higher regions of consciousness just mentioned? The latter are not even recognised or mentioned by many people in the modern era, but they <u>are</u> described in all religions.

It is fashionable these days to discount any "higher conscious-ness", even though everyone goes into a state of higher consciousness -- as dreams when asleep! Most people do not believe the scenes therein have any reality but just dismiss them as only imaginary. I will attempt to show below that we really <u>are</u> composed of the *"stuff that dreams are made of " (Shakespeare):* i.e. the atoms & molecules that compose our bodies contain smaller particles (called elementary particles, in physics), which dreams <u>are</u> made of, in a higher dimension of space.

This removes the popular notion that science and religion are separated by a non-traversable chasm: If both science and religion can <u>both</u> be shown to be descriptions of aggregations of the <u>same</u> elementary particles, in different dimensions of space, then there is then <u>no</u> chasm between them. The difference between the two then becomes only the difficulty of observation of the matter composing the world of dreams. This will then relate ordinary consciousness to higher consciousness.

Theology: A new proof of the existence of a non-physical soul, and thereby of the very probable existence of God:

The conditional proof below takes this subject away from an unprovable abstract notion, to a definite plausible contention.

For the reasons already mentioned, it is fashionable in science to ignore theology, but a scientific *Theory of Everything* requires it to be considered.

Many people, already have deeply embedded views about life and will find this paragraph very strange. But if pre-existing views can be set aside, the following logic could replace them: About 1 person in a million report detailed memories of past lives, and as there are about 7 billion people on Earth this means a lot of people are reporting such memories[**]. This (if true) is probably a direct proof of the existence of God, because:

(a) only an outside intelligence ("God") could cause someone to be re-born into an unrelated body, perhaps in another continent and another race, decades or centuries later, and:
(b) why otherwise would the soul exist as an entity, independent of the (destroyed) body, for decades or centuries?

Rebirth is very unpalatable to the Western mind, as we have been brought up to think that this is the only life we get, and the thought of having to go through the whole thing again is very unwelcome to many people. So there is an automatic (but unfounded) subjective objection to re-birth in the West. But it is no good rejecting a theory just because we do not like it personally. Rebirth is a strong indication that God exists, but His existence does not depend on it.

Under hypnosis, the "1 person in a million" ratio mentioned above, for people with memories of past lives, is increased by a factor of more than 10,000. So more than 1 person in 100 recall detailed memories of past lives, under hypnosis. An objection to hypnosis may be that imagination can play a part in the results obtained, but in many of these cases verifiable historical facts are produced, which negates that argument in those cases. These verifiable facts were previously unknown to the person producing them [1-15].

[**]See references [1-15] & many others. For a quick web reference see [17].

For example, a series of programmes in August 2007 (and earlier) by Phillip Schofield [6] on (UK) ITV1, hypnotically regressed about 25 people to recover verifiable memories of previous lives, and at least one case in particular proved (beyond reasonable doubt) that we have lived before.   This is studiously ignored in debates, but should surely be brought to the attention of all.   It is very unfortunate that copyright law prevents video copies of these programmes from being made publicly available.   [Ask the author for more information on these, at: www.4-D.org.uk/books]

There are many books written by people reporting memories of past lives and very many investigations have been done and have discovered facts which were not known before the investigation. This represents proof and cannot be dismissed [1-15].

Two objections to "past lives" memories are considered below:

It has been suggested that apparent memories of past lives could be a "DNA memory" or "cellular memory", passed on to us by our parents and their line, but this cannot explain memories of past lives in distant regions (i.e. not related to one's genealogical ancestry) and in other countries embodied as members of other unrelated races.   An inherited "cellular memory" would be expected to give memories of the lives of one's parents and grandparents, but the author is not aware of any accounts giving evidence for this.

About a year ago, a UK TV programme described cases where people had sudden character changes after receiving heart transplants. Memories which had belonged to the heart donor, spontaneously and unexpectedly appeared in the mind of the recipient!   E.g. a person with little interest in outdoor activities became a keen mountaineer after receiving a heart transplant [the paper in which this is reported is given here verbatim in Appendix 1]. Subsequent investigation revealed that the heart <u>donor</u> was a keen mountaineer. This suggests the transfer of a soul; it may appear to some to support a "cellular memory" theory but the evidence seen by the author is that <u>only</u> heart transplants can cause the memories and character of the donor to be transferred to the recipient.   Transplants of other organs do not have this effect. The following 2 points are relevant**:**

(i) Heart tissue contains some of the same type of cells as found in brain tissue, so the heart can be regarded as some kind of extension of the brain.
(ii) Other organs (e.g. a kidney) have no such cell similarity to brain cells.

If this is correct, it is evidence that mere cell transfer cannot be the cause of memory and character transfer, and the only other explanation is that it is due to transfer of what is usually called the "soul", which is attached specifically only to brain-type cells, which exist only in the brain and heart. It could be argued that "cellular memory" could still apply but is for some unknown reason only carried by brain and heart cells, but the many cases cited in Appendix 1 do not support this view.
Also, the amount of data storage space available in cells would be insufficient to store the memories and especially the consciousness reported in Appendix 1.

In relation to an inherited "cellular memory" in the above example, it is relevant that the heart recipient's parents and grandparents had no mountaineering interest, and even if this were not so, it cannot be explained why the recipient should suddenly develop a particular interest just _after_ a heart transplant (which was not present _before_ as it should have been if it were a cellular memory inherited at birth from the recipients own parents).

The _cellular memory_ theory requires that there is no involvement of any soul (and that the soul may not even exist), and that memories and inherited traits are held in brain-type cells.
The _soul memory_ theory posits that memory and consciousness reside in the soul, and brain-type cells are the link between the soul and the body.

Sometimes musically gifted parents have musically gifted children; supporters of the cellular memory theory would require that _both_ traits _and_ memories would pass on to a child, from the parents. But this clearly does not happen: no-one has their parents' _memories_ (!), and only parents' _traits_ are ever passed on (e.g. musical ability sometimes). No reason is evident from the cellular memory theory, for the transmission of traits _only_, nor of memory _only_, from parents to a child.   A soul transfer theory, does not require the parents to pass on anything at all, if an incoming already-musical soul born to them is simply attracted to them by their interest in music. So such abilities cannot be used to prove cellular memory. But primitive <u>instincts</u> can be passed on from parents by cellular memory.

The soul (assuming we have one) is said to be linked to our body via our brain and heart. So if a (living) heart is transplanted into someone else, the soul of the donor could be expected to accompany it and ensoul the new recipient, along with his/her existing soul. It seems to allow an extension of life for the transplant donor. So for transplants, explanations could be that part or all of the 'soul' of the donating person may remain with the transplanted organ.

It has been suggested in reference [16] that consciousness is 'downloaded' into each cell within the body and the transplanted organ thereby continues to download some aspect of the consciousness of the donor of that organ. But, as mentioned above, only heart transplants have been reported to have this effect, which supports a specific property of heart tissue, mentioned in (i) above, rather than a general property of cells. In the mountaineering example above, no mountaineering interest was inherited from his parents cells (as a supposed "cellular memory"), but such an interest did appear when heart cells from a mountaineer donor were implanted.

Please see Appendix 1 for compelling evidence for the continuing existence of a soul: Ten cases of heart transplant effects are reported in: "Changes in Heart Transplant Recipients that Parallel the Personalities of their Donors", by P. Pearsall, G.E.R. Schwartz, L.G. S. Russek in: *Journal of Near-Death Studies*, vol. 20, no. 3, Spring 2002. The publishers have kindly agreed to the full text being reprinted here in the present book as **Appendix 1**.

{It is also available (full text) at**:**
http://anti-matters.org/ojs/index.php/antimatters/issue/view/1/showToc
This information is gratefully acknowledged from Ulrich Mohrhoff, Sri Aurobindo International Centre of Education, Pondicherry, and from Deena O'Brien.}

On the principle of Occam's Razor (that the simplest explanation is likely to be correct: "if it looks like a duck, it is probably a duck"), the soul transfer explanation seems the most likely, especially when combined with other evidence (e.g. see Chapter 4).

Another suggested direct proof of God is from near-death experiences (NDEs) of people with heart failure who are clinically dead but are then resuscitated, and some describe remarkable experiences. However, strictly speaking, NDEs are subjective and are thus anecdotal and so can only prove something to the person involved. "Anecdotal" does not equate to "untrue", of course!

"Death is a far country, from which no-one returns"
                                                          (Shakespeare, adapted)**:**
This quotation was true in Shakespeare's time, but modern medical advances now allow people who have died from a sudden heart attack to be resuscitated in some cases.

Some sceptics say these visions may be created within the dying brain, but they studiously ignore many accounts which accurately described events occurring in the corridor outside the hospital ward, etc, to where the "dead" persons say they were able to travel "out of their body" which was lying clinically dead at the time, in the ward. These events could not possibly have been seen by the near-dead subject and thus could not have been created by or originated in his brain, which makes the "residual brain activity" explanation of consciousness untenable. They also report floating above their body lying on the bed and seeing it below them, very clearly, i.e. a clear consciousness is reported during NDEs [7-11, 15].

Other evidence for having lived before is child prodigies in fields not always shared by their parents & grandparents.

Karma ("As a man sows, so shall he reap"):

If there is a "God", He/She must be just. An unjust God is a contradiction in terms. The word God is linked to the word "good" (e.g. "God speed", and "Goodbye" which is a contraction of "God be with ye").   Logically, disasters which befall people cannot be the basis for asking why God allows disasters to injure or kill innocent people/children, which would clearly be unjust. If God exists and is just, then the only explanation remaining when all others have been discounted against these two criteria, is that the people on whom these disasters have fallen must have been responsible in their own previous lives for causing similar events to befall others. So almost no-one is an 'innocent' person, not even a

child. If this is felt to be unacceptable, then the only logical conclusion is that God does not exist (as an unjust God would be a contradiction in terms).
But if there is no God, then who causes us to be re-born, as mentioned above?
That would require the evidence of memories of past lives to be dismissed as lies. So a proof of the existence of God comes down to whether we believe or reject such memories of past lives.

Summary: The above information indicates that an external directing intelligence (traditionally called God) exists and that we have had past lives and will have future lives -- not just this one life that we are now in. There are very important implications, discussed below.

For example, a dangerously inane piece of advice placed by atheists appeared on some London buses in 2008/9: It said:
"*Enjoy yourself – There is probably no God*".
But the "*There is probably no God*" bit could be taken by some people as a licence to do anything, with impunity!

There are many diverse implications. E.g. for someone considering suicide, if suicide were done then the person is likely to be quickly re-born, but (discussed below) into worse circumstances than before, so suicide solves nothing and actually makes things worse -- there is no escape.

Even if this is not believed, the safest thing to do (on the "Precautionary Principle") is to avoid suicide, in case the above information is true! It cannot be safely believed that the many reports worldwide of previous lives' memories by many people are all lies. This would be a very hazardous assumption and it would be very unwise to base one's life actions on a belief that rebirth does not occur, nor to be an atheist on this belief. This aspect is discussed below (in a further section on Karma).

So, if only as a precaution (in case it is true), we need to change our basic way of thinking to consider what we want for our next lives after the present one!
We can't run our life "like there's no tomorrow" !

The above is not meant as a "decree" that karma must be accepted – there is freedom to reject, but consequences then occur.

Re-incarnation and Christianity:

Only at the Council of Constantinople in the 6th Century AD, the doctrine of re-incarnation was declared a heresy, but before that it was a part of Christianity. The Bible reports in several places that Christ gave (now unknown) "secret teachings" to his disciples.

The original Christian Bible gospels were "selected" from a larger group of gospels, many decades after Christ's time, and some gospels were omitted due to the personal views of those who made the selections. Some of these other gospels are now available. See the Foreword of this book.

The original Christian doctrines included rebirth:
The following quotations are from the King James Authorised version of the Bible.
Jesus asked his disciples who men thought He is and their reply is (Matthew 16: 13-14): *"And they said, some say that thou art John the Baptist; some Elias; and others Jeremias, or one of the prophets"*. Matthew 17: 10-13 is fully quoted on page 11 of the present book, so, summarising, it says: *"Jesus answered and said unto them, Elias truly shall first come, and restore all things. <u>But I say unto you, that Elias is come already, and they knew him not, but have done unto him whatsoever they listed.</u>* ((Herod had chopped his head off.)) *Likewise shall also the Son of Man suffer of them. <u>Then the disciples understood that He spake unto them of John the Baptist.</u>"* ((my underlining)) He did not rebuke them for suggesting that John might be "one of the prophets".
The statement is repeated by Mark (9: 13): *"But I say unto you, that Elias is indeed come, and they have done unto him whatsoever they listed, as it is written of him."*

The underlined text above shows that <u>Christ said Elias (Elijah) who lived centuries before, was reborn as John the Baptist, and that the idea of rebirth was taught by Christ and was familiar to the Jewish people at the time</u>. Some present-day Christians can't say Christ was wrong, but anxious to avoid the idea of re-birth, say that Elias never died and so could not have been reborn as John! But this neglects the <u>whole</u> of Chapter 1 of Luke's Gospel which clearly says John was a <u>baby</u> boy born to Zacharias and Elisabeth!

The end-product and purpose of successive re-births is explained as follows: From the Bible, Revelation 3: 12: *"For them that*

overcometh, I will make them pillars in the Temple of the Lord and they shall go _no more out_." ((my underlining – no more rebirths))
From the Old and the New Testament, and a Hermetic principle:
"_The stone which the builders rejected, I will make the Head of the Corner._" (Psalms, Mark, Luke, Acts, 1 Peter)   And:
"_...... that we may present every man perfect in Christ...._"
(Colossians 1: 28).
It should be obvious that the degree of perfection needed for the "eternal life" of Christianity cannot possibly be achieved in just one life.

Re-incarnation and Judaism:

A Jewish scriptural example is:
In the 3rd century, from the Mishnah (Talmud), Rabbi Tafan said:
"_It is not expected that you will finish the task, but you are not free to desist therefrom_".
This means that we are not expected to achieve "Salvation" or perfection in just one life (which should be obvious to anyone), but we are not free to stop trying, e.g. by committing suicide.
If we kill ourselves, we will just be re-born again!   So there is no point in suicide.
See also the Jewish Kabalah quoted just below.

Re-incarnation in Buddhism & Hinduism:

Eastern religions teach re-birth as a quite normal event. Buddhism has a vast set of scriptures, all studiously ignored in the West. Nothing in Buddhism is against Christianity.

Both Buddhism, Hinduism, and the Kabalah (a Jewish text) teach that there is a logical progression of souls through the vegetable and animal kingdoms, ending in man:

"God is unconscious in the mineral, Sleeps in the plant,
Stirs in the animal, And wakes in man."
(Kabalah).

This is a non-return underline{upward} progression. You will not be reborn as an animal or insect -- these negative possibilities are put about by those who wish to try to discredit Buddhism and the idea of re-birth. Also, regarding Hinduism -- Hinduism is monotheistic and the multiplicity is only of the many aspects of one God; this also is distorted by some in the West, to discredit Hinduism.

Hinduism is a complex of beliefs and customs comprising the main religion of India.

Buddha and Krishna:

The remarkable character of the Buddha is shown in the following verse:
"To those who revile me, I will give the protection of my most unconditional love.
And the greater their hostility, the greater shall be the measure of my love".

The following story is told of the Buddha:

A woman brought gifts to Him, two gourds.
As she approached, He said to her, "Drop it."
She dropped one gourd, and being fragile it was destroyed by the fall.
As she approached nearer, He said again to her, "Drop it."
She dropped the other gourd and it was also destroyed.
As she approached further, He said to her, "Drop it."
And in that moment, she achieved Salvation, the end of her lives on Earth.

-=-=-=-=-

"However men approach Me, even so do I welcome them,
For the path men take from every side is Mine."
      Bhagavad Gita 4: 11 (Translated by Dr A. Besant).

Our relation to animals:

Here is a Christian-background verse:

Kill not moth nor butterfly, For Judgement Day draws nigh.
          William Blake (an 18th Century Christian mystic).

Buddhism contains nothing which is against Christianity and it is a pity that it is neglected in "the West".
Here is the "equivalent" Buddhist verse:

Kill not, for pity's sake,
And lest ye harm the meanest thing
Upon its upward path.   (From: "The Light of Asia", by Edwin Arnold).

This gives a different angle and includes the Buddhist feature of compassion and also gives the notion of creatures being on an upward path, usually absent from Christianity, but hinted at in the Jewish Kabalah (see above).

Karma, a further discussion:

Every religion has the same precept that, "*As a man sows, so shall he reap* ". Eastern religions teach that this applies from one life to the next, called the Law of Karma. Action and reaction are equal and opposite!

Thus deeds done in one life will have their effects either in that life or in a future life. E.g. if someone kills someone, he himself will be killed, either later in the same life or in a future life.

Modern Christianity and Judaism have lost this "carry-over to one's next life" interpretation but they did have it originally: See the two Forewords, and the main text just above, for more details on the Council of Constantinople etc. Sins cannot be paid for after death, in purgatory: an already dead man cannot *"die by the sword"*!

The "Law of Karma" of Buddhism & Hinduism is misunderstood in the West. Its meaning was well put by Christ in six different verses, e.g. "*He who lives by the sword shall die by the sword* ". The other New Testament bible verses are given later below.

Here are some Old Testament verses:
"*Whoso sheddeth a man's blood, by man shall his blood be shed* " (Genesis 9: 6), and,
"*An eye for an eye and a tooth for a tooth*". But what most people fail to realise is that this **must** be taken along with:
"*Vengeance is Mine, saith the Lord* " (Old Testament, my underlining).
So any attempt for us to apply the first line above, ourselves, is very wrong, as it will only create bad karma for the doer (us), as it goes against the warning given in the last *italics* quotation above!
The timing of Karma can be in this life or in future lives.
Karma is often carried over from actions done in past lives:
The results of our actions (karma), which can be good, or bad, are felt:
> "*.... tomorrow, or, in a thousand years,
> For things done, or undone.*"
>   (a Buddhist verse from "The Light of Asia" by Edwin Arnold).

Hence: "*The mills of God grind slowly,
But they grind exceeding small* " (Empiricus, 160-210 AD).

If people (in general) do not realise that the Law of Karma will apply to their future lives, they can make <u>very</u> wrong decisions and take seriously wrong actions:

A terrible example of a wrong action taken, from modern times, is abortion (which often involves crushing the skull of an unborn child to death with no anaesthesia for the foetus). A woman doing this, and the surgeon, will be themselves aborted in their next life, due to the Law of Karma. It is a tragedy that <u>it never even crosses their minds</u> that this will inescapably happen to them. But if they had been taught rebirth this would surely have been different? It is doubtful that any would then proceed with an abortion.

"*Man clad in a little brief authority, doth things that do make the Angels weep*" (Shakespeare, abbreviated).

There have been 200,000 abortions per year, in the UK, over the past few years (e.g. 2006–2009) which means 1 in every 5 pregnancies ends with an abortion! In 2010, 89 UK teenagers each had three abortions, i.e. they are using it as a contraceptive method, which is a very disturbing statistic in relation to its karma.

If an unborn child is aborted before it is even born, this obviously cannot be due to any bad action yet by that unborn child in this world, and so it must be due to its actions in a previous life. So that individual must have caused an abortion for someone previously. This pre-supposes <u>only</u> that God exists, and that God is just.

If people realised that they will be re-born, their decisions would surely be very different. Another obvious example is global warming -- our generation which is causing it will be the ones who will reap its consequences, not some other people! This should concentrate our minds on preventing it.

Hypnotic regression can take people's consciousness back to the time before birth, so it should not be argued that a foetus only becomes human at the moment of birth. The moment of conception is more relevant.

The workings of karma:

The workings of karma can be seen from the events in Chapter 4.

CHAPTER 2   Theology, Karma, Re-birth

A curious indication of the workings of karma may be inferred from a comparison of Presidents Lincoln and Kennedy, if either:
(a) the same wrong actions are repeated by an individual in successive lives (i.e. lessons were not learned), <u>or,</u>
(b) there are several wrong actions to be atoned for, requiring a sequence of similar karmic events (because one can only be killed once per life!):

President <u>Lincoln</u> was shot in <u>Ford's</u> <u>Theatre</u> by a man who then ran to a <u>warehouse</u>, where he was captured.
President <u>Kennedy</u> was shot in a <u>Ford</u> <u>Lincoln</u> <u>Continental</u> car, by a man in a <u>warehouse</u> belonging to <u>Continental</u> <u>Warehouse</u> Co, who then ran to a <u>theatre</u>, where he was captured.
<u>Lincoln</u> had a secretary called <u>Kennedy</u>, who had advised him not to go to the <u>theatre</u>; <u>Kennedy</u> had a secretary called <u>Lincoln</u>, who had advised him not to go to Dallas.

Both were shot on a Friday. Both were shot in the head from behind. Both were shot in the presence of their wives. Both were shot while sitting with another couple. When both were shot, another member of their entourage was injured, but not fatally.

The assassins were born in 1839 and 1939. The assassins were both Southerners with extremist views. Both were known by their first, middle and last names. Both assassins were themselves assassinated before their trials.

The next presidents were both called Johnson and both had been vice-presidents. They were born in 1808 and 1908. Both of these successors were former senators and were Southern Democrats. Both Presidents' coffins were carried on the same caisson.

Lincoln and Kennedy had been elected to Congress in 1846 and in 1946, respectively. They were elected as President in 1860 and in 1960. Both had the legality of their elections contested. Both were directly concerned with black civil rights. Both lost a son while serving as President. Both were killed while serving as President.

Only some of the above could be coincidences! These are some of the well-documented auguries surrounding President Kennedy's death and show (beyond reasonable doubt) that events in Kennedy's life are <u>pre-ordained</u> from prior events a century earlier!

Explanation of human suffering:

Buddhism/Hinduism explain suffering well (see above), but Christianity cannot explain apparent injustices (e.g. suffering/death of children). Another (unique) problem, is Christian belief that sins are cancelled (on repentance) by Christ's suffering crucifixion; but there is a time-slip involved, as His crucifixion pre-dated our sins.

Remote viewing finds many (but not all) events are pre-ordained. Actual time-slip, implying pre-ordination of events, is evidenced by: the irrefutable Lincoln-Kennedy auguries where many events in Kennedy's life were very accurately predicted from Lincoln's life (see page 59), early Nostradamus predictions, Peter's famous "cock crowing / deny thrice" (Matthew 26 :34) and many other Biblical prophesies (see Chapter 5), Chernobl (see Chapter 5), etc.

It is thus likely that some of our own present sins were also pre-ordained (backward-linked), which logically implies that if we do not ask Christ to suffer for our sins, then His suffering would have been reduced by that amount! It is dishonourable (once the backward link is realised) to load our sins onto someone else to suffer for them (especially not onto Christ, who appeared voluntarily, to help us)! We should not ask what He can suffer for us, but what we can do for Christ. It would be more "Christian", not less, to accept full responsibility ourselves for our own sins & not to seek to transfer their karmic consequence to Christ, i.e. repentance but not transfer.

In contrast to the above Christian notion of the cancellation of sins, the Bible (New Testament) makes clear in 8 places (so it must be important) the actual effect of committing sins:

"*As a man sows, so shall he reap*", (Galatians 6: 7).
"*He who lives by the sword shall die by the sword*"
"*He who sows the wind shall reap the whirlwind*"
"*The wrongdoer shall be delivered to the jailer and shall not be let out therefrom until the uttermost farthing has been paid*";
"*For with the same measure that ye mete withal it shall be measured to you again*", (Luke 6: 38).
"*For verily I say unto you, till Heaven and Earth pass, one jot or one tittle shall in no wise pass from the law, till all be fulfilled*", (Matthew 5: 18). (("the law" = the law of karma))
"*And it is easier for Heaven and Earth to pass, than one tittle of the law to fail*", (Luke 16: 17).

"Judge not that ye be not judged. For with what judgement ye judge, ye shall be judged: and with what measure ye mete, it shall be measured unto you again", (Matthew 7: 1-2).
"Therefore all things whatsoever ye would that men should do to you, do ye even so to them: for this is the law and the prophets", (Matthew 7: 12).
"Whoso sheddeth man's blood, by man shall his blood be shed", (Old Testament - Genesis 9: 6).

These and "*the law*" referred to, are called (or include) "karma" by Buddhism & Hinduism (and perhaps "kismet" by Islam), but with the important property there that they can act across successive re-incarnations. Without this addition, they are obviously untrue, as there are cases of mass-murderers who have committed crimes against humanity, etc, and yet who have died peacefully at an old age! By ignoring re-birth, Modern Christianity has thereby reduced the above quotations to obvious untruth!

This has emptied churches because many people then conclude (from modern Christianity) that God is unjust because He "allows" apparently-unjust suffering. But the cause of the suffering cannot be seen in the present life, but lies hidden in a previous life. Thus modern Christianity has no convincing reply to suffering. People assume that God is unjust, instead of assuming "God is just" as an inviolable truth, which means logically that there must be another explanation that the suffering is just!! This unfortunately can lead to atheism and then the committing of evil actions in the dangerously mistaken belief that they can do things with impunity. This is a failure of modern Christianity. Unfortunately churches have only got a partial message to deliver (by ignoring that karma applies from one life to the next) and so they are completely unable to explain suffering.

The cause of suffering is thus obscured to us, as we only see its effect, but the cause exists in our unseen past. Each person is thereby his own absolute causer of his own sufferings!

The Purpose of Life:

From the above discussion, this is well described by the quotation:

God is unconscious in the mineral, Sleeps in the vegetable,
Stirs in the animal, And wakes in man. (Kabalah).

So there is an upward progression in which the permanent part of us (call it soul/spirit) slowly grows and improves. When all past karmas are exhausted, the final stage is that the perfected human achieves Salvation or Sainthood. In this state, no more re-births here are needed, and the soul/spirit is free. It will obviously take more than one life to achieve this high state.

From the Christian New Testament:
"*Till we all come ... unto a perfect man, unto the measure of the stature of the fullness of Christ*", (Ephesians 4: 13).
We are "*sown in weakness ... that we may be raised in power*",
(1 Corinthians 15: 43). And, as already mentioned:
"*The stone which the Builders rejected, I will make the Head of the Corner*". (Psalms, and quoted by Christ in the New Testament and also a Hermetic aphorism).

Concluding comments:

Returning to an extreme example given above, which relates to karma, a recent poll (2007) found that most UK people think that a woman should have a right to have an abortion. But there is no mention nor any notion of the dire "karmic" consequences outlined above, which people tragically seem to have no concept of.

Anyone who thinks that there is no God and who thus thinks there is no consequence of having an abortion, needs to be very certain of their belief, because the consequence of being wrong (outlined above) is gravely serious indeed, painfully destroying their own next future life, or even their next several lives.

If the Legal system allows abortions, many people think this somehow makes it "alright". But it is not "simply a medical procedure". Thinking that, just gives a false sense of security.

But as a positive comment: allowing a child to be born, that action will ensure that one receives a normal birth in one's next life/lives!

Chapter 4 (Remote Viewing of historical events) reveals the effects of karma on successive lives & what occurs if people do not learn from them. E.g. The individual known to us now as Julius Caesar was seen in a group 70,000 years ago and several members of that group have now achieved Salvation/Sainthood, but not Caesar, whom <u>recent</u> history records as still doing wicked acts.

E.g. Caesar was kidnapped by pirates in the Aegean Sea, who demanded a ransom, but did not harm him. After the ransom was paid, Caesar raised a fleet, pursued and captured the pirates. Marcus Junctus, the governor of Asia, refused to execute them as Caesar demanded, preferring to sell them as slaves, but Caesar then had them crucified on his own authority, as he had promised while in captivity - a promise the pirates had taken as a joke. As a sign of "leniency", he first had their throats cut - bad karma for him.

Later, Caesar was stabbed 23 times (Ides of March). A slow learner!

BOOKLIST for Chapter 2

1. There are too many books to cite here, but "Winged Pharaoh" by Joan Grant (Ariel Press, USA) is worth reading, written in the 1930s but still in print. And other books by the same author (same publisher).
2. Dr I. Stevenson, "Twenty Cases suggestive of Reincarnation", University Press of Virginia (1974).
3. Dr I. Stevenson, "Evidence for survival from claimed memories of former incarnations".
4. Dr L.D. Weatherhead, "The Case for Reincarnation", City Temple Literary Society, London (1960).
5. The Bloxham Tapes, Hypnotic regression tapes recounting experiences in past lives, BBC TV.
6. A series of programmes in August 2007 (and earlier) by Phillip Schofield [6] on ITV1, hypnotically regressed about 25 people to recover verifiable memories of previous lives, and one case in particular proved (beyond reasonable doubt) that we have lived before.
7. Dr R.A. Moody, "Life After Life", Bantam Books (1977) (a best seller)
8. Dr R.A. Moody, "Reflections on Life After Life", Bantam Books
9. Dr M.S. Sabom, "Recollections of Death", Corgi (1982) (on NDEs)
10. D. Scott Rogo, "The Return from Silence", Aquarian Press, UK (1989) (on NDEs)
11. D. Scott Rogo, "Life after Death", Harper-Collins, Glasgow (1992)
12. R.A. Monroe, "Journeys out of the Body" Anchor Press/Doubleday (1977)
13. R.A. Monroe, "Far Journeys", Doubleday (1985)
14. R.A. Monroe, "Ultimate Journey", Doubleday (1994)
15. DVDs on NDEs: (1) "Glimpses of Death", Presenter: Dr. P. Fenwick, BBC "QED" Series (1987; (2) BBC –2, Producer & Director K. Broome (2002).
There are very many other books on the above topics. There may be some hoaxes, such as "The Third Eye" by T. Lobsang Rampa and "The Search for Bridey Murphy", but these few can not invalidate the many other accounts.
16. B. Lipton, "The Biology of Belief".
17. http://hindustantimes.com/StoryPage/StoryPage.aspx?id=52f06560-672c-4eb5-8997-9be2dfec372b&ParentID=b7153bed-1e25-44cd-b0a7-684a93da3c57&&Headline=%E2%80%98Most+rebirth+claims+are+true%E2%80%99

CHAPTER 3

Science

◆ **Remote Viewing of molecules, atoms & elementary particles**

Small things wax exceeding mighty,
Being cunningly combined;
Furious elephants ((quarks!)) *are fastened*
With a rope of grass-blades, twined ((quark string bonds!)).
 Hitopadesha (350 B.C.)

(This chapter continues on page 65, after the 12 colour pictures)

NOTE FOR THE READER: To interleave each colour picture at its proper place in the book would have more than tripled the cost of this book, so all the 12 colour pictures are grouped here instead. The book page which each picture relates to, is given at the foot of each page.

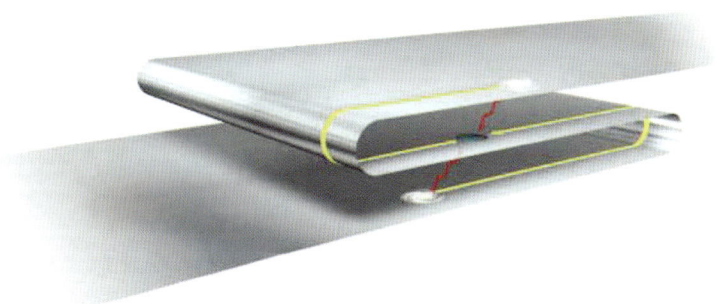

Fig. 1.1. THREE PARALLEL UNIVERSES

A 2-D sheet (membrane) analogy.

CASE 1: If the joins (shown) were not present, the universes would be completely isolated, except for gravity:
Gravity (RED line) can leak out of 3-D space & cross via 4-D space to parallel universes, but light, heat & radio waves etc cannot.

CASE 2: If joins are present (as shown, curved regions), the light etc has to travel billions of light years (YELLOW line) to get from A to B (white zones). Note: the sheets might be only 1 mm apart!!

Figure from: "The Universe's Unseen Dimensions", by N. Arkani-Hamed, S. Dimopoulos & G. Dvali, Scientific American, August 2000, page 62-69.

For page 20

Fig. 1.3. Fairy Bower photograph, by Frances in 1920. Centre: fairy looking to right, head & right arm is above drape. An enlarged section from the left, was tinted (in 1920) to better reveal the left fairy; another fairy is on the right: see also Fig. A8.5 & Fig. A8.6 in this set of colour pictures .
This photograph is considered "un-fakeable", as the wings are transparent and grass blades are seen both in front of and behind them.   See discussion of these Cottingley photographs in Appendix 8.
For a fake Cottingley photograph for comparison, see colour Figure A8.4 !
Copyright:(image): National Media Museum/SSPL.   Copyright: Christine Lynch (2009).

Fig. 1.4. Pharaoh Sesostris I

12th Dynasty, statue 4000 years old

Note the "third eye" or Eyebrow Chakra

For page 25

Fig. 1.7. Crown Chakra

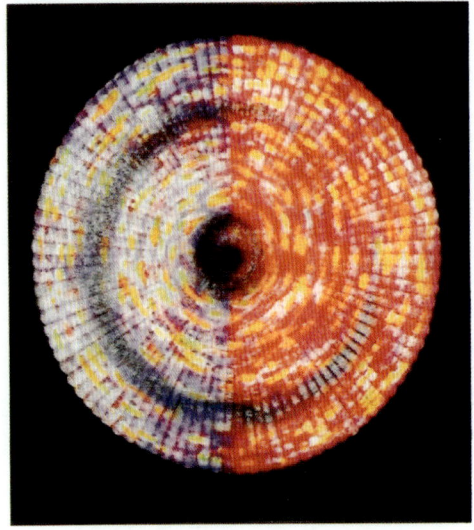

Fig. 1.8. Eyebrow Chakra
Note the division into 2 halves.

For page 28

Fig. 1.9. A blue sky

Fig. 1.10. Same sky at night

For page 31

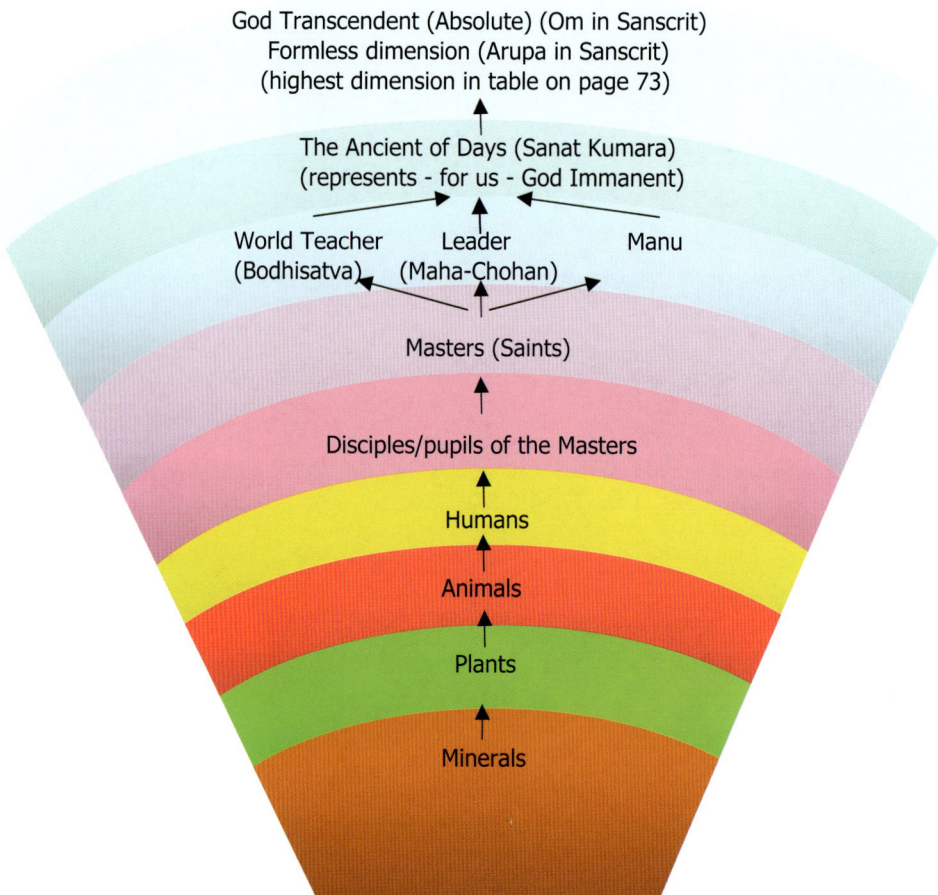

Fig. 4.1.　　　Hierarchy relevant to Table 1 on page 89

This figure is simplified, and not shown are:

(i) The non-human evolution of Arch-Angels, Angels; Elves, Fairies, & other nature spirits

(ii) A parallel 3-D evolution of pre-cursor humans living below our ground level (see page 133-137): Cf. the (fictional) "Middle Earth" of Tolkein's "Lord of the Rings", below our ground level but displaced in 4-D to another parallel 3-D space, which is solid rock in our 3-D space but is not so in their displaced 3-D space!

(iii) The "Silent Watcher"; the various Buddhas.

Notes: "Bodhisatva" means "Enlightened soul" in Sanscrit, but here it means World Teacher.   On this diagram, other Sanscrit words are: Om, Arupa, Sanat Kumara, Manu and Maha-Chohan.

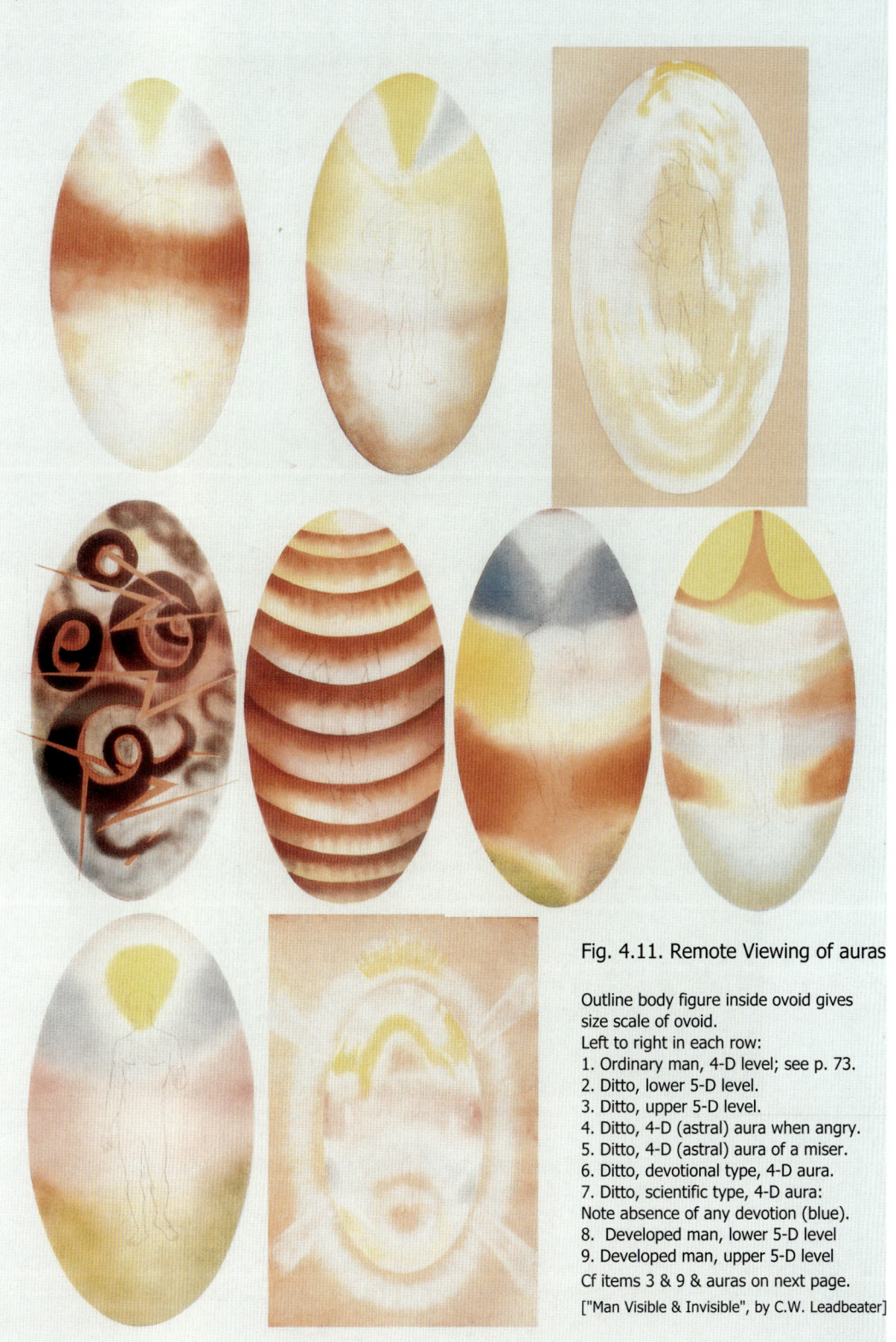

Fig. 4.11. Remote Viewing of auras

Outline body figure inside ovoid gives size scale of ovoid.
Left to right in each row:
1. Ordinary man, 4-D level; see p. 73.
2. Ditto, lower 5-D level.
3. Ditto, upper 5-D level.
4. Ditto, 4-D (astral) aura when angry.
5. Ditto, 4-D (astral) aura of a miser.
6. Ditto, devotional type, 4-D aura.
7. Ditto, scientific type, 4-D aura:
Note absence of any devotion (blue).
8. Developed man, lower 5-D level
9. Developed man, upper 5-D level
Cf items 3 & 9 & auras on next page.

["Man Visible & Invisible", by C.W. Leadbeater]

Fig. 4.12. Auras.
UPPER: Remote Viewing at "upper mental" level (upper 5-D, see table, p. 73), of an individual at stage below "Master" (see Fig. 4.1 above, & Table 1, p. 89). Aura is ~20 metres diameter; cf ratio of figure at centre to outer aura diameter, with that of a developed man (item 9 in Fig. 4.11 overleaf). [Picture from "Man Visible & Invisible" by C.W. Leadbeater)

LOWER: Buddha, at Wesak Valley, Tibet. Aura diameter for Buddha & Christ is about 2 to 3 miles, which attracts crowds from far away. [Picture copyright Heather DuPre]

Fig. 5.3.   A Coptic Icon.

Fig. 5.3 (b).   A Benedictine Icon.

Painting by Fr G.C. de la Mora, OSB, of Tepeyac Abbey, Mexico
(Copyright 2006 (c) by Prince of Peace Abbey. Used with permision)

Cf Fig. 4.3 (page 109) and the caption there.

Fig. A2.1. Admiral Stansfield Turner, Director of CIA (1977-1981) Founder of the CIA Remote Viewing Group

Fig. A8.4. Fake photograph of Frances & fairies, taken in 1917 at Cottingley. The "fairies" are paper cut-outs! See text, p.369.

See Fig. A8.5 below for a real fairy.

Image 10436087 copyright National Media Museum/SSPL: www.scienceandsociety.co.uk

Fig. A8.5. Fairy Bower photograph, taken by Frances in 1920.

This enlarged section was tinted (in 1920) to better reveal one of the fairies. The original, for inspection is colour Figure 1.3.

This picture is considered to be "un-fakeable" (in 1920), as the wings are transparent and grass blades are seen both in front of and behind (through) them.   See text.

Copyright: Christine Lynch (2009).   Appendix 8 ref. [4].

Fig. A8.6. Enlargement of the right side of Fig. 1.3 showing 3 fairies (one large one filling the whole left side above, another whose face is on the right of that one, and a third half-materialised face in between these two).

Copyright: Christine Lynch 2009, Appendix 8, ref.[4].

CHAPTER 3   Remote Viewing of atoms & elementary particles   65

Dr A. Besant   C.W. Leadbeater

Fig. 3.0. Authors of the expert-level Remote Viewing observations of elementary particles, atoms, and molecules first published in **18**95 [1].

Earlier, an organic chemist, Kekule, thought much about the structure of benzene and had a dream, which he relates as follows below. It is not Remote Viewing but an example of an <u>untrained</u> observer, whose dream dramatised his intellectual problem, and solved it. It shows the ability of the mind to create <u>subjective</u> images.

In Kekule's dream, black balls of carbon turned into black imps with forked tails that began racing around the room. Suddenly, each imp grabbed the tail of the one ahead of him, and six formed a whirling circle. One hand of each imp held a tail, and other hand a white handkerchief. They waved this to him as the group whirled by. He said that he then awoke with a start, realising that the imps were acting out the formula for benzene! As his hand grabbed the sketching pencil, the imps changed back to black balls again and the handkerchiefs had changed to hydrogen atoms: "The carbon atoms of benzene form a ring!"

But the *expert level* Remote Viewing observations of Besant & Leadbeater were of a different order, and their rigorous training prevented any subjective images, leaving only an objective item being observed. Importantly, they both had the same Remote Viewing ability, and so checked each other's observations.

BESANT & LEADBEATER's REMOTE VIEWING OBSERVATIONS [1]

There are many speculations about the origin and nature of the Universe, but this one is derived from <u>actual observations</u> made by expert-level Remote Viewing over a century ago, but which have been studiously ignored for many decades because the observations described 10 spatial dimensions, and the proton (H nucleus) as containing 3 sub-atomic particles, which were both regarded as ridiculous by science during most of the last century. But in more recent years, the existence of 10 spatial dimensions has been proposed by particle physicists, who have also found that the proton (H atom nucleus) is composed of 3 sub-atomic particles (3 Quarks).

These and other such points surely mean that it is probably worthwhile considering these <u>century-old</u> observations. This "Remote Viewing" of chemical atoms was reported by Besant & Leadbeater in a 400 page book with 230 diagrams [1]. The smallest particle observed was the elementary particle shown below in Fig. 3.1.

Here is some relevant background to their method used. It is similar to the much more recent CIA "Remote Viewing" method, on which many books have been written (referenced in Chapter 1).

Besant & Leadbeater published in **1895** their first observations of atoms & elementary particles, describing quarks (sub-atomic size particles), and 10 spatial dimensions, and they described many features of elementary particles which correspond with modern quark string theory which now also postulates 10 spatial dimensions. They were the first to use the word "string" in relation to elementary particles. "Quarks" was coined much later.

About 30 years ago the author met Dr E. Lester Smith FRS, who had met Leadbeater in the 1930s and he said that he was certain that there was no deception involved in his observations which were also verified by others.

The observations are ESP (extra-sensory perception) observations in which the observer is (mentally) able to reduce his point of perception to as small a size as required – far smaller than a quark. It is like Alice In Wonderland but without the "Drink Me" drug!

CHAPTER 3 Remote Viewing of atoms & elementary particles   67

The same method is used by the CIA remote viewing group. It is known as "Remote Viewing" by parapsychologists.

This method of ESP observation used by Besant & Leadbeater [1] is a so-called yogic SIDDHI, described by Patanjali 2400 years ago [2], of being able to indefinitely reduce the size of an ESP sensor in the mind of the observer so that objects appear increasingly large.

This very unusual method may be the only way in which the internal fine structure of quarks and their groupings may <u>ever</u> be obtained, as they are well beyond any instrumental method of observation, including electron microscopes. This method is a much cheaper alternative to costly particle accelerators (CERN etc). They describe [1] the ultimate indivisible particle (Fig. 3.1, below), the "arnoo", plural is also arnoo (*Sanskrit*, "smallest particle of matter").

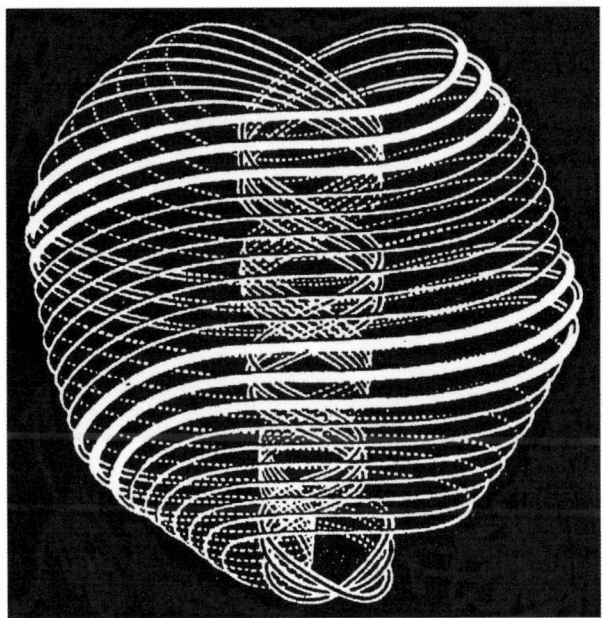

Fig.3.1. Arnoo (*Sanskrit*, smallest particle of matter): अणु

<u>3</u> arnoo are in a quark, and 3 quarks are in a proton. The arnoo is far too small to be ever detected by science. The 3 quarks in a proton can <u>just</u> be "detected", but with no indication of their internal structure as containing 3 arnoo. "Arnoo" is pronounced as "arnoo". The spelling "anu" (*pl.* anus!) used in some books should be avoided!

Fig. 3.2. Magnified detail of the arnoo shown in previous figure.

This figure shows the coiled-coil structure of the strings.
Cf. an incandescent (tungsten) coiled-coil light bulb!

Nature can be split into the Form side (using scientific instruments) & the Life side (using our minds). On the Form Side, there is no chance of using instrumental methods to observe the smallest elementary particles like quarks, which are far too small, but on the Life Side, our chakras (see Chapter 1) may allow observation.

CHAPTER 3   Remote Viewing of atoms & elementary particles

The first edition of the detailed book by Besant & Leadbeater was published in 1908 and the third edition was recently reprinted [1]. It is now also available as a free download on the internet:
www.4-D.org.uk
Their first publication on Remote Viewing of chemical structures was in 1895.   Besant & Leadbeater is referred to below as "B&L".

These observers stressed the need for very careful training in order to avoid any subjective effects intervening.

By taking their observations [1] as a starting point, the author (MGH) has derived [4] the well-known equations of Special Relativity in just one line, but without the two *ad-hoc* Principles of Relativity proposed by Einstein!   Also, Schroedinger's Equations can be simply derived in 3 lines from Besant & Leadbeater's observations [4]. This paper [4] is appended here as **Appendix 6** and is also downloadable free from the internet at www.4-D.org.uk

The present conflict between quantum theory and relativity is removed and a "Theory of Everything" is obtained: See also **Appendix 3** on General Relativity.

They [1] described some new elements, then-unknown to science, which were found by science only many years later:
-- In 1907 (B&L): Ne-22 isotope.
-- In 1909 (B&L): Promethium.
-- In 1909 (B&L): Technetium.
-- In 1932 (B&L): Astatine and Francium.

There are many other findings which were only later verified by science and this is evidence that their Remote Viewing, like that of the CIA group many years later, was genuine and accurate.

There has been some discussion about whether their observations of hydrogen were of the whole H_2 molecule or the H atom (i.e. one end of an H---H molecule). These discussions are mentioned later below, but both authors involved [7, 8] give very many reasons showing that Besant & Leadbeater's observations [1] cannot be explained in any other way than that they are genuine.   These reasons, advanced for the veracity of the observations [1], are a separate issue from the doubling problem (H_2 or H) just mentioned above. E.g. the discovery of five unknown elements/isotopes, etc.

Some points which indicate the absence of any pre-conceived ideas of what the results should be, are as follow:

(a) Their observations placed many elements which are in the same sub-group of the Periodic Table, into different categories in their classification;

(b) There are many characteristic groupings of arnoo ("arnoo" is the fundamental ultimate indivisible particle, 9 of which are in the H nucleus – explained later) which appear in chemically unrelated elements, which would not have been "expected" by them;

(c) The observers in 1895 and 1908 would expect the number of arnoo in the elements to be 9M where M is the atomic weight, not knowing about isotopes at that time. E.g. for sodium, relative to an atomic weight of H = 1.000, they would expect 22.88 x 9 = 206 but they actually observed 209 (see below). There were similar discrepancies for many other elements and so there was clearly no attempt to make the observations fit a formula that the number of arnoo = 9M. The discrepancies are not due to isotopes; e.g. for sodium there are no stable isotopes. Obviously, an isotope is what would have been actually observed. Binding energy must play a part, unknown at that time, i.e. some weight would be associated with different particular sub-grouping <u>arrangements</u> of arnoo and not just with the <u>numbers</u> of arnoo.

(d) Those elements which have <u>no</u> naturally occurring isotopes (i.e. have 100% abundance in nature) could only have been observed as the one 100% isotope, but some of the largest discrepancies occur for some of these elements. E.g. Co has a discrepancy of 26 from what would be expected from the simple formula 9M, Nb has a discrepancy of 45, Rh has 22, Tb has 54, Ho has 34 and Tm has 54. This is probably due to binding energy, i.e. the mass is not just the sum of the arnoo present but also depends slightly on the grouping arrangements of the arnoo which differ in each element.

(e) Besant & Leadbeater slowed down the rapidly moving atoms and molecules to zero velocity, essential for observing them. This means they removed their thermal (kinetic) motion, but even at absolute zero temperature there is a non-thermal motion which cannot be removed, called the **Zero Point Energy** (**ZPE**). They observed 18 arnoo in the proton (H atom nucleus), but it will be explained below that their observation of every structure as containing double the actual number of arnoo which were present, was due to the irreducible ZPE of Quantum Theory. So there are actually 9 arnoo, not 18, in the proton [7].

A consequence is that in compounds like HCl, a ZPE oscillation or vibration has one extreme (an antinode) seen as **H-Cl** and the other antinode seen as **Cl-H**, in very fast alternation, like a guitar string, so both would be seen (giving an apparent doubling).

H contains 9 arnoo grouped as three triplets at the 3 apices of a triangle **Δ**, each triplet being called a quark. So HCl was observed as: **Δ-Cl-Δ**   but because it was thought at the time that the H atom contained 18 arnoo (as 18 were observed – see below), the incorrect conclusion reached by those who interpreted the results many decades ago, was that this meant that a half-proton containing 9 arnoo was being observed on each side of a Cl atom ! See Fig. 3.3 below.

These apparent half-atoms caused the results to be set aside for many decades until it was realised [6, 7] that ZPE would inescapably cause the doubling, as just explained above, and that a proton contains only 9 arnoo, vibrating to anti-nodes like a guitar string, and thus seen double, appearing to be 18 (just like a vibrating guitar string is seen as apparently double). The proton is not 18 static arnoo.

B&L [1] describe 3D space as a continuous medium, in which small holes or bubbles move about -- this is the exact "opposite" of modern physics which regards space as being empty and containing "solid" particles.   A photographic analogy is a negative and its positive print.   Modern physics has matter and antimatter, and virtual particles which are similarly opposites of each other.

Besant & Leadbeater [1] say these holes or bubbles are caused by an energy welling up from a 4th spatial dimension which presses back the continuous medium of the 3rd dimension, forming a spherical wall.

Clairvoyant (Remote Viewing) vision [1, 3, 17] sees bright bubbles, as the smallest division of more complex groupings (atoms & molecules), observed in this black medium and these single bubbles exist in 10-dimensional space. We may tentatively identify clusters of these as photons, the smallest particles of light energy. Note: physical photons can only move in physical 3D space -- see a Scientific American (2000) article [5] for a simple exposition.

Arnoo *(Sanskrit,* 'smallest particle of matter'*)*

Arnoo are described as being cardioid in shape (Fig. 3.1), with an inrush of a force from 4-D space into the top of the "positive" type of Arnoo (source), and an outrush of force from 3-D space back to 4-D space, from the "negative" type of arnoo, which acts like a hole in space (a sink). Positive & negative are B&L's terms, but we would call it "chirality" (left or right handedness) and they may (or may not) have actual + & - electrical charges.

Besant & Leadbeater's drawing [1] of the arnoo, shows chirality; they draw details of its fine structure, being caused by an internal pressure due to a force welling-up from a 4th spatial dimension and pressing back the undifferentiated continuous medium of our 3D space, which they called "koilon" (*Greek*) or "mulaprakriti" (*Sanskrit,* 'matrix material'), which we would call a continuous structureless thermodynamic fluid; i.e. arnoo are structured holes in space.

Ten helical strings recirculate down through the core of the arnoo, each forming an endless loop of string (i.e. there are 10 endless loops of string). Along these strings are 49^6 (which is about 13,841,200,000) small spheres or bubbles. Three of the 10 strings in Fig. 3.1 have 0.57% more bubbles than the other seven and so appear to be thicker (See **Appendix 4**). The strings are like a coiled-coil electric lamp filament* but have 7 orders of coiling [six in the arnoo of the next higher (4th) dimension, five in the arnoo of the 5th dimension, etc, and none in the arnoo of the 10th dimension].

These higher dimensional arnoo would be what astronomers call the "hidden matter" or "dark matter" in the Universe. Astrophysics calculations show that over 95% of the matter in the universe is "missing" and it is currently being looked for in various ways. From

Besant & Leadbeater's observations, this hidden matter would be what is held in <u>4th to 10th dimensions</u> (see Table below) and so it appears "missing" from our viewpoint. (Note: Gravity can enter and exit the higher dimensions but photons and other electromagnetic radiation cannot, so gravity from matter hidden in higher dimensions can be detected but light etc from them cannot.)

Plane	Christian Name	Hindu Name	Kabalistic Name	Dimensions	Arnoo coil order
Adi			Kether, Atziluth	10	0
MP.Nirvanic			Daath, Atziluth	9	1
ParaNirvanic				8	2
Nirvanic				7	3
Bouddhic				6	4
Mental	Heaven	Swarga[1]	Tipheret, Beriah	5	5
Astral	Purgatory	KamaLoka[2]	Yesod, Letzirah	4	6
Physical	World	MartyaLoka	Malkuth, Asiah	3	7

*A useful analogy: A coiled-coil tungsten electric light bulb has a "**coil order**" of 2, which would correspond to ParaNirvanic in the above table!
<u>Footnotes</u>: [1]or Devachan/DevaLoka   [2]or Naraka/Paatala   MP = Mahaapara

The last column in the above table shows 7 (of the available 10) dimensions are "rolled-up" in our <u>3-D</u> world. Besant & Leadbeater reported that the smallest elementary particle, the Arnoo, has 7 orders of coiling in its whorls (see Fig. 3.1 & 3.2), but the (similar looking) Arnoo of the 4-D level has only 6 orders of coiling, which means that for a (presumed) <u>4-D</u> being, there is an infinite extent of 4-D space available. So the 4th dimension is <u>not</u> rolled up for a <u>4-D</u> being, but <u>is</u> rolled up for <u>3-D</u> beings like ourselves; so we can never see it with 3-D vision.

<u>To summarise:</u>

Anything made of 3-D arnoo (i.e. any 3-D object) has negligible extension or access into higher dimensions because they are "rolled-up" in 3-D space. Anything made of 4-D arnoo has access into 3 & 4-D but not into 5-D and beyond, etc.   3-D matter is made of atoms and molecules, but 4-D and higher matter is composed of elementary particles because electrostatic forces (which bind atoms & molecules together) fall off too steeply with distance in dimensions higher than 3 (see explanation in Chapter 1).   Elementary particles are not bound by electrostatic forces, which decrease with distance, but by quark string bonds.

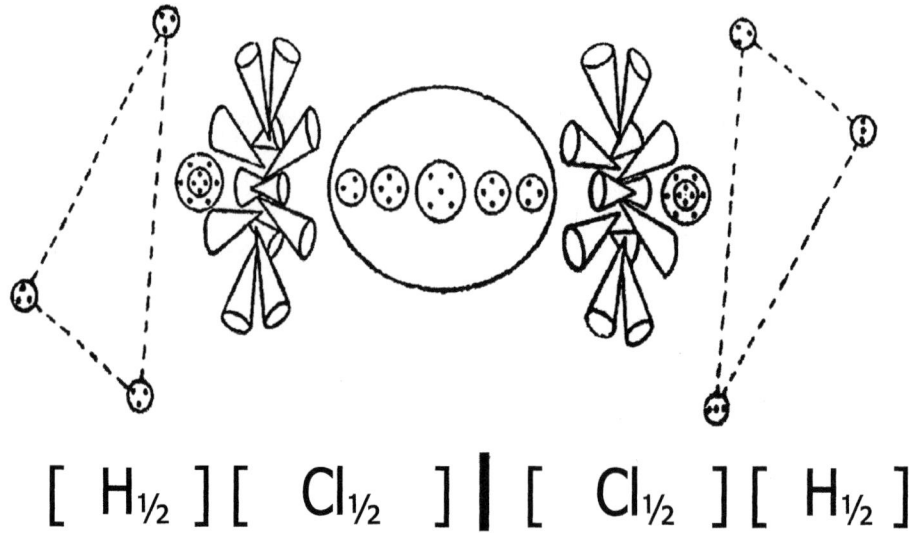

[H₁/₂] [Cl₁/₂] | [Cl₁/₂] [H₁/₂]

!! NOT TO SCALE !!

Fig. 3.3. HCl, an early interpretation some decades ago.
But the interpretation (in square brackets) is **incorrect**:
It was originally assumed that a triangle was ½ a H nucleus (proton), but actually it is a complete H nucleus.
Each triangle is a proton not a half-proton: Imagine a vertical mirror at the centre line | so that the left [H][Cl] becomes the right [Cl][H] in the mirror.
"Zero point energy" rotates or vibrates the molecule in 4-D about this "mirror", so that each half is a mirror image of the other.
Each dot is an arnoo. 3 arnoo are in a quark (small circle) and 3 quarks are in a proton (triangle) (H nucleus).

Note: The figures are not drawn to scale; B&L say that a scale figure "would require an absurdly small dot on a paper many yards square" [1].

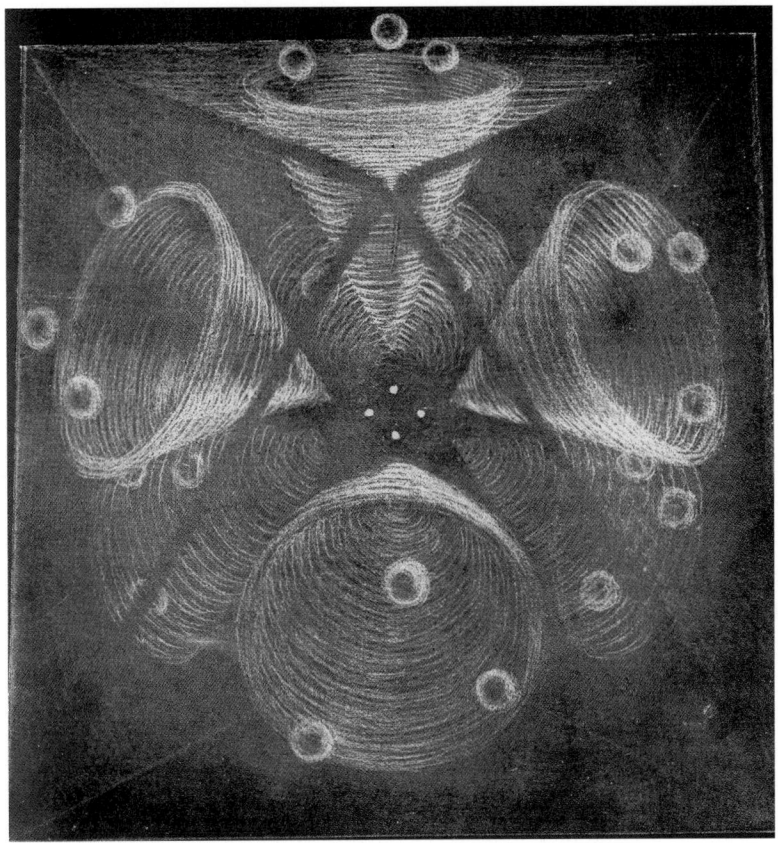

Fig. 3.4. Methane, CH_4 [1]   !! NOT TO SCALE !!

H nucleus (proton) = 3 quark spheres (the 3 small spheres hovering over each carbon funnel).

The carbon nucleus is in the funnels & at their centre.

The picture shows 8 funnels & 8 triplets of 3-quark spheres, placed octahedrally, but there are really only 4, located tetrahedrally, oscillating to appear octahedral. (C has only 4 valency).

Imagine a pyramid standing on a mirror, to appreciate an octahedron (two 4-faced pyramids fixed base-to-base).

The picture above looks down on the top pyramid only, so the lower inverted one cannot be seen, but some of its funnels are shown.

There were problems with B&L's diagrams [1], which lasted for decades because some things they described were misinterpreted, e.g. it seemed that hydrogen chloride, HCl, was $H_{1/2}Cl\ H_{1/2}$: Fig. 3.3. As explained above, there appeared to be a half-atom of H on each side of a chlorine atom!

But the author (MGH) has shown [7; ref 7 is also appended here as **Appendix 5**] that this effect is an inevitable consequence of the zero point energy (ZPE) vibration which cannot be removed by cooling the thermal motion of the HCl molecule to a stop, i.e. down to (in effect) absolute zero temperature:

B&L had to use their will (telekinesis) to hold a molecule still, to observe it, because they found that molecules were darting about and colliding at high speed, due to their thermal or kinetic energy. This "holding it still" is equivalent to cooling a molecule down, but mere cooling would not alter its internal structure.

They could slow down a molecule to zero speed, but a residual irreducible vibration exists (in Quantum Theory) called Zero Point Energy, (ZPE). This non-thermal energy cannot be removed even at absolute zero temperature (where all the thermal kinetic energy of motion is removed). If it is removed, the structure disintegrates.

Consider the following example: Methane (Fig. 3.4) was seen by B&L as an octahedron (imagine a pyramid standing on a mirror, to appreciate this 8-sided object). But it is known from chemistry that methane (CH_4) is tetrahedral, with the 4 carbon atoms at the 4 corners of a tetrahedron.

So next, imagine a tetrahedron as follows: Stand astride, stretch both arms upwards to form a V and then turn the torso 90 degrees to the right, to position the hands as far away as possible from the feet. The hands and feet then represent the four H atoms and the torso represents the C atom.

The two hands form a V shape. Now imagine the two hands oscillating down to form an inverted V and then immediately back up to form a V again, like fast flapping wings. Simultaneously imagine that the two feet (which presently are forming an inverted V) continuously oscillate upwards to form a V shape. This is

impossible to actually do (!) but the whole exercise then would produce an octahedal arrangement with the 4 H atoms appearing to be 8 H atoms, which is what Besant & Leadbeater reported seeing [1]. You would appear to have 4 arms and 4 legs!
Zero Point Energy (this oscillation) thus caused methane (CH_4) to be observed by Besant & Leadbeater as apparently having 8 half-H-atoms (explained below) located over the 8 faces of an octahedron, but chemistry gives its structure as 4 H-atoms over the 4 faces of a tetrahedron.

The H atoms are momentarily stopped at the antinodes, over the 8 faces of an octahedron, and are seen there. Cf a plucked (vibrating) guitar string, the surface detail of which can be clearly seen at its two stopped antinodes, if examined through a lens. The single string looks like two strings, due to its vibration.

The 3 quarks hovering over each funnel in Fig. 3.4 above are one H atom, which is vibrating (due to Zero-Point Energy) to a diametrically opposite funnel where it also appears, creating the appearance of an octahedron for methane. (The "flapping wings" model given above was slightly over-simplified, and it would actually be an oscillation to a diametrically opposite face of the octahedron. This does not require the proton (H nucleus) to pass through the carbon atom! This is explained in the next 4 paragraphs.)

Support for a vibratory doubling, involving an oscillation (or a rotation) to & from the 4th spatial dimension, is that: all of the left-handed & right-handed (marked + & - in figure) arnoo in one H triangle were observed to be the mirror-images of the corresponding arnoo in the other: such a chirality reversal would be expected (a 2-D to 3-D analogy is to look at a letter "**R**" on a transparent sheet from above it and then from below it (or, "**R**" will have a chirality change when rotated, via 3D, to "**Я**"). Or **d** to **b**. A rotation into 4-D space and back would be seen by us as a vibration in our 3-D space. Eagles [see in ref 7] suggested a rotation via 4-D is a viable explanation of how two antinodes can appear in 3-D space without any "blurring" between them (see below).

There is evidence of an oscillation involving such chirality (mirror image) changes because Besant & Leadbeater say*, of hydrogen, "*The 6 little bodies are arranged in two sets of 3, forming two triangles that are not interchangeable but are related to each other as object and image* ". (The "6 little bodies" are the circles in Fig. 3.5 below.) (my underlining)

* (in the 1st and 2nd editions of their book [1], on p. xiii and p.9 respectively)

As already mentioned, all molecules contain bound atoms and all atoms contain bound groupings of arnoo. All bound particles, even if brought to translational rest (0°K), must still contain Zero Point Energy, according to the well established Quantum Theory. This takes the form of a high frequency vibration or harmonic motion, the restoring force being the coulombic (electrostatic) and strong nuclear binding forces (but the quark-to-quark "string" bonds are excluded as they are not of harmonic oscillator type). This is comparable to a guitar string or a vibrating metal rod, which appear to be double due to the string or rod being repeatedly stationary at each end of its vibration trajectory. In between these extreme positions (antinodes), it is moving too fast to see.

Similarly, the number of Arnoo seen in atomic and molecular structures would be double the actual number present.

The guitar string analogy is not quite appropriate because no blurring was reported in the structures observed [1]. A better model is the escapement mechanism in a watch, but better still, is that the actual occurrence could be rotation in 4-D space (which appears as a vibration in 3-D space of the structures by spinning in 4-D space). This would seem to us to be a vibration, with only the (stationary) antinodes being seen, because the 4-D part of the motion is invisible to us. Such 4-D movements would not require the arnoo to physically (in 3-D) pass through other physical (3-D) arnoo!

See Appendix 5 for drawings of some structures.

CHAPTER 3   Remote Viewing of atoms & elementary particles   79

To summarise again:
The unexpected doubling of all B&L's structures, caused their molecules to apparently contain half-atoms, causing their book [1] to be set aside for many decades, but the author (MGH) has explained this as due to zero point energy vibrations or zitterbewegung.   The author published this in a paper [7] 27 years ago.   Fig. 3.5 (below) shows the proton as observed, and originally the double triangle was incorrectly assumed to be an H atom nucleus (proton) containing 18 arnoo.   This meant that all 6 quarks (the circles in Fig. 3.5 below) were equated (incorrectly) to an H nucleus and so if only 3 quark (circles) were seen, as in methane above, that was a half-atom of H, which is incorrect.

B&L [1] were the first to use the word "**string**", in describing the Arnoo in Fig. 3.1.   Many decades later, this word was used by scientists to describe quark-quark string bonds.

In the 1970s, elementary particle science drew diagrams of strings with quarks at their ends, like the diagrams published many decades earlier by B&L [1], but without mentioning that book.

Modern physics relatively recently described the structure of the proton as containing 3 units, called "**quarks**", but modern physics cannot "resolve" the detail of these quarks beyond being three very "fuzzy" items.   A proton is a hydrogen atom nucleus and B&L's picture of it is shown in Fig. 3.5.   This has the appearance of the Seal of Solomon (Star of David), depending on the angle of view.

Further discussion of the doubling problem:

There is no need to go to another explanation [8] which has been suggested, that for some *ad-hoc* reason two nuclei (but not explained why not more than two) had fused together when B&L looked at it, so that what they saw was a double nucleus of what it should be in the absence of their method of vision.   There are many reasons why this cannot be so [7], e.g. the valence of the molecule is unlikely to be also preserved as double during such a necessarily very energetic nucleus fusion process.

There is no reason to suppose, as has been suggested [8], that reducing the translational (kinetic) motion of an atom or molecule to zero, to observe it, would also reduce its internal ZPE to zero, as this is of a totally different type: non-thermal.   Besant &

Leadbeater saw only very clear and detailed-structured particles and nothing like the waves suggested by the Uncertainty Principle of Quantum Mechanics and the author (MGH) has suggested [4] that the mathematical appearance of waves in the Quantum Theory (Schroedinger's Equation) are actually the 4-D transits of moving particles – i.e. if a particle is oscillating between 3-D and 4-D space (as mentioned above), it is not in normal 3-D space during those transits and this will come out as imaginary in the equation of motion (Schroedinger's Equation). But this imaginary number has then been "normalised" (squared), which decision <u>creates</u> a <u>wave</u> – a wave is a mathematical abstraction only and the author (MGH) has derived Schroedinger's Equation in 3 lines on the model of a <u>particle</u> oscillating between 3-D and 4-D space [4]. De Broglie described the waves of Quantum Theory as "ondes fictives". No "quantum fluctuations" or wave effects were reported by B&L [1]. **Note**: Reference [4] is reprinted here in Appendix 6.

On this model the author (MGH) has re-interpreted the Uncertainty Principle in this light [see Appendix 6].

A suggestion has been made [8] that Besant & Leadbeater [1] saw every nucleus as doubled because if the velocity of each particle <u>within</u> the nucleus was reduced to zero ((unlikely for the ZPE – see above)) then the Uncertainty Principle would cause its position to become so uncertain ("spread out") that there would be the same chance of finding it <u>near</u> as finding it in an <u>adjacent</u> atom's nucleus. ((This ignores the report [1] that only a <u>single</u> atom was selected and slowed down for observation.)) A further *ad hoc* suggestion was then made [8] that the nuclear particle could then interact with particles in a supposed adjacent nucleus to form an entirely new structure, which would thereby be doubled, as it is now two nuclei fused together, and it is this artificial new double-structure which B&L then observed. ((The author's (MGH's) comments are in double parentheses above.))

But this ignores why triple or higher structures were never observed, and, if the position of the original particle is widely "spread out", why does it not "go back" to form the highly stable nucleus from which it first came? A whole series of double nuclei was being proposed [8] which are much <u>more</u> stable than the original stable nuclei which have evolved in stars over millions of years. And this theory requires that no original single nucleus structures at all remained to be seen. There are also many other problems with the "double nuclei" theory, which remain unanswered [7].

This all seems very unlikely to the author (MGH), who prefers the simple oscillation model suggested earlier above, where an atom or molecule is vibrating incessantly (ZPE) between two antinodes, creating the appearance (image) of a double structure, just like the antinodes of a single guitar string appear like two strings in the form: () [7].

The above does not detract from the many convincing reasons advanced [8] for the veracity of the observations [1], which is a separate issue from the doubling problem.

Symmetries are seen in all the structures observed [1], which support the doubling by oscillation model [7].

The Uncertainty Principle of Quantum Mechanics

This says in effect that the product of velocity and position is a constant, so if the velocity of a particle is reduced to zero, its location becomes infinity! The occurrence of infinities in equations is usually taken to mean that there is an error in the equation. The author's (MGH's) explanation is as follows:

Appendix 6 comments that Schroedinger's Equation (containing imaginary terms) is of the same functional form as Fick's Second Law of Diffusion and thus the "diffusivity", D, of a moving elementary particle is imaginary, meaning simply that it does not continuously exist in 3-D space. Prior to the introduction of 10-D space by quark string theory, the imaginary values of ψ in Schroedinger's Equation embarrassed physicists, who only "believed in" 3 dimensions and thus decided in effect to square ψ to force it to be real and thereby artificially created "matter waves". They called this process "normalising" ψ and it compelled ψ to conform with the then "world view" of what Nature "should be" (real, with nothing imaginary, no 4th or higher dimensions).

This understandable attitude at that time (that there are no higher dimensions) is very well illustrated by many standard textbooks which assert that "the particle must be somewhere", to "justify" effectively squaring ψ to prevent it from being imaginary (nowhere in 3-D space)! This procedure discounts the possibility that it actually could sometimes be nowhere in our 3-D space, if it oscillates or spins in and out of 4-D space. This "normalising" approach artificially creates a fractional probability (instead of an imaginary one), i.e. an "uncertainty" that a particle is present at any given location (in 3-D), which creates the notion that particles can somehow exist as waves (instead of particles) and leads to interpreting ψ as a "wave function". But it is actually an imaginary probability that the particle is present in 3-D (because it is in 4-D space!). De Broglie intuitively said that "matter waves" are "ondes fictives". The mathematics automatically consigns a dimension higher than 3 to an imaginary term.

Notes:

(1) Evidence of chirality (handedness) effects exists in out-of-the-body (OBE) literature [9]:
(i) Rooms are sometimes seen as the mirror-images of what they actually are! This is explained by analogy with 2-D: if **R** is viewed from 3-D space from above a 2-D sheet, it is seen as **R**, but if viewed from below the sheet it is seen as **Я**. So, by analogy, a 3-D room can be viewed from either 4-D "side", and thus will appear either "normal" or as its mirror image (never seen from 3-D).
(ii) Normal people (not OBE people) have reported that text seen while dreaming is often a mirror image of normal (i.e. viewed from the other side of the "sheet").

(2) The authenticated removal of part of a single crystal of vanadium carbide in laboratory conditions [10] requires the use of a 4th spatial dimension, just as the removal of an object from inside a 2-D circle (by lifting it out into 3-D) would appear miraculous to an inhabitant of a 2-D world (as it requires a transit via a 3-D spatial dimension).

(3) It is important to mention that any paper which mentions ESP or telepathy, if sent to a mainline scientific journal for publication, will be automatically rejected. Orthodox science is strongly opposed to any notion that ESP and telepathy etc exist, even though there is

overwhelming evidence for them. But one was published in Nature [11] (with considerable unjustified opposition).

(4) Also, widespread poltergeist phenomena provide overwhelming evidence of telekinesis, but it also is not "accepted" in science. It just means that these scientists have not looked at, or will not look at, the evidence. Most are frightened of damaging their reputations.

(5) As just mentioned, there are very few refereed journals that will publish un-orthodox papers but one is the Journal of Scientific Exploration. The author's article in Journal of Scientific Exploration 21 (1), 13-26 (2007) [4] is available here in Appendix 6 and also can be downloaded (free) from: www.4-D.org.uk

(6) <u>Gell-Mann's "Eightfold Way"</u>
Gell-Mann predicted 8 particles from a symmetrical model, but B&L [1] reported seeing only 1 to 7 Arnoo linked with strings of force (quark string bonds) to form composite particles (i.e. never more than 7 types of composite particle). Gell-Mann found 7 particles easily, but the 8th took over a year of searching through thousands of cloud-chamber tracks from a cyclotron, which can explain why B&L did not see it.

CHAPTER 3 Remote Viewing of atoms & elementary particles

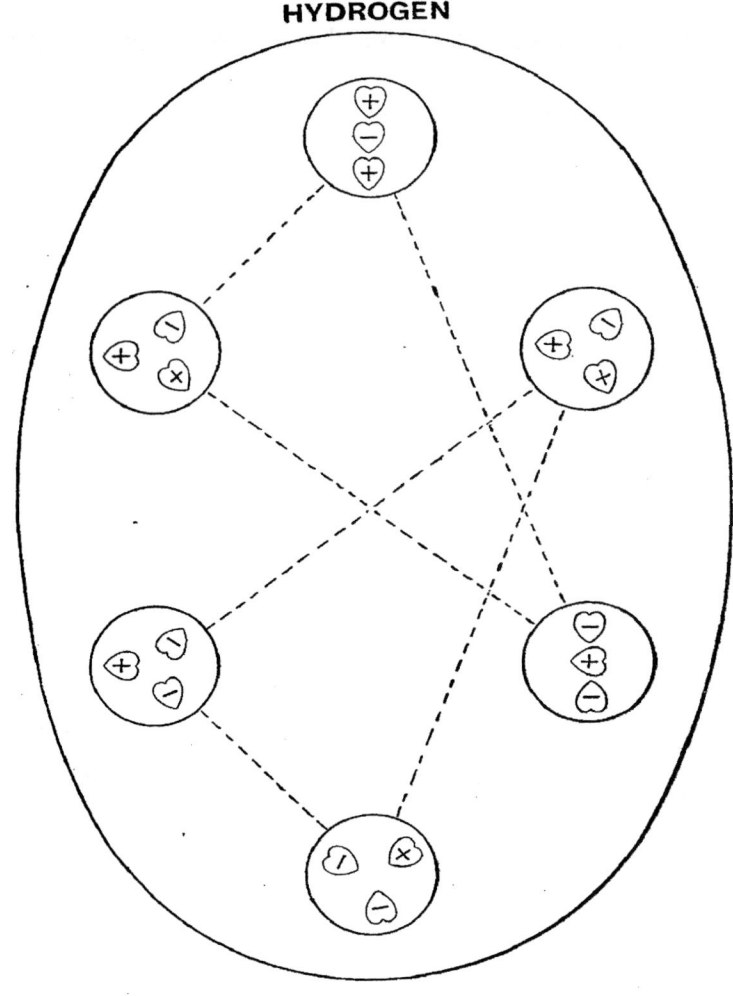

FIG. 1. HYDROGEN

Fig. 3.5. The Proton, as seen by Besant & Leadbeater [1].
Only <u>one</u> dotted triangle exists. The other is created by its fast rotation/oscillation into 4-D and back.
So proton has 9 arnoo in 3 quarks, shown in the: ◯ shapes above, not 18 arnoo.

Originally the double triangle was incorrectly assumed to be an H atom nucleus (proton). See Appendix 5 for the lines of oscillation (i.e. which quarks oscillate to those in the other triangle).

Further discussion of the micro-chemical observations of Besant & Leadbeater [1] is given in Appendix 5 here, taken from the author's website: www.4-D.org.uk

Besant & Leadbeater's book [1] contains 230 diagrams but only a few are shown here, as the book is downloadable free from the above website.

REFERENCES for Chapter 3:

NOTE: To make this paper more accessible to general readers, the references below contain general articles where available, such as those published in Scientific American.

1. A. Besant & C.W. Leadbeater, "Oc. Chemistry", 400 pp (first edition 1908, second edn. 1919, third edn. 1951 & reprinted 2001, publ. by TPS, 50 Gloucester Pl., London W1H 3HJ).   (Parts were published earlier, in 1895 in journal form: Lucifer, Nov. **1895**).
Available as a free download from**:**   www.4-D.org.uk   but for the fine detail of some diagrams the printed version is necessary.   230 diagrams.

NOTE: Refs 1, 4, 6 & 7 are available free from: **www.4-D.org.uk**

2. Patanjali, "Yoga Aphorisms", published about 400 BC.
3. G. Hodson, Clairvoyant Investigations by Geoffrey Hodson and David Lyness, 1957-1959:   **www.tphta.ws**
4. M.G. Hocking, "Linking String Theory and Membrane Theory to Quantum Mechanics and Special Relativity Equations, avoiding any Special Relativity assumptions", Journal of Scientific Exploration 21 (1), 13-26 (2007).   This is downloadable (full text, free) from Page 1 of the website: www.4-D.org.uk
It is also given in the present book as **Appendix 6.**
5. N. Arkani-Hamed, S. Dimopoulos, G. Dvali, Scientific American, page 48 (August 2000).   (article on multidimensional space)
6. M.G. Hocking, SGJ, 12, 99 and 143 (1968), publ by TS, available at Library, 50 Gloucester Pl., London, W1H 3HJ.
7. M.G. Hocking, Bull. Theo. Sci. Stdy. Gp. (India) $\underline{21}$, 53 (1983); Ibid. $\underline{22}$, 5 (1984) (**appended** to the present book **as Appendix 5**).
8. S.M. Phillips, "Anima", publ by TPH, Adyar, India & Wheaton, IL, USA (1996).

S.M. Phillips, "ESP of Quarks & Superstrings", publ by New Age Intl Publishers, Bangalore, India (1977). Also, see **Appendix 5** below (in the present book).
9. C. Greene, "Out of the Body Experiences", Inst. of Psychophys Res, Oxford (1968); R.A. Monroe "Journeys out of the Body", p. 172, Anchor Press/Doubleday, NY (1977).
10. J.B. Hasted, D.J. Bohm, E.W.Bastin, P. O'Regan and J.G. Taylor, **Nature** 254, 470 (1975).
11. Targ et al, Nature 251, 602 (1974).

See also:

M.G. Hocking website: www.4-D.org.uk

-=-=-=-=-=-=-=-=-

NOTE: References for each chapter are listed at the end of that particular chapter.
General references are listed at the end of the book.

CHAPTER 4

A documented world history

- ◆ Beginnings, 100,000 years ago
- ◆ The previous World Teacher (70,000 BC to 500 BC)
- ◆ The new World Teacher, Krishna (500 BC)
- ◆ Christ as the present World Teacher (see Table 1 below)

The contents of this chapter are the results of what is presently called "Remote Viewing", as used by the CIA for intelligence gathering, and previously used by Besant & Leadbeater [1, 2, 3] and by others. Anyone who doubts this method should see the books by those associated with the CIA Remote Viewing programme and the videos [see references 1-39 in the Chapter 1 reference list].

This type of research can no longer be discounted, and yields very valuable interesting information on world history, going back much further than previously possible by conventional methods. This chapter covers a long period before the time of Christ, going back to far beyond 70,000 BC.

It is important that the reader should have read Chapters 1, 2 & 3 before starting on this chapter, because those chapters cover the background and evidence for Remote Viewing.

If it is accepted from Chapter 2 that we require many lives to achieve the standard necessary to pass beyond this 3-D world, then it will be evident that Christ also has passed through many lives prior to His time as World Teacher, and the Bible gives an indication that He was like us before He became the Christ (The World Teacher):
St John's Gospel (16:33) says, "... *be of good cheer; I have overcome the World*" and Revelations says, "*To him that overcometh will I grant to sit with me in my throne, even as I also overcame, and am set down with my Father*" (my underlining) (King James Bible).

This also indicates that we will all become like Him at some future time. He also said that He is not God: Mark (10: 18) says, *"Why callest thou me good? There is none good but one, God "*, and, *"This is my beloved Son"* (Matthew 3: 17), and, *"I am not alone because the Father is with me"* (John 16: 32), also John 12:44, etc (King J. Bible).

The following account is taken from two books of A. Besant & C.W. Leadbeater [3, 2] which are detailed world histories as discovered by the expert level "Remote Viewing" method used by the CIA for intelligence gathering [referenced above] and by Besant & Leadbeater [1] for viewing molecules, atoms and elementary particles at high magnification described in Chapter 3, and by others.

The names of the individuals followed are obviously different for every rebirth, so permanent names (pseudonyms) were arbitrarily assigned [3] to identify individuals in successive lives, usually classical names like Roman/Greek Gods etc {e.g. Mars, Mercury; and Surya (*Sanskrit*, Sun God)}. "Surya" is Christ. Table 1 below lists these pseudonyms.

Besant & Leadbeater (referred to here as B&L) based their studies [3] on 48 successive lives of J. Krishnamurti, who was assigned the pseudonym "Alcyone". Their account below, abbreviated by the author (MGH), follows Alcyone through 22 of these 48 lives. But the aim of the present book is to retrieve information on Christ, Krishna and Buddha and so only those lives of Alcyone which involved these, are reported in this chapter. The other, omitted lives with fascinating and informative material are given in the author's (MGH's) other book [14].

This chapter and ref. [14] give a valuable insight into the workings of karma across successive lives. Karmic links cause individuals to be reborn successively into the same family groups as before.

Most of the lives below were lived before the last Ice Age ended about 10,000 years ago and before the sinking of Atlantis (see p. 91) which B&L report [2, 3] as being in 9,564 BC. In life 28 (below), iron is reported in 15,995 BC, long before the 'conventional' start of the 'Iron Age': tsunamis due to the sinking of Atlantis swept away most archaeological evidence around the world, and, took us back to the Stone Age. Also, permanent inundation due to the end of the Ice Age caused sea levels to rise 120 metres, but recent undersea archaeology has revealed ancient ruins off India and Japan. The post-ice-age period is about the limit of conventional historical knowledge (10,000 years), but it is less than 1% of the total time man has been on Earth.

CHAPTER 4   World History from 70,000 BC: 89
Buddha & Christ in their previous lives

Table 1: Pseudonyms assigned [3] to identify individuals in successive lives

(i) "Master" means one who has achieved salvation/sainthood, and so is not required to be reborn again into our world. During the 70,000 year period covered by the investigation [3, 2], many of those who are now Masters, were ordinary people like ourselves. **See colour Fig. 4.1, near p. 63.**

(ii) The term Bodhisattva, *Sanskrit* (= Christ, *English*) means World Teacher and is a Master who has achieved this status. The present World Teacher appeared <u>voluntarily</u>* as Christ in Palestine and was not <u>required</u> to be re-born, being beyond all karmas. There is a hierarchy, hinted at in Matthew 11:11 & Luke 7:28, speaking of <u>John the Baptist</u>: "He that is least in the Kingdom of Heaven is greater than <u>he</u>". Thus, e.g., Surya (the present World Teacher), Manu & Viraj are of greater rank than a Master.

PSEUDONYM	ACTUAL NAME or IDENTITY
Mahaguru	= World Teacher (Bodhisattva) of the 4th (previous) period, appearing as: **Vyasa**; **Thoth** (Hermes) (40,000 BC); **Zarathustra** (Zoroaster) (29,700 BC); **Orpheus** (6980 BC); & finally as **Buddha** (623 BC). **
Surya	= World Teacher of the 5th (present) period, called **Maitreya** (*Sanskrit*), or **Christ** (*English*), who has appeared as **Krishna** and later as **Christ**.
Manu	Vaivasvata Manu, Founder & Head of the 5th (current) period.
Viraj	Maha-Chohan, of equivalent rank to a Manu or a Bodhisattva.
Brihat	Now the Master Jesus (explained later).
Dhruva	The Master of who is now the Master "KH" (see KH below).
Jupiter	Now a Master, residing in the Nilgiri Hills.
Mars	Now the Master known as "M" (Moraya).
Mercury	Now the Master "KH" (Koot Hoomi) (previously <u>Pythagoras</u>).
Neptune	Now the Master Hilarion (previously <u>St Paul</u>; see note on p.257).
Saturn	Now a Master, called "The Venetian" in some books.
Venus	Now the Master Ragozci (<u>Comte de S. Germain</u> in the 18th century).
Vulcan	Now a Master (<u>Sir Thomas More</u> in his last life).
Alcyone	J. Krishnamurti.
Corona	<u>Julius Caesar</u>.
Herakles	Dr Annie Besant (photograph is in Chapter 3, page 65).
Sirius	C.W. Leadbeater (photograph is in Chapter 3, page 65).

SEE FULL LIST OF NAMES ON PAGE 383.   SEE COLOUR Fig. 4.1 near p. 63.
* See John 9:18.   ** ((Legend says the "Egyptian Book of the Dead" (Osirian Religion) was written by Thoth. The Avesta or Zend Avesta is thought to have been written by Zarathustra (Zoroaster). Orpheus founded Orphism (the Orphic Religion) but only fragments remain. The (relatively recent) Buddhist scriptures are extensive.))

Table 2.   The Lives of Alcyone [3].

No	Date BC	Place	Sex	Father	Mother	Brothers	Sisters	Spouse
1	70,000	Gobi	M	Mars	Mercury	Sirs,Ach,Orin	Mizar	Leo
3	60,000	Manoa	M	Manu	Surya	Cor,Sel,Veg, Hes, Cape	Herak, Ven, Brihat, Urn, Ath, Vulcan	Mercury
4	42,000	Manoa	M	Herakles	Sirius	Ura,Sel,Siwa Miz,Achil,Ald	Nep, Orion, Capri	Percy
5	40,000	Arabia	M	Jupiter	Athena	Manu, Mars, Auro	Fides, Vega	Elektra
10	29,700	Persia	M	Mars	Uranus	Jup,Mercury, Orion	Elek, Rama, Fides	Sirius
11	28,804	Persia	M	Mercury	Fides	Ura,Dhruva, Brihat, Euph	Sirius,Selene Miz, Cas, Ivy	Apollo
13	27,527	Ireland	M	Elektra	Brihat	Ath, Viola	Jup, Nep, Osir, Aquil	Mercury
16	25,528	New Zld	F	Mizar	Sirius	Eup, Osiris, Bec,Aqui, Let	Vul, Bri, Koli	Surya
17	24,700	Mexico	F	Sirius	Elektra	Os,Ra, Nep, Vega, Tel	Dhruva, Ath	Selene
19	22,662	USA	F	Mizar	Helios	Hera, Sele, Auro, Draco	Leo,Leto,Pyx And, Oak	Sirius
20	21,750	Burma	F	Brihat	Neptune	Ura, Vulcan	Mizar	Saturn
21	21,467	S. India	M	Leo	Orion	Albireo	Bee, Thesius	Herakls
22	20,574	Salem	M	Uranus	Mercury	Demeter, Elsa, Colos	Neptune, Prote, Apollo	Percy
23	19,554	China	M	Mira	Selene	Sirius, Ajax, Xan	Vega, Mizar, Gnos, Ron	Albiero
24	18,885	Manoa	M	Mars	Mercury	Uranus, Herakles	Brihat, Demet	Theseu
25	18,209	Africa	M	Leo	Achilles	Sirs, Rosa	Alethei, Polar	Helios
27	16,876	Atlantis	M	Mercury	Brihat	Achilles, Selene	Orion, Caly	Sirius, Mizar
28	15,995	Manoa	M	Mercury	Saturn	Selene, Leo, Vajra, Castor	Hera, Mizar	Albireo
29	15,402	Oudh	F	Cetna	Cancer			Scorpio
34	11,182	N. India	M	Olaf	Tolosa	Koli, Echo, Uder	Ronald	Cygnus
47	630	Benares	M	Rao	Nu			Irene
48	AD624	India	M	Ant	Irene			

Note: More detailed tables are available and detailed genealogical tables [3]. Table 1 above, includes only those lives of Alcyone which involve Surya and are covered in the present book.   22 other lives are in the author's other book [14].

(Some names above are abbreviated, e.g. Sirs = Sirius; full list is on p. 383)

CHAPTER 4   World History from 70,000 BC:
Buddha & Christ in their previous lives

Early Times

Besant & Leadbeater [2] report that the City of the Golden Gates in Atlantis was founded about 1 million years ago, about 150,000 years before the first catastrophe which flooded a large part of that lost continent. The reader who wants an account of events before this, should refer to the above book [2]. Scott-Elliot gives another account of Atlantis [5] and Donnelly's scholarly book [6] gives a large number of references on Atlantis and important conventional evidence for its existence. See also geological research showing "unexpected" sub-aerial exposure of the mid-Atlantic Ridge, published in Nature [13]; also websites, e.g.: www.atlantisquest.com/Geology.html   and: www.goldenageproject.org.uk/Survey.php

The Toltec civilisation expanded all over the world from Atlantis. Besant and Leadbeater [2, 3] followed, by Remote Viewing, a group of individuals, including themselves. They are identified down their incarnations by the names given in Table 1 above and on p. 383. As individuals progress, the interval between their successive lives was found to increase from a few hundred to about 1000 years and the group thereby remained together down the ages. Their death-birth interval in the early times of Atlantis was shorter because the experiences gathered during those early lives did not warrant a long inter-life period to work through the lessons learned.

About 600,000 BC, a group of souls from previous times [2] were born in Atlantis, including Surya (who much later became Christ). He became the Chief of the group, and Mars, his eldest son, became its foremost warrior. Mars succeeded his father as Chief, which was his first experience of ruling.

The spread of evil (explained later, below) in Atlantis led to the second great catastrophy of 200,000 BC. Atlantis which up till then had joined Africa to America, then became just two islands, Ruta and Daita. Much later the third flooding of 75,025 BC, left only one island, Poseidonis, which was the last to be flooded in 9,564 BC (the final disappearance of Atlantis). Its highest parts may be the Azores. Plato (in "Timaeus") gives the sinking date as about 9600 BC, being 9000 years before the life of Solon, and location as beyond the Pillars of Hercules (Straits of Gibraltar). For Plato, see also p. 227 & 384.

In about 100,000 BC, Alcyone was born when Corona (later Julius Caesar) was the Emperor of the City of the Golden Gates [2]; Mars was the general under him and Herakles was the wife of Mars. A major rebellion was being plotted by a man of strange and evil knowledge, known as a "Lord of the Dark Face" (in contrast with the "Lords of the Dazzling Face"). This man could control semi-human semi-animal creatures (the originals of the Greek satyrs), who seem to have been materialised from the astral (i.e. from the lower parts of the 4th dimension, Purgatory, the realm of Pan), explained later. Huge statues of himself were placed in temples, with orgies held there instead of religious ceremonies.

NOTE: The verbatim texts below (in lighter typeface) are included as world history from various periods back to 100,000 BC, obtained by expert level Remote Viewing. The observers [3, 2] selected the most interesting events from past lives in this time span. <u>100,000 years is over 3000 generations</u> and ESP was found to be much more common then and so in reading these accounts it is necessary to allow for this.

The next section below is taken verbatim from reference [2]:

<u>Black Magic in Atlantis – An Episode</u>

Alcyone is lying half asleep, half awake, on a grassy bank sloping down to a rippling brooklet. His face is perplexed, even anxious, the reflex of his troubled mind. He is the son of a wealthy and powerful family, belonging to the priesthood, the Priesthood of the Midnight Sun, vowed to the service of the gods of the Nether World, whom the priests sought in the gloom of night, in dark earth-caverns opening into passages that led down, down into unknown depths.

At this time, the great civilised nations of Atlantis had drawn into two opposed camps: the one, looking to the ancient City of the Golden Gates as their sacred metropolis, maintained the traditional worship of their race, the worship of the Sun - the Sun in the beauty of his rising, clad in the bright colours of the dawning, encircled with the radiant youths and maidens of his court; the Sun in the zenith of his glory, the blazing strength of his mid-heaven, scattering abroad his brilliant rays of life and heat; the Sun in the splendid couch of his setting, touching into rarest softest hues the clouds he left as promise of his return. The people worshipped him with choral dances, with incense and with flowers, with joyous songs, and with offerings of gold and gems, with laughter and with minstrelsy, with joyous games and sports. Over these children of the Blazing Sun the White Emperor bore rule, and his race had for long millennia held unchallenged sway. But gradually the outlying kingdoms, ruled by his lieutenants, had become independent, and they were beginning to join together into a Federation, rallying round a man who had appeared among them, a remarkable but sinister figure.

This man, Oduarpa by name, ambitious and crafty by nature, had realised that, in order to give stability to the Federation and to make head against the White Emperor, it was necessary to call to his aid the resources of the darker magic, to make compact with the denizens of the Nether World, and to establish a worship which would attract the people by its sensuous pleasures, and by the weird unholy powers it placed within the reach of its adepts. He had himself, by such compact, extended his life over an abnormal period, and, when going into battle, rendered himself impervious to spear or sword-thrust by materialising a metallic coating over his body, which turned weapons aside as would a shirt of mail. He aimed at supreme power, and was in a fair way to reach it, and he dreamed of himself as sitting crowned in the Palace of the City of the Golden Gates.

The father of our youth ((Alcyone)) was among the most intimate of his ((Oduarpa's)) friends, and privy to his most secret designs, and both hoped that the lad would devote himself to the forwarding of their ambitions. But the youth had dreams and hopes of his own, nourished silently within his own heart; he had seen in the visions of the night the stately figure of Mars, a general of the White Emperor, Corona, had gazed into his deep compelling eyes, had heard, as from afar, his words: "Alcyone, thou art mine, of my people, and surely thou shalt come to me, and know thyself as mine. Pledge not thyself to mine enemies, thou who art mine." And he had vowed himself his subject, as vassal to his lord.

Of this was Alcyone thinking, as he lay musing by the stream. For another influence was playing upon him, and his blood ran hotly in his veins. Ill-pleased at his indifference to their worship - nay, at his shrinking from it, even in its outward rites of animal sacrifice and poured out oblations of strong drink - his father and Oduarpa had conceived the plan of drawing him into the secret mysteries by the allurements of a maiden, Cygnus, dark and beauteous as the midnight sky star-studded, who loved him deeply, but had so far failed to win his young heart with her charms. Between her dusky brilliant eyes and his half-fascinated gaze would float the splendid face of his vision, and he would hear again the thrilling whisper: "Thou art mine".

Fig. 4.2. Mars = The Master Moraya ("M")

Note: The artist [4] who drew these Figures studied under the guidance of one of the main Remote Viewers, Dr A. Besant [2, 3] for many years and produced these drawings in about 1930 (when Dr Besant was still alive) as portraits from life. The pictures are the subjects as seen by Remote Viewing in about 1930 (AD!).   (Figures 4.2 to 4.8 & 4.10).

Mars visited London in 1851 (Great Exhibition, Hyde Park).

At length, however, she had so far won him persuaded to the task by her mother, a veritable witch-hag, who had told her that thus alone might she gain his love - as to obtain from him a promise that he would accompany her to the underground caves in which the magical rites were performed, which drew the denizens of the Nether World from their retreats, and gained from them the forbidden knowledge which changed the human into the animal form, thus giving opportunity for free play to the passions of the brute hidden in man, passions of lust and slaughter. Cygnus had played upon his heart with skill taught by her own passion, and had fanned his indifference into fire, not enduring, indeed, but warm while it lasted. And today the passion was hot upon him, and the power of her allurement swayed him. For she had just left him, after coaxing him to promise to meet her after sunset near the caverns where the mysteries were performed, and he was struggling between his longing to follow her, and his repulsion from the guessed-at scenes in which he would be expected to take part. The sun sank below the horizon and the sky darkened while still Alcyone lay musing; with a shudder he started to his feet, but now his mind was made up, and he turned his steps towards the rendezvous.

To his surprise a considerable company was gathered at the spot; his father was there with his priestly friends, and Cygnus with a crescent moon on her head, the sign of the bride, and a band of maidens round her, all clad in gauzy star-spangled raiment, through which the brown lithe limbs gleamed duskily; a band of youths of his own age, among whom he recognised his nearest friends, were also waiting, with spotted skins of animals for raiment, and light cymbals which they clashed as they danced round him like fauns.

"Hail, Alcyone!" they cried, "favourite of the Dark Sun, child of the Night! See where thy Moon and her Stars await thee. But first thou must win her from us, her defenders." Suddenly she was whirled away in the midst of the dancers, and vanished in the darkness of the cavern yawning wide in front, and Alcyone was seized, stripped of his garments, a skin like that of the rest thrown over him, and intoxicated, maddened, he fled in her pursuit, amid laughter and cheers: " Hey! young hunter, be swift, lest the hounds pull down thy deer!" After a few minutes Alcyone, with the shouting crowd at his heels, had raced through the outer caverns, and had reached a vast hall, blazing with crimson light.

In the midst rose a huge canopy, red in colour and studded with great carbuncles, that tossed back the light like splashes of fiery blood; beneath the canopy was a copper throne, inlaid with gold, and before it a yawning gulf, out of which flashed tongues of flame, lurid and roaring. Heavy clouds of strange incense filled the air, intoxicating, maddening.

The rush swept him onwards, and he was caught up into a wild tumultuous whirl of dancers, who shouted, yelled, sprang into the air in wild bounds, circling round the canopied throne and crying: "Oduarpa ! Oduarpa ! Come, we are craving for thee!"

A low roll of thunder crept muttering round the cavern, growing louder and louder, and ending in a tremendous clap just overhead; the flames leapt up, and amid them rose the mighty form of Oduarpa, steel-grey in his magic sheathing, stern, majestic, with his face grave, even sad, as that of a fallen Archangel, but strong with unbending pride and iron resolution.

He took his seat on the throne, where he sat throughout all that followed, silent and sombre, taking no part in the riot; he waved his hand, and the mad orgy recommenced, the wildest dancers bathing in the flames which lapped over the edges of the gulf, and tossed themselves high in air. Alcyone had caught sight of Cygnus in the midst of the youths and the girls, and he raced, mad with excitement, in her direction; she eluded him, her escort baffled him, he touched her only to see her whirled out of his reach. At last, panting, wild, he made a desperate rush, and the escort fled with screams of laughter, each youth with a girl, and he leapt on Cygnus and clasped her in his arms.

Wilder and wilder grew the revel; slaves bearing huge pitchers of strong drink appeared, accompanied by others with goblets. Madness of drink was added to madness of motion, and the lurid lights sank low into twilight of redness. The orgy which followed is better hidden than described.

But see ! out of the passage whence had emerged Oduarpa, comes a wild procession; hairy bipeds, long-armed and claw-footed, with animals' heads and manes streaming over shoulders, horrent, appalling, non-human, yet horribly human. They hold in their claw-like hands phials and boxes, and as they mingle with the wildest dancers they give these to the revellers most mad with drink and lust. These smear over their limbs the ointment in the boxes, drink the contents of the phials, and lo! they drop senseless, huddled on the ground, but from each huddled heap there springs an animal form, snarling, ravening, and vanishes from the cavern into the darkness of the outside night. ((Comments: Cf mediaeval legends of werewolves. An ointment made from henbane was reportedly used by mediaeval witches for astral projection.)) ((Note: Double parentheses indicate the author's (MGH's) comments))

The bright Gods help the wayfarers who meet these bedevilled astral materialisations, fierce and conscienceless as animals, cruel and crafty as men! But the bright Gods are sleeping, and only the hosts of the Midnight Sun, ghosts, goblins and all evil things, are abroad. The creatures return, their jaws dripping with blood, their hides draggled with filth, ere morning dawns, and, crouching on the huddled forms on the floor of the cavern, sink into them and disappear.

Such orgies as these were held from time to time, Oduarpa using them to increase his hold upon the people, and he established similar rites at many places, making himself the central figure in all, becoming a veritable object of worship, and gradually welding the people together in allegiance to himself, until he became the acknowledged Emperor. His relations with the inhabitants of the Nether World - called in latter days, as said above, the Kingdom of Pan, gave him much additional power, and he had trusted lieutenants - bound to him by their common knowledge of, and participation in, the ghastly abominations of that realm - ever prompt to carry out his commands.

He finally succeeded in assembling a very large army and began his march against the White Emperor, directing his course towards the City of the Golden Gates. He hoped to overawe and conquer, not only by fair assault of arms, but by the terror that would be spread by his hellish allies, and the ghastly transformations of the black wizards into animal forms.

He himself had a body-guard of magic animals round him, powerful desire-forms materialised into physical bodies, who guarded him and devoured any who approached him with hostile intent. When a battle was raging, and the issue doubtful, Oduarpa would suddenly loose against his foes his horde of demoniacal allies, who would rush into the fray, tearing with teeth and claws, and spread panic among the startled hosts. When his enemies broke into flight, he would send these swift demons in pursuit, and the troops of wizards would likewise take animal forms, gorging themselves on the bodies of the slain.

Thus he fought his way onwards, northward ever, till he came near the City of the Golden Gates, where the last army of the White Emperor lay embattled.

Alcyone had fought as a soldier in the army, partly under a spell, and yet awake enough to be sick at heart at his surroundings, and Cygnus, with other ladies had accompanied the camp. The day of the decisive battle dawned; the imperial army was led by the White Emperor himself, Corona, and the right wing of the army was under the command of his most trusted general, Mars. During the preceding night, Alcyone had been visited once more by his early vision, and had heard the well-loved voice: "Alcyone, thou art fighting against thy true lord, and to-morrow wilt thou meet me, face to face. Break thou then thy rebel sword and yield thee to me; thou shalt die by my side, and it shall yet be well."

And so indeed it happened. For in the fierce shock of battle, as the imperial troops were giving way, the Emperor slain, Alcyone saw, struggling gallantly against overwhelming odds, the face of his vision, the general, Mars. With a cry he sprang forward, breaking his sword in two, and catching up a spear, he threw himself at Mars' back, fiercely thrusting through a soldier who struck at Mars from behind. At that moment Oduarpa charged up, mad with fury, and struck Mars down, and with a cry that rang across the field, he summoned Cygnus, by swift spell changing her into a fierce animal, which rushed with bared fangs at Alcyone, fainting from loss of blood. But in the very act, the love which had been her life cried out from Cygnus' soul and wrought her rescue; for its strong flow changed into loving woman the form of ravening hate, and with a dying kiss on Alcyone's dying face she breathed away her life.

Herakles, the wife of Mars. was captured by Oduarpa in the assault on the City of the Golden Gates that followed and completed his victory; she indignantly repulsed his advances, and catching up a dagger stabbed at him with all her strength. The dagger slipped aside on his metallic casing, and, laughing, he struck her down, outraging her as she lay half senseless: when she recovered consciousness, he summoned his horrible animals, and they tore her into pieces and devoured her.

Oduarpa, enthroned on a pile of corpses, and surrounded by his animal and half-animal guards, was crowned Emperor of the City of the Golden Gates, assuming the desecrated title of Divine Ruler. But his triumph was not of long duration, for Vaivasvata Manu marched against him with a great army, and His mere presence put to flight the denizens of the Kingdom of Pan, while He destroyed the artificial thought-forms, created by black magic. A crushing victory scattered the army of the emperor, and he himself was shut up in a tower whither he had fled in the rout. The building was fired, and he perished miserably, literally boiled to death within his materialised metallic shell.

Vaivasvata Manu purified the City and re-established there the rule of the White Emperor, consecrating to that office a trusted servant of the Hierarchy. For a time things went on well, but slowly the evil again gathered power, and the southern centre once more grew strong; until, at last, the same Lord of the Dark Face, appearing in a new reincarnation, again fought against the White Emperor of the time, and set up his own throne against him. Then the words of doom were spoken by the Head of the Hierarchy ((The "Ancient of Days", or Sanat Kumara)), the "Great King of the Dazzling Face" - the White Emperor - sent to his brother Chiefs: "Prepare. Arise, ye men of the Good Law, and cross the land while yet dry." The "Rod of the Four" - the Kumaras - was raised ((Sanat Kumara and His 3 Disciples)).

"The hour has struck, the black night is ready." The "servants of the Great Four" warned their people, and many escaped. "Their Kings reached them in their Vimanas** and led them on to the lands of fire and metal [east and north]". Explosions of gas, floods and earthquakes destroyed Ruta and Daitya, the huge islands of Atlantis, left from the catastrophe of 200,000 BC, and only the island of Poseidonis remained, the last remnant of the once huge continent of the Atlantic. These islands perished in 75,025 BC, Poseidonis enduring to 9,564 BC when it also was whelmed beneath the ocean.

((End of verbatim text [2].))

Footnote:
** Chariots which moved in the air - the ancient aeroplanes. ((See p. 230 below))

People from Atlantis colonised many other areas: Egypt, Mesopotamia, India, North and South America [3, 2] and they reached a high standard of civilisation and developed devices which have not yet been re-invented in modern times, such as an anti-gravity drive for flying vehicles called vimanas, which are described in the great epics of India (Puranas, Samarangana Sutradhara [5, 7]; this is further discussed on page 230 below). Poison gas bombs are described as having been dropped from these air vessels during wars [2]. Powerful rockets are also described [7]. Heavy stone blocks were reported lifted by anti-gravity, which allowed the Egyptian pyramids to be easily built, more than 77,000 years ago, not the mere 5,000 years ago assumed in modern times mainly because a graffiti artist had inscribed the cartouche of Khufu (Cheops) in them! ("Conventional" Egyptology does admit that the earliest dynasties presently known, had suddenly appeared, fully formed, inexplicably.)

When a great flood occurred about 77,000 years ago, due to part of Atlantis sinking, the local people tried to climb the pyramids to escape the water, but most were drowned [3, 2]. There are very obvious horizontal water erosion lines along the Sphinx, but some would say these are due to wind erosion.

Due to the provision of universal primary education, there was an absence of poverty in ancient times. Remote viewing (by ESP) was an innate ability of most people in those times, and it greatly aided scientific development. In the life sciences, different types of grain were created by crossing different grasses. The banana was developed from a melon-like ancestor and is now found in both West Africa and Eastern South America but with no trace of its ancestor (good evidence for a now-missing linking continent where they were developed; the original normal seeds of the banana required much time to have become vestigial – the banana tree does not now propagate by seeds). One bad Atlantean experiment resulted in the production of wasps instead of bees [3,2]. "Surprising" DNA similarities link the Pharaohs to populations around Yucatan.

Two Atlantean Empire civilisations in Peru (12,000 BC) and in Chaldea (19,000 BC) are described in detail [2]. These are similar to those in Egypt and in India in their Atlantean periods.

CHAPTER 4 World History from 70,000 BC:
Buddha & Christ in their previous lives

Vaivasvata Manu led out His chosen group, selected from the civilisation in Atlantis before the major flood disaster there of 75,025 BC. He settled them first in Arabia and then at the White Island of Shamballa in the Gobi Sea (now the Gobi Desert). He planned very carefully and slowly for the development of humanity in the succeeding centuries. Centuries later, a great city at Shamballa was built, whose streets radiated out like the spokes of a wheel, now long-buried under the Gobi Desert. [3, 2]

At 70,000 BC, this community numbered about 7000 people, in villages along the southern shore of the (then) inland Gobi Sea. The Manu, as king, lived on the island and was rarely seen on the mainland, which was governed on his behalf by his son, Jupiter, in a patriarchal system where the 5 sons of Jupiter also governed for him. This system worked only because almost all were near to the high development standard of what we would call Saints (perfected beings), with a truly divine right of Kings. His eldest son, Mars, ruled one of the villages where he had a large house with his wife, Mercury. It was here that Alcyone was re-born. The house was surrounded by large trees and lawns where Mars gathered meetings when he needed to address the people. His brother was Sirius and sister was Mizar and the bond between them all has lasted down to the present day, in successive re-incarnations. Education of children was given mainly outdoors. The brothers Sirius and Alcyone fell in love with two sisters, Vega and Leo, who also reappeared together in many successive incarnations down the ages to the present day. These two sisters were then the grand-daughters of a Chief, Corona (who is, 70 millenia later, known to us as Julius Caesar). [3, 2]

When the Manu grew very old, He sent for His Chiefs, Jupiter, Corona, Mars and Vajra, forewarning them that their community would be almost exterminated by savage nomads from the north (the workings of karma) and that they should arrange to save chosen children (i.e. those whose karma allowed it) so that the group could later continue, but in bodies that would become more suitable. They sent the chosen group over to the White Island (Shamballa) for their safety. [3, 2]

Alcyone and Leo had 4 children and 2 were daughters, Surya and Brihat.
Much later, Surya became the Christ and Brihat became a perfected soul (a Saint or Master in today's language) who, as Jesus, prepared

the body which the Christ took over for His 3-year ministry in Palestine (see Chapter 5), the founding of the Christian religion. [3,2]

The Manu was reborn as a child of Saturn and Surya. Mars and Viraj followed, reborn as brother and sister. Jupiter, Selene and Corona re-appeared as the children of Elektra.

B&L give many details in their 700 page book [3], too many to reproduce here, so this Chapter is confined to 22 lives of Alcyone in which Surya (Christ) appears. For 20 other lives, see [14].

Ten thousand years later (about 60,000 BC): Surya was reborn, as his own grand-daughter, and the Manu married Her and a new race was created by this union. There were 12 children and later each of them also had 12 children, and their offspring also had 12 children (clearly not an accidental occurrence). The original group re-appeared in these and later incarnations in varying relationships to each other. Their community developed over the decades and was considered to be a divine race called the Children of the Sun, each being identified with one of the 12 signs of the zodiac. [3, 2]

The Manu organised the building of a huge capital city, which took 1000 years to complete and it lasted for 50,000 years until the cataclysm of the final sinking of Atlantis in 9,564 BC ruined it. When the city was built, it had temples and rooms used for teaching, one of which contained a series of carvings of atoms arranged in order of atomic number, with explanations: this was about 60,000 years ago.

On rare occasions the Manu was received in audience by the Ruler of the Planet, called "The Ancient of Days" in Christian theology and the "Sanat Kumara" in Hindu theology, and on one occasion the invitation was also made to the Manu's 3 elder sons, from which they had the experience of a blessing while standing in the presence of the Ruler, whose location was (and still is) on the island (now a desert) at Shamballa, but in a higher spatial dimension (see Chapter 1), from where human progress is monitored. [3, 2]

Ten thousand years passed and the community grew without outside interference. Their only neighbours were half-Atlantean and half-Lemurian (an earlier lost continent which was in the Pacific Ocean).

-=-=-=-=-=-=-=-=-=-=-=-=-=-=-

The next description [3] is 18,000 years later (42,000 BC) in a large city called Manoa, which is the big city on the shore opposite Shamballa (which was on White Island) mentioned above, in central Asia. Alcyone married Perseus and they had 6 children and their nephews included Surya, Mars and Mercury. The Manu then re-appeared as the son of Mars and Mercury. Their valley was picturesque, very wild and rugged and covered with primeval forest. There are many more details, too numerous to be repeated here.

-=-=-=-=-=-=-=-=-=-=-=-=-=-=-

An annual ceremony at Manoa is described verbatim [2] below:

At an early hour men, women and children marched in procession along the converging streets into the giant crescent, which faced the mighty bridge to the island. Rich silken cloths fluttered from windows and flagstaffs, and the roads were strewn with blossoms … . When the hall was filled to its utmost extent, except for the front, a stately group entered and filled this space, and all bowed in homage. There stood the three Manus, in their robes of office, and the Mahaguru, the Bodhisattva of the time (Vyasa) standing beside Manu Vaivasata.

Surya was close behind Vyasa, His Brother and Predecessor. Nearest to the throne were the three Kumaras, probably unseen by the crowd, supended in the air, in a semicircle, with purple and silver devas also attendant.

Continuing below is a verbatim text extract {p.279-284 in [2]}:

Then over the whole vast assembly fell an utter silence, as though men could hardly bear to breathe; and softly, sweetly, scarce seeming to break the silence, stole out an exquisite strain of music, supporting a chant, intoned by those Mightiest and Holiest who stood around the throne, an invocation to the Lord, the Ruler, to come among His own. The solemn hushed accents died into silence, and then rang out a single silvery note, as though in answer; the great golden Sun blazed out in dazzling splendour, and below it, just over the throne, flashed out a brilliant Star, its beams like lightning shooting forth above the heads of the waiting throng; and He was there, the supreme Lord of the Hierarchy, seated on the throne, more radiant than Sun and Star, which indeed seemed to draw their lustre from

Him; and all fell on their faces, hiding their eyes from the blinding glory of His Presence.

Then, in His gentleness, He softened that glory, so that all might lift their eyes, and see Him, Sanat Kumara, the 'Eternal Virgin'**, in all the beauty of His unchanging Youth, who was yet the Ancient of Days. And a deep breath of awe and wonder came from the adoring crowd, and a luminous smile, rendering the exquisite strong beauty of the Face yet more entrancing, answered their simple reverent gaze of love and worship.

** The name, translated from the Sanskrit, means Eternal Virgin, the termination showing that 'Virgin' is masculine.

Then He stretched forth His Hands towards the altar in front of Him, and fire blazed forth upon it, the flames rising high in air. And then He was gone -- the throne was empty, the Star had vanished, the golden Sun glowed but faintly, and only the Fire which He had given leapt unchanged upon the Altar. From this a glowing fragment of wood was given to the priests for the altars of the various Temples, and to each head of a household present there, and he received it in a vessel with a lid which closed above it, wherein it remained, live fire, unquenchable, till it had been carried to the altar of the home.

The processions re-formed and left the Holy Place in Silence, again passing to the Bridge and by it reaching the City. Then came an outburst of joyous singing, and hand-in-hand the people passed along, and congratulations were exchanged, and the elders blessed the youngers, and all were very glad. The sacred fire was placed on the family altar, to set alight the flame which was to be kept alive through the year, and brands lighted at it were taken to the houses of those who had not been present, for until the recurrence of the festival when another year had run its course, such fire could not be had to hallow the family shrine. After this, there was music, and feasting, and dancing, until the happy City sank to sleep.

Such was the Festival of the Sacred Fire, held on every Midsummer Day in the City of the Bridge.

Some of the people devoted themselves almost wholly to study, and reached great proficiency in esoteric science, in order to devote themselves to certain branches of the public service. They became clairvoyant, and gained control of various natural forces, learning to make thought-forms and to leave their physical bodies at will. Mindful of the melancholy results in Atlantis of occult power divorced from unselfishness and morality, the instructors in these studies chose their pupils with extreme care, and one of the lieutenants of the Manu maintained a general supervision over such classes. Some of the students, when proficient, had it as their special duty to the State to keep the different parts of the Empire in touch with each other; there were no newspapers, but they conducted what may be called a news department. News was not published as a rule, but anyone who wanted news about anyone else in any part of the Empire could go to this central office and obtain it. Thus, there were Commissioners for the various Countries, each of whom gave information about the country in his charge, obtaining it by esoteric means. Expeditions sent out on errands of peace or war were thus followed and news was given of them, as in modem days by wireless or other telegraphy.

On one occasion, when Corona was ruling a distant country, the Manu was not able to impress him with His direction; so He bade one of these trained students to leave his physical body, go astrally to Corona, and materialise himself on arrival; by this device, the message was delivered to Corona in his waking consciousness. In this way the Manu remained as the real Ruler, no matter how far the Empire extended.

Writing was done on various substances; one man was observed writing with a sharp instrument on a waxy-looking surface in an oblong case, as though he were etching; then he went over it again with a hollow pen, out of which flowed a coloured liquid which hardened as it dried, leaving the script embedded in the wax. Occasionally a man would strike out a method of his own.

Machinery was not carried to the point reached in Atlantis; it was simpler, and more of the work was done by hand. The Manu evidently did not desire the extreme luxury of Atlantis to be reproduced among His people.

From the small beginning of 60,000 BC, there gradually grew up a thickly populated kingdom, which surrounded the Gobi Sea, and obtained dominion by degrees over many neighbouring nations, including: the Turanians who had so mercilessly massacred its forefathers.

This was the root-stock of all the Aryan nations, and from it went out - from 40,000 BC, onwards - the great migrations which formed the Aryan sub-races. It remained in its cradle-land until it had sent out four of these migrations westwards, and had also sent many huge bands of conquering emigrants into India, who subdued the land and possessed it; its last remnants only left their home and joined their forerunners in India shortly before the sinking of Poseidonis in 9,564 BC, they were sent away, that they might escape the ruin wrought by that tremendous cataclysm.

From 60,000 BC, to 40,000 BC, the parent-stock grew and flourished exceedingly, reaching the zenith of its first glory at about 45,000 BC. It conquered China and Japan, peopled chiefly by Mongols - the seventh Atlantean sub-race-going northward and eastward till stopped by the cold; it also added to its Empire Formosa and Siam, which were populated by Turanians and Tlavatli - fourth and second Atlantean subraces. Then the Aryans colonised Sumatra and Java and the adjoining islands - not quite so much broken up as now; for the most part they were welcomed in these regions by the people, who looked on the fair-faced strangers as Gods, and were more inclined to worship than to fight them. An interesting remnant of one of their settlements, still left in Celebes, is a hill tribe called Toala. This island, to the east of Borneo, came under their sway, and they stretched down over what is now the Malay Peninsula and over the Philippines, the Liu-Kiu Islands, the Eastern Archipelago, and Papua, the islands on the way to Australia, and over Australia itself, which was still thickly populated with Lemurians - third Root Race.

We found Corona, about 50,000 BC, ruling over a large kingdom in these island-studded seas; he had been born in that region, and made for himself a kingdom, recognising the Manu as Overlord, and obeying any directions which he received from Him.

Over all the huge Empire with its many kingdoms, the Manu was Suzerain. Whether He was in incarnation or not, the Kings ruled in His name, and He sent directions from time to time as to the carrying on of the work.

By 40,000 BC, the Empire began to show signs of decline, and the islands and the outer provinces were asserting a barbarian independence. The Manu still occasionally incarnated, but usually directed things from higher planes. The central kingdom, however, remained splendid in civilisation, contented and quiescent, for another twenty-five thousand years and more, while activities were chiefly carried on in directions further afield, in the building up of sub-races, and in their spreading in all directions.

((End of verbatim text {p. 279-284 in [2]}.))

-=-=-=-=-=-=-=-=-=-=-=-=-=-=-=-

In 40,000 BC, the Manu decided to move his civilisation into Arabia, by setting up an army of 150,000, plus 100,000 more as wives, children etc, including most of the central core people mentioned above.   Progress was made towards Arabia through the friendly regions of Persia and Mesopotamia. During this progress, Surya was present as a child. (Refer to Tables 1 & 2 above for pseudonyms used)

In this 5th life of Alcyone, in 40,000 BC, in Arabia, Alcyone recounts a memorable meeting with Thoth (Hermes) who was on his way eastward after appearing in Egypt as the Bodhisattva of the previous period (which overlaps this one, in which Surya {Christ} is the current Bodhisattva). Egypt in those times (40,000 BC) was part of the Atlantean Empire at its height (see [2], and {p. 64 in [3]}). Surya was the young third son of Alcyone at this time and he (Surya) earnestly asked Mahaguru to allow him to serve Him forever. Mahaguru recognised the divine call as arising from services done by Surya in the distant past (which were hidden from Surya at that time, as he had not yet achieved sainthood). But Mahaguru took Surya to the Manu and asked that he (Surya) should be appointed high priest of the new religion which He (Mahaguru, as Thoth) had just founded in Egypt.   The Manu agreed. Alcyone also entered the priesthood and rejoiced to serve in it under his talented son, Surya, through whom the Mahaguru (Thoth) could speak even when physically distant. Three other sons of Alcyone, Mercury, Sirius and Selene all felt the inrush of the same sacred fire and vowed their lives to its

service. Mahaguru (Thoth) accepted them all, for He knew their past and their future and He ordained them as priests. His parting blessing to them is given below, verbatim (in quotation marks):
Turning first to Surya, and then to the rest, He said:

"Hail! my Brother through the ages; hail ! my brothers yet to be ; you shall spread God's Love and Wisdom over the world from sea to sea. Many and great shall be your difficulties and trials, yet greater still shall be your reward; for many thousands of years you must toil in preparation for the task that few can perform, but when it is achieved you shall shine as the stars in heaven, for yours is the blessing of those who turn many to righteousness. There is a spiritual dynasty whose throne is never vacant, whose splendour never fails; its members form a golden chain whose links can never be torn asunder, for they draw back the world to God from whom it came.

To that you all belong; its labours and its lustre you must share.
Happy are you among men, my Brothers of the Glorious Mystery, for through you the Light shall shine. More and more shall the Hidden Light become manifest; more and more shall the Hidden Work be done openly and be understood by men; and yours shall he the hands that raise the veil, yours the voices that shall proclaim the glad tidings to the world. Bearers of freedom and light and joy shall you be, and your names shall be holy in the ears of generations yet unborn. Farewell; in this body you will see me no more, but forget not that in spirit we are always together."

Mahaguru (Thoth) then left them and passed to Shamballa, not to be seen again for 10,000 years (to next be the first Zarathustra). [3, 2]

2,000 years later (38,000 BC) Alcyone was reborn as a female in Africa, where the Arabian Empire had then spread out in many directions. The west coast was Atlantean territory, and Arabia and Egypt divided the rest between them. [3, 2]

-=-=-=-=-=-=-=-=-=-=-=-=-=-=-

Fig. 4.3.   Surya = (later) Christ (World Teacher or Bodhisattva, Maitreya).

Drawn by Remote Viewing, from His appearance as it was in 1930 [4].

Note: It is generally accepted that no pictures of Christ which may have been made during his ministry in Palestine, have survived. Many icons and mosaics have been made by artists, but centuries later.   See two coloured icons in Fig. 5.3 (between p. 63 & 64).

See the Note under the previous (Fig. 4.2) caption.

The 9th life of Alcyone, in 30,275 BC, in Persia, found the Manu ruling with Surya and Orpheus as His sons, and grandsons Mars, Corona, Vulcan, Theodoros and Vajra.

-=-=-=-=-=-=-=-=-=-=-=-=-=-=-

The 10th life of Alcyone, also in Persia 800 years later, found Mahaguru (Thoth) reappearing as the first Zarathustra (Zoroaster) in 29,700 BC, to found a new religion (Zoroastrianism). Surya had already been born there and was Chief Priest. Mahaguru took over a body prepared for Him by Mercury, the second son of King Mars who was the King of Persia then. The brother of Mars was Corona, King of Mesopotamia. Mars had a family of 7 children, 4 sons and 3 daughters, all members of the group associated in past lives. His eldest son was Jupiter, then Mercury, Alcyone and Orion. The sisters were Elektra, Rama and Fides, while other old friends were found close by in the family of the Chief Priest, Surya, who had Sirius among his daughters. In the 500 years since the conquest of Persia, there had been much progress and the capital had become a fine, spacious well-arranged city, with some magnificent architecture. Much irrigation had converted the barren country into a fertile area.

The splendid ceremony which celebrated the occupation by the Mahaguru of the body of Mercury, is beautifully described in reference [2], a procession amidst cheering thousands, the sermon of the Mahaguru, the Rod of Power, the Fire which fell from heaven and the Blazing Star which brought the blessings of the Ruler of the World (The Ancient of Days or Sanat Kumara).

A verbatim extract {p. 309-316 in [2]} describing this is given below:

<u>The Third Sub-race, the Iranian</u>

In a few centuries this Persian civilisation dominated the whole of western Asia from the Mediterranean to the Pamirs, and from the Persian Gulf to the Sea of Aral. With certain changes its Empire lasted until about 2,200 BC.

In this long period of twenty-eight thousand years, one event stands out as of supreme importance: the coming of the Mahaguru as the first Zarathustra, the founding of the Religion of the Fire, in 29,700 BC.

The country had become fairly settled under the reigns of the Kings who had succeeded Corona, of whom Mars, the Ruler of the time - of course in a new body - was the tenth. Military rule had passed away, though occasional raids reminded the inhabitants of their turbulent neighbours on the further side of the ring of forts, now well-built and strong. It was in the main an agricultural country, though large numbers of herds and flocks were kept, and it was these which specially tempted descents from the hills.

The second son of Mars was Mercury, and his body was chosen as the vehicle for the Supreme Teacher; Surya was the Chief Priest, the Hierophant of the time, at the head of the State religion, a mixture of Nature and Star Worship, and he wielded an immense authority, chiefly because of his office, but also partly because he was of the blood royal. The fact that Mercury had been chosen to surrender his body for the use of the Mahaguru had been communicated to his father as well as to the Chief Priest, and from his childhood he had been carefully trained in view of his glorious destiny, Surya taking charge of his education, and the father co-operating in every way in his power.

The day arrived when the first public appearance of the Mahaguru was to be made; He had come from Shamballa in His subtle body, and had taken possession of the body of Mercury, and a great procession started from the Royal Palace to the chief Temple of the city. In it walked, on the right, under a golden canopy, the stately figure of the King; the jewelled canopy of the High Priest glittered on the left; and between them was carried, shoulder-high so that all might see, a golden chair, in which sat the well-known figure of the King's second son. But what was there that caused a murmur of surprise, of wonder, as he passed along? Was that really the Prince, whom they had known from childhood? Why was he carried high as the centre of the procession, while King and Hierophant walked humbly beside him? What was this new stateliness, this unknown dignity, this gaze, so piercing yet so tender, that swept across the crowd? Not thus had held himself, not thus had looked at them, the Prince who had grown up among them.

The procession swept on and entered the huge courtyard of the Temple, crowded with people in the many-coloured garments of festival days, when each wore a mantle of the colour of his ruling planet; down the sides of the steps which rose to the platform in front of the great door of the Temple were ranged the priests in long white garments, and rainbow-coloured over-robes of silk; in the midst of the platform an altar had been erected, and on it wood was piled, and fragrant gums, and incense, but no smoke arose - for the pile, to the people's surprise, was unlighted.

The procession passed on to the foot of the steps, and there all halted, save the three central figures; they ascended the steps, the King and the Hierophant placing themselves to the right and left of the altar and the Prince, who was the Mahaguru, in the centre behind it.

Then Surya, the Hierophant, spoke to the priests and to the people, telling them that He who stood there behind the altar was no longer the Prince they had known but that He was the Messenger from the Most High and from the Sons of the Fire who dwelt in the far East, whence their forefathers had come forth. That He had brought Their word to Their children, to which all should yield reverence and obedience, and he bade them listen while the great Messenger spake in Their Name. As the Head of their faith, he humbly bade Him welcome.

Then over the listening throng rang the silver voice of the Mahaguru. and none there was who could not hear it as though spoken to him alone. He told them that He had come from the Sons of Fire, the Lords of the Flame, who dwelt in the sacred City of the White Island, in far Shamballa. He brought them a revelation from Them, a symbol which should ever keep Them in their minds. He told them how the Fire was the purest of all elements and the purifier of all things, and that thereafter it should be for them the symbol of the Holiest. That the Fire was embodied in the Sun in the heavens, and burned, though hidden, in the heart of man. It was heat, it was light, it was health and strength, and in it and by it all had life and motion. And much He told them of its deep meaning, and how in all things they should see the hidden presence of the Fire.

Then He lifted up His right hand, and behold ! there shone in it a Rod, as of lightning held in bondage, yet shooting out its flashes on every side; and He pointed the Rod to the East of the Heavens, and cried some words aloud in an unknown tongue; and the heavens became one sheet of flame, and Fire fell blazing down upon the altar, and a Star shone out above His Head and seemed to bathe Him in Its radiance. And all the priests and the people fen upon their faces, and Surya and the King bowed down in homage at His feet, and the clouds of fragrant smoke from the altar veiled the three for a few moments from sight.

Then, with His hand upraised in blessing, the Mahaguru descended the steps, and He, with the King and the Hierophant, returned with the procession to the Palace whence they had come. And the people marvelled greatly and rejoiced, because the Gods of their forefathers had remembered them, and had sent them the Word of Peace. And they carried home the flowers which had rained down upon them from the sky when the Fire had passed, and kept them in their shrines as precious heirlooms for their descendants.

The Mahaguru remained for a considerable time in the city, going daily to the Temple to instruct the priests; He taught them that Fire and water were the purifiers of all else, and must never be polluted, and that even the water was purified by the Fire; that all fire was the Fire of the Sun, and was in all things and might be released as fire; that out of the Fire and out of the water all things come, for the Fire and the water were the two Spirits, Fire being life and water form. (Possibly out of this arose the later teaching of Ormuzd and Ahriman. There are passages which show that the double of Ormuzd was not originally an evil power, but rather matter, while Ormuzd was Spirit.)

The Mahaguru had round Him a quite august assemblage of Masters, and others less advanced. He left these to carry on His teaching when He departed.

His departure was as dramatic as His first preaching.

The people were gathered together to hear Him preach, as He was wont to do occasionally, and they knew not that it was for the last time. He stood, as before, on the great platform, but there was no altar. He preached, inculcating the duty of gaining knowledge and of practicing love, and bade them follow and obey Surya, whom He left in His place as Teacher. And then He told them that He was going, and He blessed them, and lifting up His arms to the eastern sky He called aloud; and out of the sky came down a whirling cloud of flame, and enwrapped Him as He stood, and then, whirling still, it shot upwards and fled eastwards, and - He was gone.

Then the people fell on their faces and cried out that He was a God, and they exulted exceedingly that He had lived among them; but the King was very sad, and mourned for His departure many days.
And Mercury, who, in his subtle body, had ever remained near Him, at His service, returned with Him to the Holy Ones, and rested for a while in peace.

After He had gone, Star-worship did not at once disappear, for the people regarded His teaching as a reform, not as a substitution, and still worshipped the Moon, and Venus, and the constellations, and the planets; but the Fire was held sacred as the image, the emblem, and the being of the Sun, and the new religion rather enfolded the old one than replaced it. Gradually the Faith of the Fire grew stronger and stronger; Star-worship retreated from Persia to Mesopotamia, where it remained the dominant faith, and took a very scientific form. Astrology there reached its zenith, and scientifically guided human affairs, both public and private. Its priests possessed much esoteric knowledge, and the wisdom of the Magi became famed throughout the East. In Persia, the Religion of the Fire triumphed, and later Prophets carried on the work of the great Zarathustra, and built up the Zoroastrian Faith and its literature; it has endured down to our own day.

The third sub-race numbered about a million souls when they settled down in Persia and Mesopotamia, and they multiplied rapidly under the favourable conditions of their new home, and also incorporated in their nation the sparse population which existed in the country when they entered it.

In the twenty-eight thousand years of the Persian Empire there were naturally many fluctuations; most of the time Persia and Mesopotamia were under separate rulers, of whom sometimes the one, sometimes the other, was nominally Overlord; sometimes the two countries were split up into smaller States, owing a kind of loose feudal allegiance to the central King. All through their history they had constantly recurring difficulties with the nomad Mongolians on one hand, and the mountaineers of Kurdistan and the Hindu Kush on the other. Sometimes the Iranians drew back for a time before these tribes; sometimes they pushed the frontier of civilisation further forward, and drove the savages back. At one period they ruled most of Asia Minor, and made temporary settlements in several of the countries bordering the Mediterranean; at one time they held Cyprus, Rhodes and Crete; but on the whole in that part of the world the Atlantean power was too strong for them, and they avoided conflict with it. At this western boundary of their kingdom powerful Scythian and Hittite confederations disputed their dominion at various points of their history; once at least they conquered Syria, but seemed to have found it a useless acquisition and soon abandoned it; and twice they embroiled themselves with Egypt, against which they could do but little.

During most of this long period they kept up a high level of civilisation, and many relics of their mighty architecture lie buried beneath desert sands. Various dynasties arose among them, and several different languages prevailed in the course of their chequered history. They avoided hostilities with India, being separated from it by a wild territory - a sort of no-man's-land; Arabia troubled them but little, for there again a useful belt of desert intervened. They were great traders, merchants, manufacturers - a much more settled people than the second sub-race, and with more definite religious ideas. The best specimens of the Parsis of the present day give a fair idea of their appearance. The present inhabitants of Persia have still much of their blood in them, though largely commingled with that of their Arab conquerors. The Kurds, the Afghans, and the Baluchis are also mainly descended from them, though with various admixtures.

((End of verbatim text {p. 309-316 in [2]}.))

Continuing from above, in the 10th life of Alcyone, Alcyone and Sirius married, with a blessing from Surya. But Orion who had also fallen in love with Sirius plotted to kill both Alcyone and Sirius. He stabbed Alcyone and abducted Sirius but had to jump into a lake with her, to escape pursuers. Alcyone and Sirius both survived. King Mars banished Orion. Mahaguru told Orion that only once in thousands of years does it come to one to help in setting up a major new religion, and that he (Orion) must tread a long and weary road before being allowed to do so again. It was not until some thousands of years later that Orion returned to the group. (Refer to Tables 1 & 2 above for the pseudonyms used)

Alcyone devoted himself to work under Mahaguru and Surya, to develop the new Religion of the Fire, which was founded by Mahaguru as the "first" Zarathustra in 29,700 BC. Later prophets continued this religion and built up the Zoroastrian Faith and its literature, which has endured to the present day.

In this time period, the group were almost all concentrated into 2 generations, and it was not until the 20th century AD that the group again re-appeared in this concentrated form. This is due to the need for a new religion requiring a strong impulse to establish it -- i.e. the imminent re-appearance of the Bodhisattva (currently Surya) is expected soon.

Alcyone died aged 85 and on his death-bed Mahaguru appeared to him in the figure of his brother Mercury – not aged like Alcyone but as young as he was on the transfer (mentioned above). Alcyone passed through the Astral (Purgatory, 4th dimension) immediately and during the 800 years of his life in the heaven world (5th dimension) Mahaguru was the principal figure in his surroundings, and in the bliss of His presence he grew like a flower eagerly opening its heart to the sun.

In his next (11th) life, Alcyone was again born in Persia, in 28,804 BC, which had then become an old and well-established civilisation. Atlantean merchants had established colonies at the eastern end of the Mediterranean and caravans went from there to Persia, taking there the wares of Atlantis. Alcyone had several adventures there, too long to relate here. See in the author's other book [14].

-=-=-=-=-=-=-=-=-=-=-=-=-=-=-=-=-=-

The 13th life of Alcyone was in Ireland in 27,527 BC, a time when the English Channel did not exist and the Thames was a tributary of the Rhine [3].

The polar ice cap extended far south from the North Pole and the present author (MGH) recalls reading somewhere that the ice cliff is said to have ended just north of where London now is, at Watford. A human skeleton called "The red lady of Pavilland", found at Gower in South Wales has been radio-carbon dated as 31,000 BC and Gower was then about 70 miles from the sea, due to the lower sea levels. This indicates that Ireland at about that time, would not have been under the ice sheet. A convenient reference for "The red lady of Pavilland" carbon dating is Wikipedia.

The original population was described [3] as smaller and darker with broader Mongolian faces than the invaders from the south. The natives accepted the domination of the white invaders (who had arrived with little opposition some decades earlier), believing them to be a semi-divine race, who were kindly in their bearing. Their leaders were, as usual, a King (Mars) and a High Priest (Surya); King Mars had married Vesta who was a cousin of Surya.
((Recall, "*And did those feet, in ancient time, walk upon England's mountains green*" – William Blake; there was no Irish Sea then, so England and Ireland were one land.))

Surya was led to his own consort by a strange recurring vision. In the race, visions were common and considered important. Surya was only 10 years old when it first occurred. The full description is given verbatim below, with a very few author's (MGH's) comments in double parentheses as ((comment)):

Surya was 10 years old when a recurring vison first came to him, or rather when he first clearly remembered it. In his sleep it seemed to him that he was floating high in the air, looking down on a city of marvellous beauty, a city larger than he had ever seen with his physical eyes or imagined as possible in his physical brain; a city on the shore of a great lake, in which near the shore was an island covered with glorious white buildings which seemed to his entranced gaze like the very courts of heaven itself. ((This would have been Manoa in central Asia – see earlier lives.))

Yet not to the wondrous island was he drawn, but to a large, low rambling house, which stood in an extensive park of its own a little way outside the city. And in that park he saw a little girl ((Dhruva, named later in text)) perhaps eight years old - a little girl of rare beauty, whom he somehow knew quite well and loved with an intensity of affection which astonished him. She stood all alone at the edge of a tank with massive stone walls, watching the sporting of some bright-hued fish that dwelt therein; and even as he floated low to see her face more clearly she leaned over too far, and fell with a cry of fright into the water. Obviously she could not swim; there was no-one there to help her, and the wall rose sheer and smooth several feet above her head; but before she could sink a second time Surya somehow found himself in the water beside her, holding her up, and trying to swim with her across the tank to some steps which ran down to the level of the water. It was a tremendous strain upon him, for she threw her arms round his neck and impeded his motions; indeed, she all but drowned him, for his strength was going from him when after a last despairing effort he felt his feet touch the steps. Somehow they staggered up them, and threw themselves upon the grass; and the girl, who had not yet unwound her arms from about his neck, looked deeply into his eyes and gave him a long, loving kiss. And then - he woke in his own bed in far-away Ireland with that kiss still upon his lips, and his clothes all dripping with the water of that Central-Asian tank !

So excited was he by the adventure, and so certain that it was a real occurrence and no mere vision, that he rushed at once to the room of his father and mother and waked them to hear his story, showing his dripping garments in proof of it. They were much amazed, and could not comprehend how such a thing might be; yet they did not disbelieve, for among their race there were traditions of rare events not quite unlike this - of priests who had had the power of appearing and disappearing mysteriously, of showing themselves at a distance from their sleeping bodies, and sometimes even of striking or of saving men who, were physically far away. And Surya's mother was already predisposed to believe wondrous things of this noble and fearless son of hers; so like another Mother in later history, she kept all these things and pondered them in her heart. But Surya wondered greatly how he knew that girl so well, and loved her so intensely, and even then as a little boy he vowed that to her and to no other should his life be devoted - that she and no other should be his wife when he grew up to manhood.

The memory of his strange adventure remained fresh and clear-cut in his mind; and as he had some skill in drawing, he drew several portraits of the little girl, and also made a drawing of the tank and the house which he had seen. He had no idea in what part of the world these places were situated, nor could his father the Chief Priest help him in discovering this, for though the priests were the principal depositories of the knowledge of the nation, geography was not a strong point among them.

But though he did not know where she lived, he thoroughly believed in the physical existence of the heroine of the story, and resolved that when he became a man he would find her. He was a boy of many day-dreams, and she always played a large part in them. He liked much to be alone, and often spent hours quite contentedly walking or lying in the sunlight, and telling himself interminable stories in which he and she passed through all sorts of stirring adventures. By thus constantly dwelling upon her perfections he naturally fanned the flame of his love, and at last he resolved to make a mighty effort to leave his body and reach her once more by definite materialisation. He had long before questioned his father as to the possibility of doing this; but the High Priest had dissuaded him from attempting it, saying that such power could only be attained by a long and severe training which could be safely done only by an adult of great strength of will, and not by a boy of tender years.

But at last his yearning became too strong to bear; and so one night, after an earnest prayer to the Sun-Deity, he cast himself upon his bed and entered into the great endeavour, determined to succeed or die in the essaying. After long strain it seemed to him that something snapped and at once he was free from the body and floating in the air. Startled at first, he quickly steadied himself and as he fixed his will once more strongly upon his objective he began to move with great rapidity. He retained enough self-command to notice the direction of his flight, orienting himself by the stars, as he had been taught to do in the physical world. The journey seemed to him a long one, and before its rushing ended the stars which had been just rising upon his horizon when he started where well beyond the zenith, showing that he must have swept round a quarter of the circumference of the globe. And then - to him all unexpectedly - he came out into rosy dawn, and saw by its sweet light the city and the island that he knew so well.

Quickly he found the long low house, the garden and the tank; beside the latter he alighted, and stood wondering what to do, yet willing strongly that his love might come to him. And so, surely enough, she presently did, for she came running through the garden and dancing lightly over the grass, followed more soberly at a little distance by a stately yet kindly lady who was evidently her mother. His girl-friend had grown taller and more beautiful, and when she caught sight of him she stopped for a moment, startled, and then rushed towards him with a cry of glad recognition and threw her arms round his neck. With a wild outflow of long pent-up feeling he held her to his breast, amply rewarded now for the weary waiting of the last four years; and it seemed to him that earth could hold nothing more of bliss for him, if but that moment might be prolonged for ever. But all too soon it passed, for her mother came up and stood looking at the children with an expression of intense though by no means unfriendly amazement. Releasing him from her embrace, but still holding him by the hand, the little girl excitedly poured out a torrent of information in a language entirely unintelligible to him, and the smiling mother drew him into her arms and kissed him warmly. He spoke to her in terms of respectful salutation, such as he had been taught to use to the great ladies of his own land, but it was evident that his new friends could no more understand his language than he had comprehended theirs. The mother spoke to him with several different intonations, probably trying various languages, but none of them conveyed anything to him; and seeing this, she took him by the hand and led him towards the house, her daughter clinging closely to his arm on the other side.

While full of the deepest happiness, Surya was acutely conscious that he was attired only in a single night-robe, while his companions wore garments of rich materials which, though quite unlike any he had ever seen, were obviously their ordinary costume. But he was fortunately consoled by the thought that, as they must regard him as a foreigner from some unknown country, they might suppose the customs of that country in the way of dress to be simpler than their own. The house into which they brought him was more sumptuously furnished than those to which he was accustomed, and, when presently they took him into a room where food was served, he found both provision and the mode of eating strange to him. He was an observant boy, and by covertly watching the methods of his entertainers he was able to get through the meal creditably, and he found the victuals palatable, though their flavours were entirely new to him. Just as breakfast was finished, a tall commanding-looking man entered, and was effusively greeted by the little girl, who at once presented her boy friend to him. He first placed his hand on Surya's head as though in blessing, and then held him by the shoulders and gazed long and earnestly into his eyes with a piercing look that seemed to read into his soul. The scrutiny was satisfactory, for he drew him to his breast, enfolded him in a warm embrace, and then again blessed him. He also spoke to him in several languages, but in none which he could comprehend; and after listening to a long story excitedly told by the girl with occasional confirmatory interjections from the mother, he smiled kindly upon Surya and left the room.

The little girl then drew him out into the garden, guided him to an exquisitely carved stone seat, sat down beside him and began to try to establish some sort of communication with him. First she pointed to herself and recited several times a word, which he took to be her name, and she seemed much pleased when he repeated it after her. Then she pointed to him, and evidently asked his name he spoke it, and after several trials she was able to say it accurately. Then she began to point to various objects, evidently giving him names of them in her tongue, and he picked them up quickly, although the intonation of the language was quite different from his own. Many other words she made him learn, at whose meaning he could only guess but in the course of two or three hours he had accumulated quite a number of detached words and several little phrases about the signification of which he was by no means certain. Presently the mother came out to them; and when she heard what they were doing, she joined in the attempt to explain. Suddenly, while all three were deeply interested in his efforts to pronounce some unusually difficult word, an extraordinary feeling overwhelmed him; he sunk into a few moments of curious whirling rushing unconsciousness, and awakened out of it to a sense of weakness and lassitude such as he had never before known. He found himself lying on his own bed at home in Ireland, with his own mother bending over him, evidently much perturbed at his condition.

It was some minutes before he was able to speak, and then he asked faintly where the little girl was. At first no-one understood him, but presently his mother realised that he must be referring to what they had called his dream. He was anxious to tell his story, yet felt too weak to talk; seeing that, his mother soothed him, and in a little while got him to sleep again; but if during that sleep he returned to his friends in the garden, he had no recollection of such return when he awoke. Clearly his violent and persistent efforts had overstrained some part of the brain-mechanism, for it was several months before he completely recovered, and his father and mother insisted that he should promise never again to risk his life and his the attempt to force his way where it was manifestly not natural that he should go. He promised, though reluctantly, but declared his unalterable conviction that his young love really existed, and his intention to search the world for her. He carefully wrote down as well as he could the words and sentences that he had learnt, and asked all the learned people he encountered whether they recognised them; but none ever did.

Three years later, however, there came into that land a traveller of unknown race, who did not understand the language of the country; and because none could converse with him, they brought him to the Chief Priest as the most erudite of their people, hoping that he might be able to communicate with him. The Chief Priest was helpless; but Surya, who happened to be present, thought that he recognised the intonation, and tried upon the stranger some of his well-remembered words and phrases. The traveller's face brightened immediately, and he began to speak rapidly in the very tongue of Surya's friends. Of course, Surya could not follow him, but he obtained leave from his father to receive the stranger as a guest, and devoted many hours each day to working hard with him until each knew a good deal of the other's language, and they were able to exchange ideas.

He gathered that far to the south, on the shores of another sea, were many who spoke that other tongue; and because men of his race had not infrequently travelled to the Mediterranean, and had even settled there, he hoped by going there to find someone who knew perfectly both that language and his own. So he asked his father's permission to take that journey; but his father suggested that he should wait a year, until he had fully entered the priesthood. He assented to this, but did not forget his resolve; and so in due course he found his way to a certain great southern city, where he had no difficulty in obtaining a teacher who could do what he wanted.

Now for the first time he acquired some definite information about the country of his experience; he met with men who knew the city and the island which he described so minutely, and were able to give him some idea of its direction and its distance - both of which agreed very closely with the results of such calculation as he had been able to make from his childish observations of the stars. But he told no one the details of those strange early visions or visits, keeping the memory of them to himself as a sacred thing. Only before he returned home he learnt the Manoan language so that he could speak it like his own, in preparation for the visit which he intended to make to Central Asia.

His father and his mother were reconciled to his making this long journey, though the latter begged him not to go quite yet, but to postpone it for a few years. The date of his departure was eventually determined by yet another vision, though it was of a different kind from the others. This time he found himself not in the garden but in the house, and in an inner room of it which he had not previously visited. He had had no special intention of going to Manoa that night (though the thought was always in his mind); nor had he any recollection of the journey; simply he found himself watching and listening to a conversation between his beloved (now a tall and beautiful woman) and her mother, and his newly-acquired familiarity with the language enabled him to understand every word. He gathered that they were discussing an offer of marriage which had been made by some suitor of high rank, who was evidently considered specially eligible. The mother was half-heartedly pressing his suit, or at least enlarging upon its advantages; but the daughter would have none of it, and declared that she had no wish to marry. After the affair had been presented from various points of view, and the young lady still remained uninterested, her mother remarked :

" My dear daughter, I know exactly what you are feeling; you have never lost the memory of your spectre-suitor, and you cannot bear the idea of unfaithfulness to him. I sympathise deeply, yet I also feel that we have absolutely no certainty that he really existed, that he still lives, that he on his part is faithful to you. Even if he lives, even if he still loves you, he may have been forced into a marriage in his own country; we know nothing of its customs; we do not even know where it is. Is it well to sacrifice your life to what may after all have been only some strange kind of unusually vivid dream? You know that your father and I wish to see you settled, and you will never have a better offer than this." The daughter admitted that her heart was entirely devoted to her spectre-boy, and said quite frankly that though she did not know whether she would ever see him again, she would rather submit to perpetual spinsterhood than marry anyone else, for she felt that the boy she had twice so strangely seen was her only true mate. Her mother acknowledged that the dictates of her own instinct agreed entirely with her daughter's decision, though on the physical plane such a course could not be defended as sensible.

"If only he would come to us again," she said, "we could perhaps discover something more about him, so that we might have a comprehensible reason to give for at least asking for a delay."

Surya heard all this, and burned with eagerness to manifest himself; but he remembered his promise to his mother, and so was torn between two duties. Suddenly it occurred to him to wonder why he was obviously invisible to his friends, though he could himself hear and see quite clearly. Without understanding the detail of the matter, he saw that the circumstances of his presence were somehow different, and he instinctively felt that even if he had been free to make the same effort as before, it could not have been successful. So he turned his attention in another direction. He had lately been studying what we should now call mesmerism, and so it came naturally to him to try to turn to account his newly acquired knowledge. He exerted all his strength to impress upon the mind of the girl the fact of his presence and in a few moments he saw that he was succeeding. She started, turned towards him, and peered into the shadows in the corner where he stood. He redoubled his efforts, throwing his whole soul into his fiery glance, and directly afterwards she uttered a loud cry:
"Mother, he is here! Do you not see him?"
She rushed towards him, but her outstretched arms passed through him, and she cried:
"He is but a spectre indeed; I cannot touch him; alas he must be dead!"
With all his strength he impressed upon her the reply:
"Not dead, but living! Within a year I shall come to claim you."

And she heard and understood, and eagerly repeated his words to her mother. Then he turned the current of his will upon the mother, and for a moment she saw him too; then the strain told upon him, and he vanished from their sight. But he was still able to watch long enough to see them fall into each other's arms, weeping tears of joy, and to hear them speak of his noble appearance, and say that he had more than fulfilled the promise of his boyhood. Then he returned to his body, woke up in great excitement and high resolve, and as soon as it was light went to his father and mother and told them what he had seen and heard. They agreed with him that his destiny was manifest, and that the will of the Sun-God had clearly declared itself in this matter. Indeed, his father publicly related the circumstances at one of the great religious gatherings as a gracious indication of the interest of the Deity in his worshippers, and he sent his son upon his long journey with equipment worthy of his rank.

It seems evident that on his first visit to Manoa as a little boy, he was at first in his astral body in the usual way, and probably materialised himself by drawing what was needed from the surrounding ether; it may be that his intense desire to help was sufficient to enable him to perform that feat, or it may be that he was specially assisted by some passer-by, or by some Great One who was watching his struggle. The fact that when he awoke his physical garments were wet, seems to suggest that he borrowed matter from his own etheric double; yet we have no instance of such rapid action at such a distance. On the second occasion it is clear that he tore away much of the matter of his own etheric double, and thereby injured himself so that it took him weeks to recover. This however enabled him to maintain the materialisation for a much longer time than is usual, to eat and drink, and to repeat clearly the words which were spoken to him. On his third visit he did not materialise at all, but mesmerised the mother and daughter into believing that they saw him.

With such methods of physical transit as were then available, it took him almost a year to reach the city of Manoa, but when he arrived he soon found his way to the house and garden which he knew so well; and a very curious sensation it was to stand physically where before he had been only astrally. Enquiry in the neighbourhood had obtained for him the name of the lady of the house, so he boldly asked for her. When he was ushered into her presence, she recognised him immediately, and welcomed him with profound joy and many exclamations of wonder. Her daughter was instantly sent for, and when she entered the room she sprang into his arms with a glad cry of triumph and love. He was at once on the footing of a friend of the family, or rather of an honoured member of it; and he lost no time in enquiring about their side of the amazing story of their previous meetings.

It agreed exactly with his own recollection in every particular; but naturally they had also to tell of the shock of stupefaction with which they had seen him vanish on his first and second visits. They had doubted that he was a real living man, though only the daughter had been unshakably certain that she would one day meet him in the flesh.

Presently the father came in, and Surya was introduced to him; indeed, it was then that for the first time he really explained who he was, and from what country he came, for before they had all been too busy discussing his previous appearances to do anything else than take him for granted. His account of himself was eminently satisfactory, though his prospective mother-in-law looked very sober when she understood how far away from Manoa her daughter's new home would be. Surya was careful to explain that in Ireland there was less of luxury than in Manoa, and that their life there was lived chiefly in the open air; but of all that his lady-love reconned less than nothing, caring for naught else now that she had at last found the lover who for so many years had been to her half-myth, and yet at the same time the most vivid fact in her consciousness. Naturally she had filled up by her imagination the numerous gaps which inevitably existed in her knowledge of him; and she was surprised to find in how many cases she had guessed exactly right, so that eventually they began to see that some sort of clairvoyance or intuition had guided her when she thought she was giving rein to her maiden fancies.

There had been so manifestly an intervention of divine power in their wondrous story that it never even occurred to the parents to object to the departure of their daughter to a far-away and unknown country; but they did plead for some delay, and eventually it was decided that the marriage should take place immediately, but that the newly-wedded pair should reside in the bride's old home for a year, especially in the hope that the first child might be born under that roof. Surya gladly agreed to this, and despatched one of his suite to return to Ireland and bear to his mother news of his safe arrival, his marriage and his plans, and ask her to be ready in a year's time to welcome her daughter-in-law. The twelve months passed quickly, and before they were over the hopes of the elders were fulfilled, for a noble son was born - our old friend Elektra.

When the time came the farewells were said, and the young couple, with their newly-born baby, started on their way into what was to all intents and purposes a new world to the bride; yet so perfect was her love that she faced it without a qualm. The journey was prosperous, and a right royal welcome was accorded to the happy pair - literally royal, for Surya's parents had told the romantic story, and the King of the country had been greatly interested in it, and invited the travellers to pay him a visit. This was done, and he received them with every mark of favour, and would have had them stay long at his court; but Surya wished to get his wife home again quickly, to put her under the care of his mother. Soon Elektra had a little sister - Mizar, whom he had loved so well long ages ago, whom he was to love no less in the life now before them.

Thus it will be seen that Aryan blood was introduced into the family of the Chief Priest; and they further intermingled with the royal blood of their own country, for the King continued his friendship towards those whom he felt to be favoured of the Deity. He drew them into closer relations with him; his eldest son in due course married Mizar, while two of his daughters wedded sons of Surya, Elektra himself taking to wife Brihat ((who was millennia later, Jesus, who prepared his body for Christ - Surya)), and Rama espousing Vulcan. Elektra and Brihat had three sons and four daughters, and the eldest of their family was our hero Alcyone, who was thus born directly into the succession to the position of High Priest, and had furthermore the advantage of a close alliance with the family of the reigning monarch.

The work of the priesthood was very interesting, for it comprised not only the religious teaching of the people but the education of the children. All children in the kingdom learnt to read and write a curious rounded script, but hardly any of them except the Priests made much use of this accomplishment in later life. They had books written on rolls of parchment, consisting chiefly of epic poems and ascriptions of glory to the sun, which they worshipped as the source of all life and the symbol or manifestation of the Deity. Daily hymns were chanted to him at sunrise and sunset, and at certain seasons of the year special festivals were celebrated in his honour.

Elektra was a wise father, and contrived to retain the full confidence of his little boy, so that they were always very happy together .
Alcyone was a great favourite also with his grandfather Surya and his grandmother Dhruva, and he loved nothing better than to sit at the feet of the latter while she told him wonderful stories of the city where his father was born, of its wide streets and its magnificent buildings, and above all of the marvellous beauty and sanctity of the mighty temples, built who knows how long ago, by the hands of giants and godlike men of old upon the mysterious White Island.

" Why have we no such temples here, grandfather? " he asked Surya one day. And the great Priest answered: "My boy, each race has its own customs, and its own ways of worshipping God; and so long as they acknowledge Him, it matters but little how. We have no temples because our wise forefathers have taught us that our God is everywhere, and that we need not set apart one time or one place more than another in which to serve Him who should be always in our hearts, so that every grove or field or house is to us a temple of His service, and every day a holy day upon which to do Him honour. We think that the trees and the sky which He has made are grander than any human work, and so we make them the pillars and the roof of our temple. For the same reason we have few ceremonies, because we think that our whole life should be one long ceremony of devotion His service. You do not remember how, soon after you were born, you were carried up the hill in the early dawn to the great altar-stone near the summit, and laid upon it to await the morning kiss of our Lord the Sun, and how, as the first glad beam of rosy light fell upon you, I blessed you in His name and offered to Him as a sacrifice the life-long devotion of your strength to His service, and of your body as channel for His love. And if you so choose, later on there will be yet another ceremony which will dedicate you in a new sense to a still fuller service, when you become a Priest like me and like your father."

Alcyone was satisfied; but he nevertheless resolved that as soon he grew old enough he would travel to far-away Central Asia, and visit the great city with which his fate seemed so strangely linked.

CHAPTER 4 World History from 70,000 BC:
Buddha & Christ in their previous lives

This resolve he duly carried out, for he made that journey, bearing gifts from King Mars of Ireland to the Emperor of Manoa, and he spent two years in the city which long centuries before he had helped so much in building. Perhaps it was this latter fact, or perhaps it was only the many stories which he had heard about, it, which caused him to feel that nothing there was really strange to him, but that he was almost as much at home as upon his own hill-sides. His great grandmother was still alive, and delighted to see him, and to show him the tank from which his grandfather saved his grandmother, the room in which his father was born, and all such mementoes of earlier days as very old people delight in. She was much pleased with him, and heaped upon him gifts of great value, so that he returned home after two years' stay in Manoa a far richer man than he had been on his arrival.

When he (Alcyone) reached home, it was he who had tales to tell to his grandmother Dhruva - tales of the country which forty years before she had left for the sake of love, yet had never forgotten even for a day.

Soon after his return he (Alcyone) married his cousin Mercury, with whom he had been in love ever since her birth, or at least since the day when, himself a tiny boy, he had been taken up the hill by his mother to see the consecration of the infant daughter of his uncle Rama. Not long after this came the ceremony of his own consecration and initiation into the full mysteries of the priesthood – an occasion of deep import, the memory of which abode with him through the rest of his life. The scene was, as ever, the great prehistoric altar-stone near the summit of the mountain which their repeated ceremonies had made so sacred; and the supreme moment was, as before, the falling of the first sunbeam of a new day upon the brow of the candidate, crowned with roses and lilies, to typify at the same time the love of God which he must preach, and the purity of the life which he must lead. The ceremony was performed by his grandfather Surya, and in the course of it he delivered the following exhortation:

"This is an important occasion in your life - perhaps the *most* important in this life, because it admits you to the brotherhood of those whose duty it is to keep alight the fire of devotion in the hearts of the people, and to hold up before them the shining light of a good example. See to it that you never falter in these duties, that you exercise worthily the power which I have this day entrusted to you. Remember always that this life is but one of many lives – one step on a vast staircase, leading up to the portal of the Temple of our Lord the Sun.

When at last all the steps are trodden, when you shall enter the glorious portal, a splendid destiny lies before you. Servant of the servants of God shall you be, to help them on their way to Him, to guide their feet into the path of peace and happiness.

But for an office so magnificent the preparation is arduous. For many lives in the past you have lived among us, among the Kings and Priests of the earth who are your true spiritual kin, in order that their spirit might permeate you, that you might become one in heart and mind with them; for a few lives yet you will do this, but before the end there must be times of trial, lives in which you stand alone and away from us, lives spent in lower walks of life and among those who are less evolved; for only so can final debts be paid, only so can uttermost sympathy be developed, only so can be gained the power which enables a Prince of Life and Death to pour out his own life in final self-sacrifice for the saving and the blessing of the world. For ever shines our Lord the Sun; keep your mind ever fixed on Him, and learn to see Him through the darkest earth-born clouds, so that His reflection in you may be ever steadfast, and in you His people may find an ever-open gate through which they may reach His feet; so that through you they be saved from their sin and sorrow and ignorance, through you the little streamlets of their lives may reach at last the shoreless sea of His infinity, the ocean of eternal bliss which is the life of God."

Alcyone and Mercury had nine children - all of them characters whom we have met many times before. His first was Sirius – a daughter this time; but his eldest son was Corona. In due time Surya passed away, and Elektra became the Chief Priest; and at about the same period Mars also died, and Viraj succeeded to the throne, thus making Alcyone's aunt Mizar queen of the country. Now that our hero was next in succession to the office of High Priest, he frequently acted for his father, and ranked next to him in power and importance.

The residence of the Chief Priest was not at the capital, so the civil and religious centres of the country were not the same - much as, in England, Canterbury ((today)) is really the seat of the ecclesiastical head of the Church, though London is the capital of the country. There was, however, no suggestion of rivalry between the two powers, as each had its own sphere, with which the other did not interfere.

The spot where the capital city stood in those days is not now identifiable, for it has been whelmed beneath the sea in the changes which took place at the time of the sinking of Poseidonis ((sea level rose ~120 metres at the end of the Ice Age, 10,000 years ago)); but the mountain where Surya officiated still remains, and is now known as Slieve-na-mon, in Tipperary. The Priests of the Sun knew much of magic, and were well acquainted with the various orders of the nature spirits, as well as the greater Angels; and it was Surya himself who first gave to Slieve-na-mon the sacred character which it bears even to the present day. The arrangements as they exist there now were made by the Priests of the Tuatha-de-Danaan just before the Milesian conquest; but it is to Surya that the inception of the great scheme is due, for he first conceived the idea of establishing in the country a number of centres from and through which power might be radiated. Elektra and Alcyone understood these plans; and each in his turn carried on the magnetisation, and handed on the tradition to his successors.

The life of the times was spacious and leisurely, for there was plenty of room in the land and everyone had plenty of time, and so it often happened that such Priests as felt so disposed climbed the hill and sat in meditation near the altar-stone. The common folk came there but rarely, though sometimes one who had some trouble, or some difficult problem to solve, would sit alone in that sacred spot and wait for an inspiration, taking what came into his mind on such an occasion the response of the oracle, as suggested by the guardian spirits of the place. This custom is eminently characteristic of the whole attitude of these people. Their entire life was permeated with the knowledge that close around them and in intimate relations with them was another world, unseen yet ever present and always to be taken into account in every word and action. Indeed, that world was hardly regarded as unseen, so frequently did some token of its presence obtrude itself upon the physical senses.

The dead were not considered as absent, but as present in a slightly different way; it was fully recognised that many of them remained very closely in touch with mundane affairs, and were for some time after death deeply interested in the health of their friends, the progress of their crops, the well-being of their horses and cattle. The living did not fear the dead, but regarded them with a certain reverence, possessing new powers and having in some respects a wider outlook.

Sometimes people invoked a departed relation, but it was considered a dangerous and selfish act, and was discouraged by the Priests, who taught that when the dead could speak, and wished to speak, they would try to do so, and that when they did not, it was rash and presumptuous of the living to thrust petty earthly concerns upon them.

Nevertheless, manifestations of some sort from the departed were by no means uncommon; and, as the race was on the whole distinctly psychic, there were many who constantly received strong impressions as to the wishes of the dead, and these were almost invariably carried out.

The existence of angels and nature spirits was universally accepted – indeed, to most of the people it was a matter of first-hand knowledge, for such beings were often seen and all sorts of strange adventures with them were on record. I ((sic, B&L)) have mentioned that though everyone knew how to read and write, but little use was made of these accomplishments. To a large extent their place was taken by story telling, which was elaborated to a degree of which under modern conditions we have no conception – elaborated until it became both a custom and a science. They had no such things as balls or garden-parties, but instead of them they had what can only be described as orgies of story-telling. The neighbours met somewhere or other for this purpose every night, usually taking the houses of the district in turn, and the party settled down round the fire and composed themselves to listen or to narrate. There was a vast store of legend and of supposed history - mainly the personal adventures of certain great heroes - and another huge department of accounts of angelic or fairy intervention; all these were recognised and accepted tales, which had to be told according to tradition, from which no departure would be tolerated, and the persons who knew most of these, and had a reputation for reciting them dramatically, were sure of an enthusiastic reception anywhere. Besides these classics, there were constantly new narrations of present-day adventures and happenings - stories which had their vogue, and then either died out and were forgotten, or took their place among the received body of such romances.

Alcyone himself had some experiences of that kind, having seen the fairies at their gambols more than once; but the great fairy story of the family was a visit paid to some sort of underworld by his youngest daughter Yajna. When the child was about seven years old she disappeared one day, and though the distracted family searched the whole hill they could find no trace of her. Wild beasts, though rare, had not been entirely eliminated, and the first fear was that she had fallen a victim to some of them. But there was no evidence for this theory, and no such creatures had been seen in the neighbourhood for years, so presently suspicion took another turn, and it began to be whispered that perhaps the fairies had taken her, as she was an especially beautiful child, and it was known that in the past such children had been coveted and captured by nature-spirits. Her father immediately employed certain arts of conjuration with which he was acquainted, and soon obtained confirmation of this surmise, and a promise that his daughter should be returned to him unharmed if he would seek her in a dell which was indicated by his informant. He promptly repaired to the appointed place, and found the little girl asleep under a tree.

When aroused, she told a strange tale. When wandering on the hill, quite near her own home, she had come upon a little hollow in the hillside, which she had never seen before, and had found in it the entrance to a cave. She had hesitated whether to go farther, because of the darkness; but while she stood looking, a handsome boy came out of the cave, and with a deep bow invited her to enter. She was flattered by the deference with which he seemed to regard her and asked him who he was, and where he lived. He replied that the cave was the entrance to his home and that he would gladly show her the beautiful gardens which were but a little way within. She wondered much, but curiosity triumphed, and she put her hand trustingly in that of her guide, and let him lead her into the darkness. He seemed to be able to see quite well, and led her unhesitatingly forward; and after walking for a few minutes they came, quite suddenly and round a corner, upon a hall so vast that it was as though they were again in the open air. Yajna had no recollection of seeing the sky, but had the impression of a pleasant warm light like sunlight. They seemed to be in a garden, full of the loveliest flowers and trees, yet none of the flowers or trees were exactly like any which she had ever seen before.

The boy led her forward through the garden, and presently they came upon a number of other children, who seemed to be playing some sort of game, in which both she and her guide joined; but she was never able to explain quite what the game was, except that it was not like any played on earth. The merry party played and danced for hours without the slightest feeling of fatigue, and varied their proceedings by wandering hand in hand among the gorgeous vegetation, and on one occasion plunging into a crystal lake and splashing about deliciously warm water. Yajna was deliriously happy, and earnestly wished that her brothers and sisters and friends could share her enjoyment; indeed, she asked her boy friend whether she might come again and bring them all with her. He laughed joyously, and said that they would be heartily welcome if they could find the way - a cryptic utterance which Yajna did not understand, but she asked no more, lest she should seem rude. Nevertheless, in the midst of all her play curious little twinges of longing for her mother obtruded themselves into her mind - doubtless the result of the anxious thoughts of Mercury while the search was going on.

Suddenly there came to them through the garden a shining form to whom the playing children paid great deference; he spoke earnestly to the boy who had befriended Yajna, and then passed rapidly away. The boy called to Yajna, and told her that her father wanted her, and that he would take her to him. She ran to him at once, and he led her away from the garden, and up a curious stairway, which led them out among the roots of a great tree, and so into the old familiar world of daily life. But somehow that world seemed strangely dull, and the very sunlight itself looked pale after the golden light of the cave. The boy asked her to sit down beside him on the ground, and when she did so, he put his hands upon her shoulders and looked long into her eyes. His gaze was kind though compelling and under it she found herself sinking into sleep. Her last remembrance was that he stooped forward and kissed her as she sank to rest, and after that she knew no more until her father's touch awoke her.

She made repeated efforts to find the entrance to the cave, and the head of the stairway which came out among the roots of the tree, but could never come across the least trace of either, though she and her father and her uncle Naga spent many hours in the search. She was much impressed by what had happened to her, and tried again and again to get back into that beautiful underworld, but without success.

CHAPTER 4 World History from 70,000 BC:
Buddha & Christ in their previous lives

One day Naga sat meditatively upon the hill-side alone, and presently fell asleep in the sunshine. When he woke he found standing near him a radiant young man who looked upon him benignly; and it was somehow impressed upon him that this was the shining form of which his niece Yajna had spoken. He accosted the man, and asked if this were so, and the visitor smiled assent. Naga continued :
"My little niece was so strongly attracted to the boy who led her the garden, and it makes her sad not to see him again; cannot this be arranged? May they not meet and play sometimes as they did on that occasion?" The young man answered: "Tell her that just as she loves that boy, so does he love her, and desires earnestly to see her; yet it is better that they should not meet, for they are of different worlds, and it is not meant that these worlds should intermingle too freely. If she came to us she would be lost to you; and she has work to do in your world. Believe me, things are best as they are. The boy will continue to love her and watch over her unseen. See, I will call him".

In a moment a handsome boy stood beside him. Naga held out his arms to him, and he came forward and gravely allowed him to embrace him; his look was full of longing, but he spoke no word. Naga kissed him on the forehead, saying :
"Take that as a greeting from her who loves you".

Then in a moment the figures were gone, and Naga tried to persuade himself that it had been but a dream. Yet he knew well enough that it was nothing of the kind; and Acyone and Yajna realised it too as soon as he told them the story. Many times Yajna dreamed of her boy friend, and often unexpected and inexplicable help was given to her in sundry childish difficulties; and she always attributed such help to his watchfulness. She clung tenaciously to his memory, and always said as a child that she meant to find him and marry him; but as she grew up the impression gradually wore off, and she finally married Muni - though she said that she did so only because he reminded her of her fairy boy more than anyone else.

Alcyone lived as usual to a ripe old age, loved and reverenced by all the thousands who knew him.

((This is the end of the verbatim text, {p. 177-195 in [3]}.))

Comment: The story of Yajna above is similar to several others given [3], in different time epochs. E.g. Alcyone's 35th Life contains a much more detailed account of one – a visit lasting 2 weeks [14]. The en-trance (literally an "en-trancing" experience!) to that fairy world seems similar to that of "Alice in Wonderland" and to the well-known story, "The Lion, The Witch and the Wardrobe" where a region is entered via a wardrobe, and is possibly a "portal" to the astral or dream world, i.e. to 4 dimensional space. E.g., quoting from above, "*Yajna had no recollection of seeing the sky, but had the impression of a pleasant warm light like sunlight.*" And when she returned, the "normal" world seemed "*strangely dull, and the very sunlight itself looked pale after the golden light of the cave.*" This seems to be confirmed by another quote from above, "*One day Naga sat meditatively upon the hill-side alone, and presently fell asleep in the sunshine. When he 'woke' he found, standing near him, a radiant young man who looked upon him benignly; and it was somehow impressed upon him that this was the shining form of which his niece Yajna had spoken.*" This apparent waking was probably the so-called "false waking" in lucid dream studies where the dreamer thinks he has woken but is really still in the dream. But the difficulty with the astral (4-D) interpretation, is, what happened to Yajna's physical body during the many hours that she disappeared!? Modern folk stories tell of children who similarly disappeared and only returned years later … .

A better explanation is that there are parallel 3-D worlds (a "multi-verse", but limited here to within our own planet) and it is possible somehow to enter one of these in one's normal physical 3-D body. Alcyone's 35th life [3, 14] mentions that they ate local food there when their supplies ran out, on a pre-planned expedition. But it is difficult to explain what the "pleasant warm light" (no sunlight) underground is due to, if not what is called the "astral light" which illuminates everyone's dreams and makes 4-D objects self-luminous but is not light from our normal 3-D world sources. As already mentioned in Chapter 1, physical atoms would be unstable and cannot exist in 4-D "Astral" space, so the physical body must be present in a 3-D parallel world but Yajna's visual senses must have also involved astral (4-D) vision – i.e. just like a "day-dream". The episode occurred inside solid rock, i.e. it must have been in a parallel 3-D world, in which "our" solid rock would be absent. This "other world" is similar to the fictional Middle Earth, of Tolkein's "Lord of the

Rings", or the brick wall "portal" at King's Cross Station in the Harry Potter books by J.K. Rowling! See also references [12] & [14].

In Chapter 5 it is speculated that Christ may have used a parallel 3-D world to escape from a hostile crowd ("*He passed through their midst...*") (Bible, Luke 4: 28-30 and also John 8: 58-59).

Some other underground people are mentioned elsewhere in the books [3, 14]. Speculating, portals are rare locations where one 3-D world touches another, but no 4-D linking corridor could be used because 3-D objects would disintegrate in 4-D space (see Chapter 1). Further speculating, some crop circles are completely inexpilcable (not all are hoaxes) and there is a small possibility that they are portals into our world, left by inhabitants of a parallel 3-D world!

Dhruva was the girl whom Surya married and because of their very strong attraction, the author (MGH) speculates as follows:

She (Dhruva) may perhaps be identified, 30,000 years[**] later, as Mary Magdalene. The Gospel of Philip, a Gnostic text from the third century AD, declares of Mary Magdalene: "She is the one the Saviour loved more than all the disciples". Likewise, the Gospel of Mary (from the early second century AD) suggests that Jesus entrusted Mary Magdalene to instruct the disciples on his religious teachings. She was also the one who reported the resurrection to the others.

Dhruva is listed in a unique way by B&L [3] in Table 1 above. Permission would have been obtained from the individuals concerned, to publish the books [2, 3] and maybe it was agreed not to say who Dhruva was in later times.

-=-=-=-=-=-=-=-=-=-=-=-=-=-=-=-=-=-

In Alcyone's 16th life, in New Zealand in 25,528 BC, the ruler and High Priest was Viraj. His eldest son was Surya, who was drawn to the priestly side of the royal duties and so Viraj gave him that aspect to look after. Surya married Alcyone who was the daughter of the younger brother of King Viraj. She (Alcyone) shared with Surya the priestly duties.

[**] or 1000 generations ! (taking 1 "generation" to be 30 years)

Surya and Alcyone's children were twins, Mars and Mercury. Mars was drawn to the kingly side and Mercury to the religious side with his mother.

Surya was greatly impressed by a vision seen by Alcyone - so much so that he took the words out of her mouth and spoke himself as though inspired: ((verbatim below))

"You and I, my wife, and these flowers of our tree, have a wondrous destiny before us. As you follow me now, so shall you and they follow me in that glorious future. Some of these who now call you mother shall pass in advance of you, and shall be my more immediate helpers in the work which I have to do. And when your share in that work comes, others of these your children shall stand around you as helpers and disciples. So the members of this our family shall not be separated as so often happens; again and again shall they be born together, so that it becomes a permanent family whose members shall meet in fraternal affection through the ages that are yet to come."

So when Viraj was gathered to his fathers it was Mars, not Surya, who was proclaimed King in his stead; and it was not long after he came to the throne before it was found necessary to take further control of that part of the island inhabited by the Turano-Lemurians. These latter had an obscene form of religion, which, among other unpleasantness, involved occasional human sacrifices - usually sacrifices of especially beautiful children. These were sometimes selected from among their own families, but more frequently one of their tribes made a raid upon another in the hope of finding suitable victims. On one occasion, however, it was decided by the priests of this unpleasant form of worship that an unusually choice sacrifice was required, because an unknown infectious disease had broken out among their people. So the priests met in conclave and decided that, as ordinary methods had proved ineffectual in turning aside the wrath of their deities, a white child should be captured and sacrificed.

Their only hope of obtaining such a prize was through some of those of their tribe who were in close touch with the ruling race. There had been a certain amount of intermarriage between the races, although this was discouraged by the authorities, and it was from some of these mulatto families that the most powerful and the most scheming of the priests were drawn. Among them were found just at this period two with whom we are acquainted, Lacey and Tripos.

Aided by a woman named Cancer, they resolved to steal a child from the white settlements, and after much lurking and watching they contrived to carry off Phra, one of the grandsons of Surya and Alcyone. When he was missed, the boy's father, Naga, rescued him.

((End of verbatim quote {[3], page 221]}.))

-=-=-=-=-=-=-=-=-=-=-=-=-=-=-=-=-=-

Alcyone's 17th life was as a female in Mexico in 24,700 BC, a cult of human sacrifice had grown up and been thwarted, but Spica's son had been rescued from the power of Scorpio, his mind was clouded and the evil influence was still strong upon him. ((verbatim [3])):

She heard from one of those who had been monks, who was therefore acquainted with the nefarious mesmeric powers of Scorpio, that one who had once come under his control could never break away from it again, but must inevitably pass through the various stages of degradation which ended in vampirism. Much horrified at this, she carried her case once more before her father the King, but he had to confess himself powerless in this matter, knowing nothing as how to deal with it. He spoke with great kindness to his daughter, and showed much sympathy and sorrow, but yet he knew not what to advise. At last he turned suddenly to Alcyone and said to her:

"Daughter, through you there came to us the advice which has saved for me my kingdom, and has freed it for ever from the powers of evil. Can it be that in this case also you can come to our assistance, and rid this poor suffering boy of evil, even as you have done for the country as a whole?"

Then the power seized Alcyone once more, and she arose and said: "O King, the power of evil is terrible indeed and to oppose it may well mean the rendering asunder of body and soul. Yet it must be opposed, even though the victim die, because if we do not oppose it, then is he lost not for this time only but for all time, for never again can he free himself from the downward course of the vampire. I cannot tell what the result may be, yet must I set him free even though in doing so I may destroy his body."

So she turned upon her shrinking nephew, and raised her hands in the air over his head, and cried aloud: "In the Name of the Great Father of all, let this curse depart from thee!"

The boy uttered a terrible cry and fell to the ground as one dead. He lay in a trance of unconsciousness for many days, but at least he did not die, and after a long time consciousness returned to him, and he called faintly for his mother. Weak and ill he was indeed yet she knew that she had her son back again from the dead, for now he knew her and clung to her as of old. Presently he slowly recovered, yet the shock had been so terrible that all through his life he remained nervous and easily disturbed. Indeed, for many lives and through thousands of years something of the effect of that terrible psychic convulsion was still to be seen. For the evil high priest had seized upon the very soul of him, and had made for it a link with that whose name must be spoken. And the breaking of such a link is a feat which but few can accomplish, yet in this case it has been done by the power and love of Alcyone - and of Surya who worked through her, though not then in physical incarnation.

Alcyone's life passed on in great love and happiness. She married her cousin Selene, and her eldest son was Herakles, in very truth a friend of many lives. Among her ten children were two who have now attained, and others for whose near attainment we may hope. Her life was one long benediction to those around her, for she remained to see her grandchildren and great-grandchildren, even to the age of ninety years. And the good work in which she had so large a share remained as a monument after her, for never again while that Toltec race occupied the ancient land of Mexico were sacrifices re-established. Long after that race had been destroyed by the flood which accompanied the sinking of Poseidonis, it was re-peopled by a half-savage race who, having in themselves much of cruelty and greed, psychometrised the ancient stones, and revived to some extent the ancient horrors, but for twenty thousand years and more the work of Alcyone and Surya bore its fruit.

((End of verbatim extract from {p. 238 – 239 in [3]}.))

-=-=-=-=-=-=-=-=-=-=-=-=-=-

Alcyone's 19th life was in North America on the east side of the Rocky Mountains, in 22,662 BC, and the description is given verbatim below. (Refer to Tables 1 & 2 on p. 89 & 90 for pseudonyms used)

King Mars was once again ruling in North America, but this time on the other (east) side of the Rocky Mountains, in the great Tlavatli kingdom of Toyocatli, which we previously described as including all the Southern States of the Union. With him, naturally, came the band of Servers; and among them we find our heroine Alcyone as the eldest daughter of Mizar aud Helios, who were exceedingly kind, tender and devoted parents. Her father Mizar was a man of great wealth, as he not only owned vast flocks and herds, but had also on his estate a good deal of alluvial gold, which was washed out of some gravel on the banks of a rapid stream in a hilly region. These flocks were not, however, goats or sheep exactly as we know them now, but more resembled the gnu. The commonest animal was a kind of heavily-built long-haired goat, with head, neck and horns not unlike those of a miniature ox. The hill country round the gulf seems to have been of quite a different outline in those days. The river now known as the Mississippi cut across the State now bearing that name, instead of flowing round it in a curve between that State and Louisiana as at present. The Gulf of Mexico was less in size than now, and its configuration was quite different. ((Note: sea levels rose 120 metres at the end of the last Ice Age, 10,000 years ago.))

In a beautiful grove not far from Alcyone's home stood a magnificent temple, built in the form of a five-pointed star, in the angles of which were stairways which led up to the central ceremonial chamber.
Over this chamber was a large dome, coloured blue on the inside. On the inside wall just below the dome was a frieze about three feet high of some metal which looked like silver, inlaid with symbols and hieroglyphics. In the upper part of the dome hung seven silver bells, heavy and large enough to give clear deep tones, resonant and beautiful. Beneath the temple itself were crypts in which were kept instruments of gold and silver studded with precious stones, which were used in ceremonial worship on special and secret occasions.

The central hall under the dome was circular, and its walls were decorated with rare stones inlaid in symbolical forms; its whole appearance gave one the idea of Byzantine architecture. In it all the sacramental and festal ceremonies were conducted. On the second floor of the temple, in the points of the star, were the rooms of the priests; one of the windows in each room looked into the central hall, and sometimes minor services and ceremonies were conducted by the priests from their rooms through these large openings.

It is here that we find the first scene of importance in the life of Alcyone, the occasion being that of her presentation or consecration, which took place at the age of six months. Over this ceremony Mercury presided, assisted physically by three other priests, Osiris, Venus and Brihat, and astrally by the Mahaguru, who hovered above the altar, visible only to those who were clairvoyant. This group is a most interesting one to contemplate, and it can hardly be considered a mere coincidence that those who later represented four separate forms of the Great Mysteries should have been there together at this time. The ceremony of the consecration of Alcyone appears to have been largely astrological.

The colour used on the altar was electric blue, the colour ascribed to the planet Uranus, which was in the ascendant at the moment of the child's birth. The influence of this planet also accounts to some extent for the latent possibilities of psychic development, which came info manifestation later in her life. During the consecration ceremony an angel appeared, and into his guardianship the child was given, with the approval of the Mahaguru, who directed from the higher planes the work of Mercury. The Mahaguru was the Founder of the religion of this people, and He appeared in order to make a link between the child and the over-shadowing angel. He seemed to take possession of this, the first-born child of the family, and stretched out His arms over it with words to the effect that He took this ego into his care, not for this time only, but for the future.

Venus was in charge of the astrological part of the ceremony; he had cast the child's horoscope and arranged the necessary details in accordance with the planetary aspects in it, though it was Mercury who performed the actual ceremonies of the consecration. The child was placed upon a smaller altar, made of metal and highly magnetised; this stood in front of the principal altar and the rites were intended to make a magnetic link between the child, the angel and the Mahaguru, and also to inhibit any lower disturbing influences. During the ceremony the seven silver bells in the dome chimed three short musical phrases, the priests chanting in unison with them, as they stood each in the centre of one of the sides of the great square altar facing towards it. During the ceremony the little Alcyone wore a magnificently embroidered robe, made by her mother

CHAPTER 4 World History from 70,000 BC: 143
Buddha & Christ in their previous lives

Fig. 4.4. <u>Mercury</u> = (later) Master Koot Hoomi ("KH")

Drawn by Remote Viewing, from His appearance as it was in 1930 [4].

Studied at Oxford in 1850.

Note: The differences in appearance of ordinary people is said to be mainly due to their degree of advancement and karmic effects. But for the perfected Men shown in Figures 4.2 to 4.8 and in Fig. 4.10, these factors have dropped away and the Genesis (1:26) verse: "God said, Let us make man in our image, after our Likeness", applies without any constraints.

Helios, who often also enjoyed the privilege of embroidering the priests' robes and some of the decorations of the Temple. On the child's robe was worked a large swan as a centre-piece (probably the Kalahamsa) and there was a border of curved swastikas. The temple itself was attached to a great central temple far away in Atlantis, over which Surya presided as High Priest, assisted by Jupiter and Saturn.

The people were a light-brown race, belonging to the Tlavatli subdivision of the fourth Root Race; and about two years after the ceremony described, we find Alcyone a little toddling whitish-brown baby, wearing golden anklets which were really her mother's bracelets; as the baby enjoyed playing with them when on the mother's arm, she put them on the little ankles, and they would often fall off as the child walked.

On this occasion our old friend Sirius was the son of the priest Brihat, and his first sight of Alcyone was at that consecration ceremony.
Although he was only about three years of age, he had been brought by his parents to witness this dedication ceremony, which was an exceptionally brilliant one, as Alcyone's parents, being wealthy people, spent a great deal of money on decorations and processions. The grandeur of it greatly impressed him, and he at once fell in love with the baby, declaring his intention of marrying her when he became a man. When he was a few years older and again expressed the same sentiments, his parents advised him to put the thought out of his mind, since they were comparatively poor and Alcyone's parents were very rich.

The two families lived on opposite sides of the river, which at this point was about a mile wide. Sirius did not share the view of his parents that difference of circumstances should be a barrier to his love, and when he was about twelve and Alcyone about 9 years old, we find him having himself ferried across the river in order to pay his sweetheart a visit. He brought her a piece of sugar-cane which she refused to eat alone, so they compromised matters by taking alternate bites of it, as they sat together under the shade of a wall. Sirius could not forget Alcyone, and contrived to continue visiting her; presently he swam across the river daily for this purpose, even though the current was very swift and it took courage to accomplish it. As no one knew where he went on these occasions, he acquired the reputation of being a strange boy who took long wandering walks all alone. While swimming across the river on one of these visits he was attacked by an alligator, but contrived to kill it by stabbing it under the foreleg wlth a knife which he had carried for several days, because he had seen a similar reptile shortly before. Acyone's brother Herakles became an intimate friend of Sirius, and being some years younger rather worshipped the older boy, and was glad to carry letters for him to Alcyone, thus considerably helping on this juvenile love affair.

Years went by, and the children grew up into youth and maid, but still remained faithful to one another. The young lady's parents of course knew all about it by this time, but they did not look with much favour upon the impecunious suitor, especially as an opportunity offered Alcyone to become the bride of Vajra, who was the son of King Mars and heir to his throne. Alcyone, while admitting that it would be very pleasant thing some day to become a Queen, still would not give up her love for Sirius and wished to marry him. When a final decision had to be taken in the matter of marriage, and she was pressed by her parents to accept Vajra, she wept bitterly and was deeply distressed and dejected. Her mother's tender heart could not bear this, and her father too was deeply moved, so she had her way at last and was permitted to accept Sirius. All being settled, Helios wished to make a settlement of a sum of money upon the two, and to carry things out gracefully and generously. Sirius and his father were proud and found it hard to accept this, but it was finally arranged. Helios and Mizar made the best of things, and considered themselves fortunate that their daughter had chosen the son of one so honoured in the temple as was Brihat.

The parents on both sides having made the final arrangements, the marriage of the happy young lovers took place in most gorgeous state in the temple, and the ceremony was performed by the high priest Mercury, aided by Brihat, the father of Sirius, and by his uncles Osiris and Yajna. Alcyone looked most beautiful in a white robe, and here again the skilled handiwork of Helios showed itself, as the dress was profusely embroidered with gold and jewels.

Mercury, handsome as a Greek God, recited the marriage service in a most impressive and dignified manner, and threw much cordial personal feeling into the words which he had to repeat, for he had known and loved both bride and bridegroom since their childhood.

The central feature of this marriage ceremony seems to have been a sort of eucharist. The celebrant invoked the Mahaguru, and then handed the sacramental cup to Sirius, who passed it on to Alcyone; she drank some of its contents and handed it back to him, and then he in turn drank. The cup and the liquid had been highly magnetised, so that all earthly influence was removed from it, and only that of the Mahaguru left paramount. The husband and wife, after receiving the blessing of the Mahaguru, were bound together with ropes of roses and walked hand in hand round the altar, bowing in turn before each of the priests who were taking part in the ceremony. After this circumambulation they were seated side by side in a sort of palanquin, which was drawn up into the air by ropes and left swinging high above the heads of the people while further blessings were chanted. This was to symbolise their new relation to each other, that they were now alone together and apart from the rest of the world, and also that they could rise together to planes higher than either would be able to reach apart, and that thus they could work together for a higher good. Then they were once more lowered to the floor, and received a final blessing from the priests preparatory to leaving the temple.

Many handsome presents were given to them, and it is noteworthy that all these were brought to the temple to be magnetised by the priests. Among them was a huge golden bowl from Helios, which was wrought in the form of a lotus. Some beautiful chased silver swinging lamps were given by Mizar, and were filled with sweet scented oil, which perfumed the whole temple. At various points during the ceremony the bells in the domes sounded soft muffled tones, but as it finished they rang out joyfully.

In this family alone we find a considerable gathering of the clan, for in addition to the little children of Helios there were sixteen born to Sirius and Alcyone, all of them egos well-known to us. If we include the children of the King, and those of Vajra and Herakles, who are also numerous, we find nearly all our *dramatis personae*. The children of most of these families were taught by the Priests of the temple, and some of the sons became inmates of it.

Besides the sixteen children of Alcyone and Sirius, they also adopted the orphan Olaf, because Mercury was deeply interested in him.

Somewhat strained relations existed just at this time between the court of Mars and the authorities of the great Temple, chiefly owing to a number of small misunderstandings intentionally created by two young priests of bad character, Thetis and Scorpio, who cherished a bitter grudge against the King because he had been compelled to banish their father Cancer for a series of heinous crimes which he had committed at the instigation of a stronger ruffian than himself. These two young fellows contrived somehow to become aware of a conspiracy against the King, and joined themselves to it, intending either to use it or to betray it, as might best suit their own machinations. They decided to request an audience from the King, and, if he granted it, to endeavour to utilise the occasion to assassinate him. There was a certain important functionary (Castor) in the King's household, among whose duties it was to arrange audiences for him; so these two young scoundrels wrote a letter to this man asking for an appointment, and hinting that they could betray a dark conspiracy against the King, and could also show that the Temple authorities were trying to undermine his power.

In going up the steps of the palace the functionary accidentally dropped the letter, and Herakles happened to pick it up. Herakles was now an intimate friend of Vajra and in consequence was much at the palace. He was on his way to Sirius at the time, and when he read the contents of this letter he had so odd a feeling of danger that he showed it to Sirius, and discussed the matter with him. Sirius at once consulted his wife Alcyone, who proceeded to psychometrise it, and saw the plot in the minds of the scoundrels. In order to confirm what she saw, they took the letter to Helios, who was also psychic. She agreed as to the plot, and they felt that they aught to take some action, but since some high authorities of the Temple had been accused of treachery to the King, and this was mentioned in the letter, it was a serious question what to do with it.

It was finally decided to say nothing for the moment to the King, but Herakles went to the functionary to whom the letter had been written. The latter had been seeking for it everywhere before reporting its contents to the King. So Herakles told him what he feared, and together they arranged that the ruffians should have the desired audience, but that they themselves should be present and also have in readiness a strong guard.

The would-be murderers presented themselves, and as they were rising from the usual prostration, Thetis thrust his hand into the front of his robe and grasped a dagger. Herakles, who was close to the King's side, saw the action and guessed its meaning, so he sprang forward just in time to seize the man's wrist as he raised the dagger and was about to leap upon the King. Both the villains were quickly overpowered and imprisoned, and shortly afterwards they were banished from the kingdom. The law condemned them to be buried alive, but the Monarch commuted their sentence to banishment, because he said that, wicked as their action had been, and worthless as they themselves appeared to be, their treachery had been dictated by a perverted idea of filial affection and family honour.

The King was grateful to Herakles for having thus saved his life, and when he heard the part that Alcyone and Helios had played in the affair, he called them before him and publicly thanked them.

The entire family, including that of Sirius, was much advanced in royal and public favour. Herakles was honoured by receiving the Kings daughter (Bee) in marriage, and was appointed as ruler over the large province in which the family of Sirius lived. Vajra was made ruler over the province in which Mizar and Helios lived, and as only the river separated these two provinces there was much happy social intercourse between all these families, the Court and the temple Priests. After the attempted assassination of the King, it became known at once that the rumour that the Priests of the temple had tried to undermine the power of the King had no foundation whatever. Mars sent for the Chief Priest Mercury, who came to the palace with Herakles and Vajra. A wonderfully clear understanding was at once established between the Priest and the Monarch, and harmony was restored between the Court and the Temple; so much so, that when later the King abdicated in favour of his son Vajra, he took up his permanent residence with the Priests in order to live a life of devotion.

Various expeditions were sent out from time to time by the King and one of them was given into the charge of Vajra and Herakles. They were sent to make a sort of treaty with the ruler of the kingdom where they had lived a thousand years before, and bore rich gifts with them. On the way, near where New Mexico now is, they were attacked by savage tribes similar in type to Pueblo Indians, who captured them and then sent to Mars for a large ransom. But instead of a ransom the King sent Sirius with a large army of trusted men to rescue the captives. This they succeeded in doing, the army engaging the Indians in front of their village while Sirius entered it from the rear and easily rescued Vajra and Herakles, who were borne home amid great rejoicing. Herakles had learned the Indian language while a prisoner among them. Some time after their return a second expedition was sent, to the same kingdom, which reached its destination and returned safely; but this time the King would not permit Sirius, Vajra or Herakles to go. Another expedition was sent towards the north-west, as a rumour had come of great silver and gold mines in that direction. It was successful and returned with much treasure and large numbers of sparkling gold stones, such as those now found in Arizona, and also great quantities of other gems of various kinds.

During the expedition of Sirius to rescue Vajra and Herakles, a rather interesting experience occurred in the family of his son Demeter who had married Elsa and settled in a house in the suburbs of the city. They soon found that there were other previous tenants who paid no rent, for the house was haunted in the most extraordinary way, and they were much disturbed by all sorts of unwelcome manifestations. Noises were heard, doors opened and shut unexpectedly, and they were frequently troubled by heavy footsteps, although no visible bodies were to be found on investigation. There was also a deep feeling of sadness about the place, and sometimes spasms of acute but inexplicable fear seized upon them both. The manifestations appeared to centre themselves round a certain room, though no part of the house was entirely free from them. The constant pressure of this psychic trouble quite wore out both Demeter and his wife. It was the wife who was first actually seized upon by the haunting entities, but, in endeavouring to protect her, Demeter himself became partially obsessed and after that ((the possession)) had once happened, quite long periods of time elapsed in many cases during which he had no accurate knowledge of what had occurred or what he had been doing. Both he and his wife were quite worn out with this, and as an addition to the family was impending, the mother of Demeter (Alcyone) felt that some decided steps must be taken. She determined to go herself to the house and spend a night alone in the room which seemed to be the central point of the disturbance, in order to try to discover exactly what was the matter, and to see if there were any possible way of dealing with the subject.

Demeter and his wife strongly urged that they should be allowed to remain with her, but she insisted on being alone, saying that she could not be responsible for anybody but herself. When everything was quite quiet, she covered the light and sat waiting. For a long time nothing happened, but at last there came three heavy dull knocks or blows, such as might be made by a large slow-moving object. Cold chills ran down Alcyone's spine, and an overmastering sensation of fear came over her. She shook this off, hastily uncovered the light, and stood looking expectantly towards the place from which the knocks appeared to come, reciting mantras by which she expected to call in the aid of deities. All at once she felt a cold breath on the back of her neck. She spun round and then something tapped her lightly the back. Again she span round but could see nothing there, and as she was thus looking into space something brushed her ankle. Looking down she saw a horrible object on the floor; it was like a large worm, perhaps four feet in length, but somewhat cigar-shaped, covered with hair, black, coarse, short and bristly; it had a sort of rudimentary face, with no features but a big red hole which took the place of a mouth, and the whole gave out a horrible and most sickening odour, as of something that had been long dead. It writhed along, and came curling round her leg, and as she reached down to tear it off, it fastened on her hand like a vampire, and then began to coil round her body. Just then she saw her son Demeter approaching, looking like one drowned, with horribly distorted features - lead-coloured, greenish, and bloated - and with a baleful deadly fire in his eyes, lambent and unholy. At first she thought he was coming to defend her; the horrible worm was just getting at her throat, and she called to Demeter to help her. But he came towards her in a curious stooping, crouching manner, his fingers clutching the air, and instead of helping her he seized her by the throat. With all her strength of will she called upon Sirius (who was absent on the expedition more than a thousand miles away) and he at once came astrally in answer; he seized the beast with one hand and Demeter with the other, tore them apart, dashed the beast to the floor and stamped upon it, till it was nothing more than a jelly; then he shook Demeter into wakefulness, and was gone as suddenly as he came. Demeter looked at his mother in a dazed sort of way, and said again and again:
"What is it ? What is it ? What is it?"

A great weakness overpowered him, which did not pass away for a long time, but he was never again obsessed. Alcyone's hair was white on one side where the beast had struck her, and for days afterwards she could not get rid of the horrible odour. The incident made a deep impression on her mind, and whenever she thought of it, it made her physically sick. For years she could not bear the sight of any creature that writhed, and she nearly fainted one day when a harmless cat happened to curve itself round her ankle, although it was a year after her adventure; and for a long time even the sight of a small worm would cause her to grow pale and weak.

When Alcyone had called Sirius to help her, he and others were sitting round a camp-fire, and at once he fell back in a trance. He plainly heard his wife's call, and somehow found himself in a room which he did not know. Seeing his wife in dreadful danger, he rushed to her aid, endowed with superhuman strength; when he had rescued her in the manner described, he seemed to lose consciousness, and when it returned his friends of the camp were sprinkling water on his face. He felt quite weak after this, and was not, fully himself for several days, so his exertion had evidently been a great strain upon him.

Alcyone went to Mercury and told him her story, asking him what, could have been the cause of all these strange happenings; he looked into the matter and unearthed the fact that on the spot where Demeter lived there had been long ago a centre for a peculiarly obscene form of early magic. Its devotees used to provide at their seances a bath of human blood, and huge scorpion-like creatures materialised and stalked round it, squirting out, a poison which seared everything which came near them.

Among these creatures was the unpleasant object that attacked Alcyone, and as it had been starving for a long time it was proportionately ferocious. These elementals were expressions of a certain form of evil thought, deliberately intensified and materialised by magical ceremonies, and, being ensouled by familiar spirits of a particularly obscene kind, they were exceedingly dangerous. By those who made them, they were called '*sendings*', because they could be sent, to anybody whom the magician hated, to materialise in his bed-room, to sit on his breast in the night and spit venom on him. An entity of an evolution lower than the physical used to be put into such a thing, and enabled to hold it together.

In the year 22,605 BC, when Sirius was about sixty years old, the King prepared an expedition to a certain holy city in Yucatan, which was about to be visited by Surya, the Head Priest of the great Atlantean religion, and Alcyone, Sirius, Mizar, Helios, Mercury, Uranus and many others set forth, starting in the late summer and travelling southward round the Gulf. At first, they used carts, but, after a time they had to leave the great, main rock road and abandon the carts, using their mule-like horses or mustangs both for pack and riding. The main rock roads were really remnants of a previous age. When Atlantis was at the height of its glory, wide roads of solid rock were formed radiating in all directions from the Great City of the Golden Gates, stretching over hill and dale for thousands of miles; and these were crossed by a network of local roads, which, however, were not so made or kept.

On one occasion our party fell into difficulties in trying to cross a river. At a later point in the journey they met a caravan of merchants who were using a curious camel-like sort of animal, resembling a big llama. It was some type between the two; the Atlanteans had been fond of experimenting in the crossing of animals. Once our travellers came to a deep canyon, and though it was less than fifty yards across they had to travel thirty miles round to reach the opposite side of it. When about half way on their journey, they met another caravan, of which all the people were in a dying condition, because the savages had poisoned the water of the stream from which they had drunk. Mercury magnetised the people and neutralised the poison, thus saving them all. They now bent their course towards the east, and then a little to the north, and soon a guide met them, a curious aboriginal man, who had been sent from Yucatan for the purpose of showing them the way. The people in the city were aware of the approach of pilgrims, at least, of this particular caravan, and a procession met them at the gate.

Mars, Mercury and the Priests at once repaired to the great temple of which Saturn was the Chief Priest, where they found some kind of initiation ceremony taking place. The number of people admitted to this was of course limited, but both Sirius and Alcyone were allowed to be present. There was a sort of golden throne, magnificently decorated; it had lion arms and a flight of nine steps leading up to it with carved animals on either side, something of the Egyptian style of work. Surya sat upon this throne, and received the people as they were presented to him, exchanging with each of them certain signs.

Each Priest, as he appeared before Surya, gave him the same secret salutation, which is one of those still used in the White Lodge at the present day. Surya sent out streams of blessing - or perhaps they were sent through him. Afterwards the huge brazen gates of the temple were thrown open, and the rest of the party came in, and Surya came down from his throne to speak with them, saluting them with the most friendly words. One remarkable fact that was observed is that <u>he must have known even then the name</u> which Alcyone would choose on his admission to the Sangha <u>twenty-eight incarnations later</u>, in the life in which he met the Lord Buddha, because he distinctly referred to it. ((see p. 225; author's (MGH's) underlining!))

Our friends attended also another great gathering on an occasion when Surya spoke to the assembled people. Even then he preached the doctrine of love, which is so characteristically his own, telling all the pilgrims the emphasis that must be laid upon that quality.

"Love is life", he said, "the only life that is real. A man who ceases to love is already dead. All conditions in life are to be judged fortunate or unfortunate according to the opportunities that they offer for love. Love will come under the most unlikely circumstances, if men will but allow it to come. Without this all other qualifications are only as water lost in the sand."

Our band of pilgrims stayed in the city for about two months and then started for home. On the journey they ran short of water and could find no source of supply, but the Priests located a spring by means of some sort of divining twig. While they were still on the way Helios died, to the great sorrow of her friends and relations. Mizar could not bear to leave the body to decay in the wilderness, and was grief-stricken because they had not the usual acid which it was the custom to inject into the corpse to burn it up at once. In compassion for Mizar, Mercury placed his hand on the body and disintegrated it by some means, as though by sending a current of consuming heat through it. Alcyone, being psychic, felt no separation from her mother, and so through her, Helios was just as much in touch with the family as ever, accompanied them on their journey in her astral body.

Sirius died at the age of sixty-four, but both he and Helios continued for a long time to keep up the closest relations with Alcyone, lingering intentionally on the higher levels of the astral world in order to do so. Her children and her brother Herakles looked after her thoroughly well as far as her physical wants were concerned. She occupied herself for the last twenty years of her life in writing a great book on religious subjects. It was in four parts, or volumes, with curiously epigrammatic and untranslatable titles. The nearest we can come to rendering them into English is: "Whence? Why? Whither? Beyond?" Mercury ordered that when this work was finished it should be preserved in the crypt of the Temple; but some centuries later, in consequence of the danger of invasion, it was removed to the other Temple in Yucatan. A copy of it was made by Alcyone herself for the Chief Priest Surya, which she sent to him in Atlantis; it now rests in the secret museum of the Great White Lodge.

Ajax had married Erato, and had a son (Melete) who was about five years old when the following curious incident happened. One day he was not to be found, and his mother, half mad with anxiety, went to Alcyone the grandmother, who tried in every conceivable way to find him, even to the sending of a servant down a well by means of a rope to see if he had fallen into it. At last, all physical resources having failed them, Alcyone sat down, determined to look for him psychically. She was successful in discovering where he was, and she told the father to take his sword and come with her at once to save the child. She led the way to an old half-ruined hut, to which a savage woman had carried off the boy, with the intention of sacrificing him in a black magical ceremony. Her intention was to make his intestines into strings for a musical instrument to be used for demoniacal invocations. The woman was resting with the child at this hut, in the course of her journey to a dark shrine, which lay further in the forest. By means of a magical potion she had put the child to sleep, so that she could carry him more conveniently, and was just about to start on her way when Alcyone and the father arrived. At first they threatened to kill the woman, but after a time relented, telling her, however, that if she came near their house again she would meet with certain death.

Alcyone lived to be 84 and in the year 22,578 BC this eventful life closed and Alcyone passed away, loved and respected by all who had known her.

((end of verbatim text {p. 259-272 in [3]}.))

-=-=-=-=-=-=-=-=-=-=-=-=-=-=-

In Alcyone's 20th life, near Chittagong (then in west Burma), in 21,759 BC, the description is again given verbatim, below:

Alcyone was born again in a female body in the year 21,759 BC, not far from where Chittagong now stands. She was the daughter of Brihat and Neptune, and was one of a family of four. Her elder brother was Uranus and her younger sister was Mizar, but both of these died young: Uranus at the age of eighteen and Mizar, in childbirth, at the age of fifteen. There was also a younger brother, Vulcan, who was taught from boyhood by the Priests in the temple.

The father Brihat seems to have been both ruler and Priest of a small community or kingdom. Astrology was a prominent factor in the religious ideas of the day, and Alcyone's horoscope was cast with great elaboration. It destined her to a marriage with Saturn, who was a distant relation, and it foretold that she should bear a child of remarkable power and holiness, and directed that all her early life should be arranged as a preparation for this coming event. The instructions were obeyed, and she was specially instructed by the Priests with a view to this.

Her childhood was a happy one. We see her as a graceful, beautiful child, with long, streaming black hair. The only mode of dressing the hair was to catch it back from the face with golden clasps, in which were mounted most magnificent diamonds, so large that they looked like brilliant stars against her dark locks. The hair was washed daily and anointed with magnetised oil, which was supposed to stimulate the intellectual faculties. She was carefully secluded from all possible trouble or difficulty. Her only sorrow was the death of her elder brother Uranus, to whom she was profoundly attached.

At the age of fifteen she was duly married to Saturn with great pomp, and a year later a noble boy was born, Surya. There was great rejoicing over this event, and every care was taken of the child of promise. Alcyone was very sensitive and impressionable, and when the child was about to come to her she had a wonderful dream in which she saw a bright star leave the sky and enter her. This dream caused her to be considered a holy person. She was also clairvoyantly conscious of the presence of the ego when it attached itself to her.

Everything seemed to promise for her a long and brilliant life under the most favourable conditions; yet all these expectations were disappointed, for her life was abruptly terminated at the age of seventeen by an accident in which she voluntarily sacrificed herself in order to save her child. The circumstances were as follows:

Alcyone's house formed part of a great suite of buildings erected round a sort of square which was within the palace of the King. A slave-woman, who was changing the water in a glass vase containing gold-fish, was called away on some other business, and set the vase down on a table in the full rays of the sun. The glass acted as a lens, and the sun-rays, streaming through it, converged on some neighbouring wood-work and set it on fire. The house was built entirely of wood, richly gilded, and the flames spread like lightning in every direction, blazing up like a furnace. Alcyone was, at the moment, at some little distance off, but as the servants rushed out in every direction, shouting and screaming, her attention was attracted, and she flew, fleet as a deer, towards the burning house.

The baby had been left with his nurse in an upper room, but she had gone out, confiding her charge to some fellow-servants. These fled downwards on the alarm or fire, forgetting the baby, and the terrified nurse, rushing for the child, fell back at the sight of the blazing staircase; which was the only way to the nursery. Wringing her hands, she screamed out: "The child! the child!", but dared not face the roaring flames which "barred the road". "My boy?" gasped Alcyone, and as the woman pointed upwards, shrieking, Alcyone pushed her away and sprang up through the sea of fire. Several of the stairs had already fallen, leaving only in some places the supporting wooden bars not yet burned through, though blazing.

Desperately she plunged on, climbing, slipping, leaping across the gaps through which the flames, flaring upwards, caught her garments and scorched her flesh. Surely no human strength would suffice to carry her to the top! But mother's love is omnipotent, and, in less time than it takes to tell it, she reached the room where the baby lay. Smoke was pouring into it, and she wrapped an unburnt fragment of clothing across her mouth and crawled along the floor.

The babe, crowing at the dancing flames, stretched out chubby arms to his mother, and, catching him up, she pressed his face into her bosom and fled downwards with her boy close wrapped in her arms. Again she crossed that burning torrent, her body nude, her hair blazing, the diamonds dropping from it, flashing back the flames. Somehow she reached the bottom; the open air, and fell prostrate outside, shielding the babe even as she fell. He was unhurt, but she was dying, and in less than an hour she breathed her last. More out of her body than in it, too terribly injured to retain feeling, she was scarce conscious of suffering, and her last smile seemed to be reflected on the freed astral form, as it bent over the rescued boy. Is it not the karma she made, by dying for Surya then, being reaped in the present opportunity given to Alcyone to serve the Blessed One again ?

After its mother's death the child was taken in charge by his aunt, Viraj (Saturn's sister), who was even then an advanced ego, and has since become an important member of the Esoteric Hierarchy. She was psychic, and through her Alcyone was still able to help and care for the child. The aunt never allowed any of the servants to touch the baby, and swung him herself in the garden in a sort of cradle hung up between the trees. There, in the quiet grove, Alcyone would speak to her from the astral world about the child, who was thus brought up altogether in a holy atmosphere and soon became a wonder, at the age of seven delivering teaching in the temple, so that people from all quarters came to hear him.

It seems as though from time to time the members of the present Hierarchy of Adepts were born together in different countries to assist the founding of a new religion, or of a magnetized centre. We see them also spreading the religion and sending expeditions to other distant centres, as in the previous life in North America, where an expedition was sent to Yucatan. In the present one, some twenty-five years after Alcyone's death, we see Surya sending one north to the city of Salwan. Some of the party lost their lives from the hardships endured; and among these was Alcyone's younger brother Vulcan, at the age of about 35.

((End of verbatim text {p. 279–281 in [3]}.))

-=-=-=-=-=-=-=-=-=-=-=-=-=-=-

Alcyone's 21st life, was in South India, in 21,540 BC and the full description is again given verbatim below:

The next appearance of our band of Servers is in South India, and on this occasion Orion, who had been wandering in outer deserts for some thousands of years, finds his way back into the group - but in a very peculiar way. He ((Orion)) was born in 21,540 BC as a girl in one of the hill tribes of the Nilgiris, a clever, good-looking and unscrupulous young person. She had no intention of living the life of the hill-tribes, so she engaged herself to serve a noble Tamil lady, and was appointed to attend upon her daughter Iota, to whom she speedily made herself indispensable. In this Tamil family was an heirloom - an enormous emerald credited with magical powers. It had been magnetised in Atlantis by one of the Lords of the Dark Face, and it was supposed to earn for its possessor whatever he most desired, but it always brought misfortune in the end, and those who used it became tools of the original magnetizer. Iota persuaded her father to give her this wonderful stone, and by its means to arrange for her a marriage with a neighbouring King. By the power of the stone this plan was carried through, and the King sent an escort to bring his bride to him. Iota took with her three attendants, Orion being one of them, and on the journey Orion contrived to murder her young mistress, and then impersonated her. Her plot succeeded and she was duly married to King Theodoros, to whom she made a good and clever wife. It is to the credit of Orion that in her new surroundings she did not forget her brother Egeria, whom she had dearly loved, but sent secretly for him, had him educated as befitted her new station, and eventually married him to one of the ladies of the court, though never openly acknowledging him as her brother.

Ten years later, for reasons of State, the King took a second wife Nu, a princess of a neighbouring house. After all these years the murder of Iota at last came to light, and Orion's true status was discovered.
Her husband was indignant at the outrage on his pride, and promptly condemned her to death. When thrown into prison she invoked the Lord of the Emerald, and he appeared to her and ordered her to throw the emerald out of the window to Sigma, one of the little children of the second wife Nu, who was playing outside. As soon as this was done he ordered Orion to take poison, and as she left her body the little girl Sigma fell down dead in the courtyard, and Orion

was pushed into her body. When they went to lead out the queen to execution she was of course found dead in her cell.

In Sigma's body she ((Orion)) was contracted in marriage to Leo, the prince of a neighbouring kingdom, in what is now the Telugu country, and after they were married she induced her young husband to bring about his father's abdication, so that she herself might be queen of the country. Alcyone was born in the year 21,467 BC as their eldest son, and there were four other children. When Alcyone was eleven years old his mother Orion fell ill of some internal disease which was found to be incurable. As soon as she knew that death was drawing near, again appealed to the Lord of the Emerald, who told her that he would help her once more to take another body, but that it must be that of her daughter Theseus, whom she loved dearly. For some time she refused this. But at last increasing suffering drove her to accept it. So she drowned Theseus, hung the emerald round the child's neck and then threw herself into the water and sank. When she recovered consciousness she ((Orion)) was in the body of Theseus, and so, instead of being Alcyone's mother, she was now, as far as outward appearance went, his sister.

The politics of the time were complicated and troubled, and Alcyone, though anxious to do his duty, was more interested in his studies than in affairs of state. He learned whatever was customary for boys of his class and time, and was proficient in riding, shooting, swimming, and the various sports of the race. When he came of age he married Herakles, who was the daughter of a neighbouring Raja, and they were happy together in their religious studies. The Priest Mercury was a neighbour and close friend.

In order to save the King Leo from certain defeat at the hands of a coalition of neighbouring States, Alcyone's mother Orion had induced Leo to place it under the suzerainty of the Atlantean Emperor, Jupiter, and there was much discontent among the people about this. A few years after, when Orion had had to change bodies and could therefore no longer direct Leo's policy, the discontent broke out into open rebellion, and Leo was defeated and killed. Sirius (the son of Gimel, an Atlantean noble) was sent over from Atlantis by Jupiter to be Governor of the kingdom, which was thus made a province of the vast Atlantean Empire.

Sirius made friends with Alcyone and Orion, at first perhaps from motives of policy, but the friendship quickly ripened into real affection. He fell in love with Orion, and demanded her hand from her brother Alcyone, who gladly gave it, and a close tie united the two families, and also that of the priest Mercury. This made the government of the province an easy matter, as the official heads of both the parties in it were now so thoroughly united. In fact the three families were almost like one, and made a kind of little society of their own, in which all sorts of interesting problems were discussed.

We find that the language commonly used then in India was not Sanskrit, and ceremonies usually began with the word' Tau', not with 'Aum'. The doctrines of Reincarnation and Karma were commonly known to the people. The Teacher (Mercury) knew of the Great Ones behind all who sometimes helped. Some of the expressions which are familiar to us now were in use then also, as for example: "I am THAT". Mercury told the people that of all the qualities that they could develop, of all the qualifications they could possess, the most important was the power to recognise that all was THAT.

"You cut down a tree", he said, "THAT is the life of the tree; dig up a stone, THAT is what holds the particles of the stone together; THAT is the life of the sun, THAT is in the clouds, in the roaring of the sea, in the rainbow, in the glory of the mountain", and so on. These words are taken from a discourse of Mercury on death. In a book from which he read to the people there were well-known phrases, such as:

"One thing is the right, while the sweet is another; these two tie a man to objects apart. Of the twain, it is well for who taketh the right one; who chooseth the sweet, goes wide of the aim. The right and the sweet come to a mortal; the wise sifts the two and sets them apart. For, right unto sweet the wise man preferreth; the fool taketh sweet to hold and retain".
(Katha Upanishad; words in Mead's translation). The wording in Mercury's book was not actually identical, but it was clearly the same set of verses.

There was another saying:
"If one is killed and I am the slain, yet I am also the sword of the slayer, and none slays or is slain, because all are one.
There is no first nor last, no life nor death, because all are one in Him".

The books which Mercury used did not come from the Aryans; this book from which he read (evidently the original of the Katha Upanishad) was written in the City of the Golden Gates by one who was a member of the Brotherhood. It belonged to a great collection, and had been handed down through many centuries. The Nachiketas story had not yet been connected with it.

In one Temple there were no images at all. The religion was not sun-worship - at least not exclusively; rather a worship of the powers of nature. Outside the Temple, there was a large bull in stone, facing the Temple and looking in. Inside there was a curious arrangement - a depression instead of a raised altar. Two or three steps led up to a great low square platform, paved with beautiful tiles, and then there was a depression in the centre with a railing round it. People threw flowers into the depression, in the middle of which was a slab, which was specially holy; it had some markings on it, but we could make nothing of them.

In another Temple there were many images which were set in niches in its back wall. The people here wore a different dress from those in the former Temple, and there were men who were distinctly priests, which was not the case in the other. The images sat cross-legged, and had not more than the natural number of arms. This was the old form of Jainism, presumably, and the images Tirthankars. Some images were naked; others, which had a looped garment over them, were probably regarded as dressed, or perhaps a conventional symbol was intended.

In another Temple a long way to the north, there was already a lingam. Up there the Trimurti was fully recognised, though the names were not those used now. In one cave-temple there was a gigantic face carved out of the rock which was three faces in one, though it was so arranged that only one face could be clearly seen at a time. There was a great Temple in South India which also contained a Trimurti. We tried to discover the meaning of the name attached to it, to see what idea was connected with it in the mind of its priests and we found that one priest thought of it as: "He whose life flows through all", while another had the idea that the three persons were: "He who opens the gates, He who guides the stream, and He who closes the gates". We saw no specimens here of the many-armed images which are so plentiful at the present time.

The priests had strong ideas about a 'Lake of Light', which was also Death and Life and Love; all streams led into the Lake of Light, whencesoever they seemed to begin. There were traces also of the theory that all that we see is illusion, but the only Reality is the Lake of Light. "We live in the Lake of Light and do not know it. We think of ourselves as separated, but we are each a drop in the Lake". The priests seemed to be perpetually urging the people to get behind the illusion of the senses, and to realise that THAT was the real Presence behind all, and that the separated forms were the separated drops: "When they fall in again they are all one", they said, "and it is we ourselves who make all the sorrow and trouble".

They had a prayer to the "Lords who are in the Light, who consist of the Light".

What is written above represents something of what was taught to the people, but in this small and strictly private family circle, Mercury was willing to go a little further, and expound the true meaning of the symbols, and give far more information about the Lake of Light and the Lords who are the Light. He told them of a great Teacher who might be invoked by certain prayers and ceremonies, whose blessing might be called down upon them if they asked for it earnestly and with pure heart. They invoked Him at their meetings, and a response always came, and on two special occasions He even showed Himself. This Great One was He whom we know as the Mahaguru, and His special connexion with this group was that He had, in a previous birth, founded their religion and arranged that He would, as its Founder, respond to certain invocations made under proper conditions by its true adherents. He threw into the mind of Mercury the solution of their problems and the answers to their questions on religious matters, and twice certain personal directions were given to them, though this was a rare occurrence. The priest Mercury had married Ulysses. The pleasant interaction between the families and their study of the questions which so deeply interested them went on harmoniously for many years, and the first break in the party occurred in the year 21,423 BC, when Orion confessed her black magic to Mercury and to Sirius, threw the magic emerald into the sea, and retired to an ascetic life, with a view to atoning for her deeds. She handed over her children to the care of her friend Helios, and four years later the latter married Albireo, a younger brother of Alcyone.

The children of these families all grew up together, and naturally fell in love with one another, so that when they became men there was a good deal of intermarriage between them. Achilles took to wife Mizar, while Uranus married Vega, and Hector Selene.

Aldebaran, however, caused much trouble to the family through becoming involved with and marrying a woman of bad character (Gamma), who ruined his life, and left him a miserable wreck when she finally abandoned him, and ran away with Pollux, who was a rich but dissolute merchant.

Vajra also was a source of anxiety to his mother Herakles, because he developed a wandering disposition and became a great traveller in search of knowledge and experience. He, however, wrote a brilliant account of his travels, which was read over and over again by the family group, and practically learnt by heart by the younger members. Alcyone was so interested in some of its glowing descriptions that he actually went on no less than three difficult and dangerous journeys in order to see the places of which his son had given so attractive an account. In the course of these he met with various adventures, the most serious being that he was once captured by robbers and held for ransom, though he contrived to escape by disguising himself as a woman. In another case he was carried off his feet while trying to wade across a swollen river, and was swept down more than a mile, and nearly drowned. He also accompanied Sirius on several of the latter's official tours through the province; indeed, Sirius, delegated many of his powers to him, being anxious to show the people what thorough accord existed between the Atlantean power and their old royal family. The tie between these two men was singularly close, and, though of different races, they seemed always to understand each other perfectly. Sirius, who was patriotic, told Alcyone much of the glories of Poseidonis and the City of the Golden Gates, and fired him with great enthusiasm about it and an intense desire to see it, which bore fruit much later in life.

Herakles died in 21,396 BC, at the age of seventy, and Sirius, to whom she had been a particularly close friend, mourned her loss quite as much as Alcyone, and accorded her the most gorgeous obsequies. This left Alcyone much alone, and he clung more than ever to his friend Sirius, who fully returned his affection, so that the two old men were like brothers. For thirty years Sirius had been visiting regularly every month his wife Orion, who was living as an ascetic; and when she died in 21,392 BC, he felt himself unable to stay any longer in India, and applied for leave to resign his Governorship and return to Poseidonis. Alcyone, though seventy-five years of age, immediately announced his intention to accompany him, and actually did so.

The two septuagenarians had a prosperous voyage, and Alcyone found the splendours of the capital even greater than he had expected. Few of those whom Sirius had known forty-four years ago were still living to greet him. The Emperor Jupiter was long ago dead and his son Mars reigned in his stead; he received the two old men with great honour, and gave them honorary posts at his court, distinguishing them with many marks of favour. He must have felt drawn to them, for he set his court astrologers to calculate the particulars of their connection with him, and was informed that both had worked with him more than once in the past, and that both were destined to serve him in some mighty work far in the future, when nearly a quarter of a lakh of years ((25,000 years)) had been added to the roll of time. None of them then understood this prophecy, but it is evident that it will be fulfilled in the Californian community (USA) about 2,750 **AD**.

Vajra, who had accompanied his father, soon took a prominent position under the Emperor and enjoyed his fullest confidence. Sirius and Alcyone lived together in the same house as brothers for ten years, and both died in 21,382 BC -- hale and hearty to the last. During these ten years they jointly prepared a book upon Southern India, which was highly esteemed, and was regarded for centuries in Poseidonis as the classical work on its subject. It was in two volumes, one treating of the different races and their customs, and the other of the various religions - the latter embodying much of the teaching given to them long before by the priest Mercury.

((end of verbatim text {p. 283-289 in [3]}.))

-=-=-=-=-=-=-=-=-=-=-=-=-=-

In Alcyone's 22nd life, in 20,574 BC, in South India again, two interesting episodes were as follows. Jupiter allowed Alcyone to see some secret ceremonies which resembled the Eleusinian Mysteries, which stimulated his psychic faculties and gave him a vision of his (dead) mother and allowed him to communicate with her. On another occasion, Alcyone visited a shrine at Lakhimpur near the Brahmaputra river, which was in the charge of Lyra, a priest from Tibet who was founding a new religion under the direction of Mahaguru. Lyra much later was reborn as the philosopher Lao Tsu (author of "Tao Te Ching"). Another temple visited by Alcyone was near Brahmakund, where many shrines had been consecrated personally by Mahaguru 20,000 years earlier. Many other interesting events of this 22nd life [3, 14] are too numerous to repeat here.

-=-=-=-=-=-=-=-=-=-=-=-=-=-=-

In 19,877 BC, Surya inspires a group who were reforming the religion in Tibet, under the direction of Jupiter. Surya was not born there but inspired the group from the Astral (4th dimensional) level. 500 years later, in the Amazon valley, in 19,380 BC, the same sub-group were reforming the religion there; at this time Surya was born and married Jupiter's daughter Naga. Surya generated a religion lasting some thousands of years. This developed into a major Peruvian civilisation.

-=-=-=-=-=-=-=-=-=-=-=-=-=-=-

During Alcyone's 24th life, in 18,885 BC, in Manoa (at Shamballa, in central Asia), Mars was inspired by Manu to lead a group away from the constant warlike raids they were experiencing. This was to be the greatest migration in history, going to a "promised land" prepared for him in India, where the ruler, Viraj welcomed them.
(Refer to Tables 1 & 2 on p. 89 & 90 for the pseudonyms used)

Fig. 4.5.   <u>Viraj</u> = The Maha-Chohan.

Drawn by Remote Viewing, from His appearance as it was in 1930 [4].

Note: The differences in appearance of ordinary people is said to be mainly due to their degree of advancement and karmic effects. But for the perfected Men shown in Figures 4.2 to 4.10 and in Fig. 4.10, these factors have dropped away and the Genesis (1:26) verse: "God said, Let us make man in our image, after our Likeness", applies without any constraints.

CHAPTER 4 World History from 70,000 BC:
Buddha & Christ in their previous lives

During this time, Alcyone had a series of visions, memories of past lives, given verbatim below:

These visions were of varied character, some of them connected with lives which we have already investigated, but others which are at present unknown to us. In many of these scenes his father and mother appeared, and he always recognised them, under whatever veil of race or sex they might be hidden. Sometimes, when a rare wave of confidence swept over him, he would describe these visions to his mother, making them marvellously picturesque and life-like. He called them his picture-stories, and he would say:

"Mother, in this story you are a priest in the temple" or "In this you are my mother, just as now", or again, "In this you are my little baby, and I carry you in my arms". Whenever he said these things his mother felt herself identified with the figure in the vision, and her memory was as it were awakened by his. She remembered now that when she was herself a child she used to have similar recollections, though as she grew older they faded from her mind; and she realized that her son was seeing what she used to see. In one of his most splendid visions - that which he liked best of all - neither his father nor mother appeared, but he saw himself as a young girl filled with intense love and determination, rushing through raging flame and suffocating smoke to rescue a child who was the hope of the world - a memory of the 20th life (west Burma) three thousand years before. But he had also other memories in which his parents bore no part, and some of these were far less desirable.

One curious set of visions which came now and then appeared to image some ceremonies of the darker magic, evidently from a remote past. They were indescribably weird, yet thrilling, and they excited a feeling of inexpressible horror and loathing which was yet somehow mingled with a kind of savage ecstasy. There was about them a distinct sense of something radically unholy and evil - something from which Alcyone's present nature shrank with terror and disgust, while he was yet keenly conscious that there had been a time in the far-distant past when it had filled him with a fierce joy - when he had somehow been able to revel in what now he utterly abhorred. He disliked these visions intensely, yet occasionally they asserted themselves, and when one had commenced he seemed compelled to play his part in it to the end. Of these he had never been able to speak to his mother, though she had twice noticed the prostration which followed them, for he came out of them in a condition of profuse perspiration and utter nervous exhaustion. But he said only that his dreams had been terrifying, but that he could not describe them.

It is not easy to recover the actual subject matter of these evil visions, but they evidently reflected some of the wild orgies of the darker worship as practised in Atlantis – something of the same order as the alleged witches' sabbath of the Middle Ages - a kind of riotously sensual adoration of some strange personification of evil belonging to an existence which humanity has now altogether transcended. Its devotees appear among other things to have been able by the use of some potion or unguent to assume animal forms at will and to levitate these transformed physical bodies. In looking back involuntarily upon unholy revels Alcyone always saw himself with a partner – always the same partner; and he knew that it was for the love of that partner that he had thrown himself into this cult of evil, that her seduction had drawn him into it and taught him to enjoy it. Yet even amidst his horror he knew that she had herself no evil purpose in doing this - that it was because she loved him that she shared with him what for the time made her happy, that in reality she would have died rather than harm him, and that it was only her ignorance which permitted her to be used as a lure by malicious powers behind it. These unpleasant visions came to the boy but rarely, and they would not have merited such detailed mention but for the fact that a few years later they were shown to have a close connection with one of the recurrent characters in our story. ((The above is given in more detail near the start of this chapter ("Early Times", page 91).))

Some time before the birth of Alcyone a certain Mongolian chieftain had come to take refuge in the kingdom of Mars. This chieftain was the younger brother of a reigning chief who was (apparently not undeservedly) decidedly unpopular with his people. The younger, on the contrary, was universally liked, and there was a conspiracy, though entirely without the young man's knowledge, to dethrone the elder brother and set him up in his stead. This was discovered and suppressed, but as it was impossible to persuade the elder brother that the younger had not been privy to it, he had to flee for his life, and it was in this way that he came to seek refuge with Mars. He and two or three friends who had escaped with him proved harmless and indeed desirable members of Aryan tribe, so they settled down and were accepted without further question.

They had brought their wives and children with them; so they formed a kind of minor community within the tribe, living amongst it but not intermarrying with it. This young chieftain (Taurus) had several children, but the only one that comes into our story is Cygnus, a daughter who was about the same age as Alcyone, with whom she fell violently in love. They played together often as children, but along with many others, and it does not seem that Alcyone specially differentiated her from the rest, though he was always affectionate to all. As they grew older, the boys and girls drew more and more apart in their games, and so he saw less of her, but she never for a moment forgot him.

When she was seventeen her father married her to Aries, who was the son of one of his companions. He was much older than she was, and she had no affection for him, but her wishes were not consulted in the matter; it was entirely an affair of policy. Her husband was not a bad man, and was never unkind to her, but he was absorbed in his studies and had no attention to spare for his young wife, whom he regarded rather as part of the necessary furniture of a home than as a sentient being who might possibly have claims upon him.

For a long time she fretted silently against this, being all the time madly in love with Alcyone, and seeing him only occasionally and casually. At last there came a time when he was sent ahead of the main body on a dangerous scouting expedition; hearing of this and fearing that he might be killed, she seems to have been reduced to desperation, and she fled from her husband, dressed herself in male attire, and joined the small band of men whom he was taking on this perilous expedition. Alcyone succeeded in carrying out the instructions of Mars, but only at the cost of the loss of many of his men, and among others Cygnus was fatally wounded and her sex discovered.

She was carried before Alcyone, and when he recognised her she asked to be left alone with him for a few moments before her death. Then she told him of her love and her reason for thus following him; he was much surprised and very much regretted that he had not known of her affection before. As he stood beside her his mind was persistently haunted by the most vivid presentment of his old vision of the wild orgies of Atlantean magic, and like a glare of lightening it burst upon him that Cygnus was identical with the female companion of that strange old witchcraft. He was so struck by this revelation that his manner showed it, and she, who had known something in childhood of the visionary side of his nature, at once divined that he was seeing something non-physical, and set her remaining strength to see it too. She had not been at all psychic during life, but now as death approached, the veil was to some extent broken through by her earnest effort, and as she seized his hand the vision which he saw opened before her eyes also. She was horror-stricken at his evident horror, but at the same time in a way delighted also, for she said :

"At least you loved me then, and though through ignorance I led you into evil, I swear that in the future I will atone for this and regain your love by loyal and ungrudging service to the uttermost".

Saying this she died, and Alcyone mourned over her, regretting that he had not known of her love for him, for had he done so, he might have prevented her untimely end. When opportunity offered he told the story of this strange experience to his mother, and she agreed with him that without doubt his visions did represent the events of previous incarnations, and that she, his father, his sister, his elder brothers and Cygnus had really borne in those lives the parts which the visions assigned to them. ((The story of which this particular incident brought the recollection, is given earlier in this chapter (page 91) and in [2].))

The strong influence of his mother Mercury over Alcyone seemed to increase rather than decrease as the years rolled on, and though the visions of his childhood now visited him but rarely, he still remained impressible as far as she was concerned, and frequently caught her thought even when at a distance from her. For example, on one occasion when her sons were out on a scouting expedition clearing the way through the hills for the main body of the caravan, she became aware through a dream of an ambush into which Herakles and his party were danger of falling. The whole scene was so vividly before her eyes, and the natural features of the country so deeply engraved on her mind, that she could not but feel sure that the danger was a real one. She called before her some natives of the hill-country who happened to be in the camp, described minutely to them the place which she had seen, and asked if they recognized it. They immediately replied that they knew it well ... ((and was able to warn the group. The text continues on ...

End of verbatim text quotation of p. 327–330 in [3].))

The native population as they moved south in India were Atlantean. Viraj ruled north India and Saturn ruled the south. In the south, Surya was the established High Priest, (who was overshadowed by Mahaguru, who was not physically present) who advised the ruler, Viraj, to welcome the newcomers (Mars and his group). Mars devolved his duties to his son Herakles and his younger son Alcyone took on the duties of Deputy High Priest, following on from ... .

Alcyone was also able to receive guidance from Mahaguru. Alcyone died at age 79, but he noted that Surya had hardly aged at all, which is a characteristic of sainthood; in modern scientific terminology, Surya was able to control the telomere cells, which control ageing.

AB (Herakles) & CWL (Sirius) add a note in their book [3] that the present-day Brahmanas of south India, commonly called the "dark Caucasians" are descended from the tribe whose arrival in India is described in their book (a brief summary only is given here above), but from their long stay in the tropics they have now become much darker than their ancestors (simply by natural evolution – survival is increased for darker protected individuals where sunlight is strong).

-=-=-=-=-=-=-=-=-=-=-=-=-=-

Alcyone's 25th life, in 18,209 BC, was in the region now called Algeria & Morocco, but which was then an island. The present day Sahara Desert was then a sea. The people were Atlantean Semite, similar in appearance to today's Arabs. The civilisation was advanced, with parks and gardens and fountains fed by aqueducts similar to the much later Roman ones. The roads and gardens were beautifully kept, in contrast to today's in North Africa! There was a University with a high reputation, which conferred degrees in Divinity, Mathematics, Literature and Rhetoric (debating and lecturing).

Fig. 4.6. <u>Saturn</u> = later a Master ("The Venetian").

Drawn by Remote Viewing, from His appearance as it was in 1930 [4].

Was the painter Paul Veronese in a previous life.

Note: The differences in appearance of ordinary people is said to be mainly due to their degree of advancement and karmic effects. But for the perfected Men shown in Figures 4.2 to 4.8 and in Fig. 4.10, these factors have dropped away and the Genesis (1:26) verse: "God said, Let us make man in our image, after our Likeness", applies without any constraints.

CHAPTER 4   World History from 70,000 BC:
Buddha & Christ in their previous lives

The University made Alcyone a staff member and he advanced so rapidly that he became Head of the University at age 30, by far the youngest who had ever held that post. He arranged for a new campus out in the country and new buildings, with transport to the town by a novel tramway operated by water power from a 300 foot cliff which a river went over, using a bucket-chain arrangement to drive a rope which pulled many carriages (full details are in the book [3]). Water wheels were later used.

Alcyone invited lecturers from Atlantis and built a large library and exchanged manuscripts with libraries in Egypt, Atlantis and India. He thus came into relation with the large library in south India which he had founded 600 years earlier when working there as deputy to Surya. His eldest daughter, Mercury, took interest and pride in the University. Brihat, a learned and holy man lived a hermitic life, but came out to teach philosophy and divinity at the University.

The next text below is given verbatim [3]:

Brihat had some reputation also as a healer though it appears not to have been so much his own doing as that Surya sometimes sent power through him and effected cures in that way. This was done once with regard to Alcyone himself, after an unfortunate accident which occurred at the University. Alcyone's second son, Aldeb, had taken up keenly the study of the chemistry of the period, having travelled as far as Egypt in order to obtain additional information from the professors there. He had made several important and useful discoveries, and was always engaged in experiments, often of the most daring character, in which his sister Mercury also took much interest.

One day when Alcyone had been invited to the laboratory to inspect the results of some new process, a serious explosion took place, stunning both Mercury and Aldeb, and setting on fire the garments of the former. Alcyone displayed great personal bravery in this emergency, rushing forward and beating out the flames with his hands, and dragging the body of Mercury out of a pool of blazing liquid, thereby unquestionably saving her life. He was badly burned himself in doing this, and it was a consequence of this that he was taken to Brihat. The latter passed his hands lightly over the wounds and blisters, applied to them some sort of oil which especially magnetized, and then deftly enveloped them in bandages, telling Alcyone not to touch them for a certain time and promising that when at the end of that time he removed them he should find the wounds healed, which proved to be the case. It is noteworthy that Brihat always used the name of Surya in his magnetizations, and that he invoked him when operating upon Alcyone, saying: "I cure him in thy name and for thy work". Owing to Alcyone's prompt action Mercury was but slightly injured, but Aldeb, who had been nearest to the retort, was much hurt by the force of the explosion, though hardly burnt at all.

Alcyone was so much interested by Brihat's procedure that he afterwards went to him to learn the art of mesmeric healing, and practised it among his own students with considerable success. Once Brihat himself fell ill, and was sedulously nursed by Helios.

On yet another occasion Brihat's semi-esoteric influence came usefully into the family life. During one of the vacations of the University an attack was made upon a village in the neighbourhood by African pirates from the southern shore of the Sahara Sea. Brihat by some means or other became cognisant of the impending attack - from his eyrie on the hill-top he may have seen the fleet of boats approaching - and he managed by means of thought transference to warn Alcyone of the danger. Leo, Alcyone, and Herakles, representing thus three generations, happened to be within reach, and they all at once hurried down to the village and organised the inhabitants to resist the raid.

The villagers were ill-armed and unaccustomed to fighting, and if caught unawares would undoubtedly have fallen an easy prey to the savage marauders. But having three gentlemen to lead and encourage them, and to make a definite plan of defence for them, they were able to do much better. Our heroes thought it best not to attempt to oppose the landing of the enemy, but succeeded in decoying them into an ambush in which large numbers of them were slaughtered.

Mizar, the youngest son of Sirius, happened to be staying out there with Alcyone for his holidays, with two boy friends. These boys had of course been left behind when the news arrived, and strictly enjoined to keep out of harm's way. But equally of course they desired to see something of the fighting, and stole down after their elders, and while they were watching from a distance Leo's arrangements for the defence, a brilliant idea suddenly dawned upon Mizar, which he instantly communicated to his companions. The pirates ran their boats up on the shore, made them fast and left them while they charged into the village to pillage and murder. The boys ran unobserved round the back of the village, rushed to these boats and set them on fire, helping the conflagration by pouring into them a quantity of pitch which they obtained from the yard of a neighbouring boat-builder.

The pirates had not dreamt of any serious opposition and had left their craft entirely undefended, so the boys had a clear field of action, and in a surprisingly short space of time, by working with feverish energy, they had the entire fleet of boats blazing merrily, and whenever they could not get the flames at once to seize upon some part of the vessels they stove in their sides with an axe and cut away such rigging as they could easily reach. In this they were assisted by another of our characters - Boreas, who was a boy servant to Mizar. Fortunately for themselves they contrived to get away just before some of the pirates, disgusted with their unexpectedly warm reception, came trooping back to the beach and realised that they were cut off. This discovery made them fight with redoubled savagery, but Leo's plans were so well laid, and he was so ably seconded by the younger men, that they were able to keep the pirates at bay until the arrival of Sirius with a large armed force from the city - for immediately on receipt of the first warning pf danger Alcyone had sent a messenger to him for military assistance. The pirates were then ruthlessly exterminated. ((author's (MGH's) comment: How does the karma of this work? Could it mean that those who killed the pirates would be killed in future themselves <u>only if</u> they were <u>pirates</u> attacking someone, but not otherwise?!
The mills of God grind slowly, but they grind very exceedingly small?!))

The younger branches of the family intermarried to some extent, Vega taking Bee to wife, and Bella joining with Aqua. The childish association of Cygnus with Sirius and Alcyone led to her falling seriously in love with the latter when they grew older. Though she had never previously shown her love openly, his marriage with Helios was a great blow to her, and she went and reproached him bitterly for forgetting her, as she put it. He was much concerned about the affair, and spoke gently and kindly to her, though he was in no way shaken in his devotion to his wife. Cygnus could not forget him, and refused several eligible offers because of this; but after some years she at last yielded to the oft-repeated solicitations of an old suitor, married him and lived a sober and happy life. Her brother Algol married Psyche, which was considered an exceedingly good match for him.

Perfect understanding always subsisted between Sirius and Alcyone, and when the former died at the age of sixty-nine Alcyone felt that he had lost himself as well as his brother. But he soon realised that nothing was really lost, for each night he dreamt vividly of Sirius and during the two years which he survived it may truly be said that he lived through the days only for the sake of the nights. Up to the last, however, he retained the keenest interest in his University, and it was his greatest joy to see how thoroughly his son Herakles entered into his feelings, and how eagerly he carried on his work. Finally Alcyone passed away peacefully during sleep, at the age of 71, leaving behind him as a monument a University the renown of which lasted some two thousand years, until the civilisation wore itself out, and was overrun by barbarous tribes. We find another of our characters Phocea, acting as a clerk in the office of the University .

((end of verbatim text, {p. 348-351 in [3]}))

-=-=-=-=-=-=-=-=-=-=-=-=-=-

The 27th life of Alcyone was in 16,876 BC, in the south-central part of Poseidonis (the last then remaining island of Atlantis). Mars was the Toltec emperor then and the people were merchants and sailors, resembling the (later) Etruscans and Phonecians. The High Priest was Surya, a man of saintly life and great wisdom, revered through the whole Atlantean Empire. Herakles, the son of Mars married Saturn and he thus became the effective ruler of the country. The sons of Herakles were Mercury and Venus, who married Brihat and Osiris. Alcyone was the eldest son of Mercury and Sirius and Mizar were the daughters of Venus. Surya considered Alcyone to be a child of great promise and gave much time to his education.

Herakles being the son of the Emperor, was not of the priestly line and so not eligible to succeed Surya. Alcyone, who was eligible, had heard stories of sailors' adventures and so he wanted to be a sailor, but his grandfather Surya told him the life of a sailors and merchants

CHAPTER 4   World History from 70,000 BC: 177
Buddha & Christ in their previous lives

Fig. 4-7. <u>Venus</u> = later the Master Ragozci;
formerly The Comte de S. Germain of the 18th century.

Drawn by Remote Viewing, from His appearance as it was in 1930 [4].

Note: The differences in appearance of ordinary people is said to be mainly due to their degree of advancement and karmic effects. But for the perfected Men shown in Figures 4.2 to 4.8 and in Fig. 4.10, these factors have dropped away and the Genesis (1:26) verse: "God said, Let us make man in our image, after our Likeness", applies without any constraints.

was mainly one of self-interest, but that of a priest was altruistic: the former worked for this life only, but the latter worked for a higher life and for all eternity.

Surya saw him again shortly after and Alcyone told him that he would choose the temple life. Surya blessed him, and said (verbatim below from here on):

"You have chosen wisely, as I knew you would. I have prayed much for you, and last night, as I was praying, the past and the future opened before my eyes, and I know what has been and what shall be. Just as today you saved another life at the risk of your own, so long ago did you save my life, even mine, at the cost of your own; and once more in the future you may give up your life for me if you will, and through that sacrifice all the kingdoms of the world shall be blessed".

The boy looked up at Surya in wonder and awe, for the old man's face was transfigured as he spoke, and it seemed as though mighty flames were playing round him; and though Alcyone could not then fully understand what he meant, he never forgot the impression which it made upon him. He was duly admitted into the Temple, and was very happy in his life there, for though the studies were arduous they were well arranged, and were made interesting to the postulants.

Surya, wishing perhaps to show the boy that in the priestly life also one might have travel and adventure, offered him the opportunity of accompanying his father Mercury and some other priests upon a mission to a great library and University in Northern Africa. Naturally Alcyone accepted with the greatest joy, and the voyage was a never-failing wonder and delight to him. It was long and slow, but not too long for him; indeed his excitement and interest when land came in sight were somewhat tempered by the regret which he felt at leaving the vessel, every sailor in which was a personal friend to him.

As they sailed along the coast a curious feeling came over him that he had seen it all before, and it grew so strong that he amused himself by telling the sailors what would come in sight beyond each headland as they came to it and the remarkable thing was that he was always right. He described in detail the city which was their port of disembarkation long before they reached it and the sailors who knew it said that his description of the hills and valleys and the position of buildings was marvellously accurate, but that what he said as to the shape and size of the buildings themselves and the extension of the town was almost all of it wrong. When at last they came in sight of it his feelings were of the most mixed description; he recognised instantly all the physical features of the place, but the town was enormously larger than in his opinion it ought to be and the buildings seemed all different. He was strangely excited at this astounding half recognition of everything and constantly questioned his father about it but at first Mercury could only say that he must have travelled on in advance of the ship in his eagerness and seen these things in a vision.

Presently, when it became evident that the city which he knew was much smaller, it occurred to his father that they might be presence of the phenomenon of a memory from a past incarnation and when they landed he became almost sure of this because when Alcyone described how, according to his idea, the various streets ought to run or the buildings to stand, in several cases the inhabitants said:

"Yes, there is a tradition that it used to be like that". Then they were carried out to the University on a curious hydraulic rock tramway he became still more excited and described exactly how it used to work and the form of the old cars, which had for centuries been superseded by another type; and when they reached the University itself he was quite unable to contain himself, for he declared that he knew every walk in the garden, and dragged his father about to show it all to him. Presently his fullness of memory reawakened that of his father, and Mercury also began to see things as they used to be and to recollect events as well as scenes of a far-away past. Then father and son were able to compare notes, and to realise that in those old days they had been, not father and son, but father and daughter, and that the relative positions had been reversed.
Then Alcyone said to his father:

"You are an advanced priest of the Temple, and I am only a beginner; how could I remember all this before you did ? "
Mercury replied: " It is just because your body is younger than mine that it is easier for you to remember; I have changed sex too, and so have an entirely different outlook on life, while you have not. Besides, this University was your life-work, and so it was impressed more strongly upon your mind than upon mine".

They talked over all that old time together, and marvelled greatly as they recalled incident after incident of the earlier life, and went from building to building, noting the changes. Most of all, perhaps, they were interested in the library, where they found some of the very books in which they used to read - some even that they had copied with their own hands.

Among other recollections, the language of that country came back to them, but of course as it used to be spoken fifteen hundred years before, so that to those who heard them it sounded archaic and almost unintelligible; indeed, the professor of ancient languages was the only man with whom they could converse quite freely. The University staff were greatly interested in this wondrous phenomenon, and they had an amusing argument with a professor of history, who insisted that their memory of various events most be wrong because it did not agree with his books.

Alcyone found with great glee a statue of himself in that earlier incarnation, and after much persuasion he induced the authorities to inscribe on its pedestal his present name, and a record of the fact that he was a reincarnation of the founder, and the date on which he had visited the University. From this it will be seen that after a searching enquiry the claims of our two travellers were admitted, and this unusual occurrence aroused a vast amount of interest, and was noised abroad in many neighbouring countries, for the University was widely known and had a great reputation.

After their work in connection with the library was completed, they started on their homeward voyage. The ruler of the country sent for them, and desired to persuade them to stay in his realm, but Mercury respectfully declined the invitation, alleging as excuse that he had undertaken in Poseidonis duties belonging to his present incarnation, and that he must return to fulfill them.

Their voyage home was accomplished without serious mishap, though a heavy storm carried them far out of their course and gave them some new experiences. The vessel this time called in passing at the great City of the Golden Gates, and Alcyone was much impressed with its architectural splendour, though Mercury felt its moral atmosphere to be foul and degraded. Of course they took this opportunity to pay a visit to Mars, who received them with great kindness, and kept them with him for two months.

By force of example and by stern repression of evil tendencies, Mars had kept his court at least outwardly decent; but he was well aware that the Toltec civilisation was even then decadent, and that a strong party among his subjects scarcely veiled their impatience of the restrictions which he imposed upon them. He felt that the outlook for the Empire was a gloomy one, and congratulated his descendants that their lot was cast in a part of the continent in which, though the inhabitants were often materialistic and avaricious, they were at least much freer from the darker magic and from what was called 'refined' forms of sensuality. Even Alcyone, young though he was, felt that there was something wrong with the place, despite its magnificence, and was glad when the time came for them to pursue their journey.

Mars was deeply interested in the account of the remarkable recovery of memory on the part of both father and son at the North African University. He had no recollections of that nature himself, but said that in dreams he frequently found himself leading vast hosts through stupendous mountain ranges, and that he had speculated as to whether those might not be memories of actual achievements in some previous birth. As Alcyone sat and listened to all this, it seemed to him that he too could see those towering peaks and those slow-moving multitudes, with his great-grandfather riding at their head, and his vision added many details which Mars would certainly have recognised if Alcyone had not been far too shy to venture upon describing them in the presence of the Emperor. He *did* describe them afterwards to his father, but, as we know, Mercury had not been in the emigration to which they referred, and so they awakened no memory for him.

When at last they reached their native city, the aged Surya welcomed Alcyone warmly, and rejoiced to hear of his visions of the past.
The report of these, which had preceded him, caused him to be regarded in the Temple as the most promising of its neophytes, and it was universally felt that he had a great future before him. One person at least reckoned upon that, and determined if possible to share it, and that was Phocea, the girl who had so nearly drawn him away from entering the Temple several years before. She had tried to attract him then; she tried with maturer arts to attract him now.

But by this time he was trebly armed against her wiles, for immediately on returning from his voyage he had met his cousin Sirius, and at once felt so strong an attraction for her that he determined offhand to marry her at the earliest possible moment.

She thoroughly reciprocated his feelings, and was just as eager for instant marriage as he was, but the parents on both sides did not quite understand such a violent case of 'love at first sight', and insisted kindly but firmly on a delay of at least a year. The young people unwillingly consented to this, because they could not help it, but this intervening period was one of severe trial to both of them, and this became so evident to the discerning eyes of Brihat that she contrived to get it shortened by almost half, to the great relief of the lovers.

Surya himself performed the marriage ceremony, though it was but rarely that he took any personal part in the services, usually giving only his benediction to vast crowds from a lofty opening in the facade of the Temple, much as the Pope used to do at Rome. This marriage was indeed his last appearance at any public function, and only a few months later Alcyone and his wife were summoned to his bedside to receive his farewell message. He said to Alcyone:

"Now I stand on the threshold of another world, and mine eyes can pierce the veil which hangs between this and that. I tell you that there lies before you much of tribulation, for all that has been evil in your past must descend upon you now speedily, in order that its effects may be expiated, and you may be free. In your next birth will pay something of your debt by a death of violence and after that you will return amidst surroundings of darkness and evil; yet if, through that, you can see the light and tear away the veil which blinds you, your reward shall be great. You shall follow in my footsteps, and shall fall at the feet of Him whom I also worship. Yes, and she also", (turning to Sirius), "she also shall follow me, and your father shall lead you, for you be all of one great Race - the Race of those who help the world. And now I go down into what men call death; but though I seem to leave you, yet in truth I leave you not, for neither death nor birth can separate the members of that Race - those who take upon them the vow that can never be broken. So take courage to meet the storm, for after the storm the Sun shall shine - the Sun that never sets".

A few days later Surya breathed his last, but Alcyone never forgot him through all his long life, and he often saw him in dreams and received blessing and help from him. So Mercury took charge of the great Temple in his stead, and strove to carry on everything as Surya's wisdom had ordered it, his father Herakles cooperating in every way as the head of the temporal government.

The daughters of Venus had been a closely united family; indeed their feelings were so nearly identical, that Sirius and Mizar were both in love with Alcyone, as well as with one another. When he married the former, the latter, incapable of any feeling of jealousy, loved both husband and wife just as dearly as before, and they so strongly reciprocated the affection that they invited Mizar to live with them.

She joyously accepted, and no one could have been a more loyal and loving coadjutor than she was to Sirius during all the years that followed. A more piteous case was that of Helios, a niece of Osiris, who had been left an orphan at an early age, and consequently adopted by her uncle Venus. She had grown up with the family, and was so much one with it that she followed the example of the two elder girls in falling in love with Alcyone, and was quite heart-broken when he carried them both off, since she could not well offer to join his new household. She did, however, later come on long visits to the family, and in course of time accepted Alcyone's younger brother Achilles, thus remaining in close touch with all those whom she loved so well.

The authorities of the North African University had never forgotten their reincarnated founder, the little boy who had told them so marvellous a story and exhibited such vivid enthusiasm. The tale had caught the popular imagination and been repeated in every home in the land, and when, some twelve years after his visit, the headship of the University fell vacant with no obvious successor, and somebody set on foot the idea that the post should be offered to the original founder, there was a tremendous outburst of enthusiasm over the whole country, and the ruler in consequence sent so pressing an invitation and made so generous an offer that Alcyone felt it would be churlish to refuse. Though he had now a wife and three children he consented to expatriate himself, and set up a home for them in a foreign land.

He was received in Africa with a perfect ovation; he landed at the capital city, by the special request of the ruler, and after being feted there for some time made a triumphal progress through the country to his ancient home. He was able to arrange to inhabit the very same suite of rooms or halls in which he had lived fourteen hundred years before, and he even had furniture constructed on archaic models, and endeavoured to reproduce as far as he could the exact appearance of the place in that previous life. The recollection of his earnest efforts then was a never-failing wonder and joy to him now, and he had such an opportunity as is given to few to see the permanent results of his own work after many generations. He threw himself into the University work with a vigour and enthusiasm which fourteen hundred years had not diminished, and his wife Sirius and his sister-in-law Mizar (who of course had accompanied them) co-operated with equal zeal.

Infected by his eagerness, both Sirius and Mizar began to remember something of that remote past, but they never attained to anything approaching his perfect familiarity with the older time. Vesta, who at that time was the youngest child seemed as thoroughly at home in it all as his father, but Bella, though he also had been equally intimately associated with it all in that other life, had no memory of it whatever.

Alcyone soon found that to establish a University and arrange it all just as one wished was one thing, but that to administer it when all its customs had the weight of a thousand years of tradition behind them was quite another. Still, he was happy in his work, and he managed everything with such tact that no outcry was made against various reforms which he contrived by degrees to institute. He kept up a constant correspondence with his father Mercury, this being indeed one of the stipulations which the latter had made before giving his consent to his acceptance of the headship of the University. He had also made it a condition that his son should return whenever he had urgent need of him, or whenever he felt his own strength beginning to fail.

Some comparatively uneventful years of hard work followed, his children growing up around him. Though they had married so young, he and his wife were exceedingly happy together, and as closely united as when they were twin brothers in the same country in that other life. While Alcyone was working in Africa, his great-grandfather Mars passed away in the City of the Golden Gates, and his grandfather Herakles was called to assume the imperial purple. Venus then took the place of Herakles as temporal chieftain of the Akkads, since his elder brother Mercury was already in charge of the Temple work.

Herakles found that the position of Emperor was no sinecure, for he did his best to carry on government on his father's lines, though the opposition of the party who demanded greater licence in morals grew ever stronger and more restive. Various conspiracies were unmasked and suppressed, yet new ones were ever coming to light, and it seemed that the hostility between the few who wished to retain the semblance of decent living and the majority who cared little for such things must soon break out into open war. Under these circumstances Herakles found the government of such an Empire a weary and thankless task, and often wished himself back again in the steady-going mercantile oligarchy.

Although the North African University was at that time probably the most famous in the world, the education of the poorer classes in that country was entirely neglected. This matter did not seem to occur at all to the upper classes, but it was brought prominently before Alcyone and Sirius by the fact that an especially faithful servant of theirs, who was really almost a friend, had an exceptionally bright-looking little boy (Boreas) to whom Alcyone's children took a great fancy. It was on enquiring about the education of this boy (in consequence of some remarks made by his own sons) that Alcyone first realised that there was absolutely no provision of this sort for the poorer classes.

He arranged easily enough for the teaching of that particular child by a private tutor, and in due course admitted him as a free pupil of the University; but the incident suggested to him that there might well be many more equally bright children among the poor, for whom no such possibilities presented themselves. He and Sirius discussed the matter for a long time together, and finally worked out a tentative scheme, to the carrying out of which they resolved to devote some of the large income of the University.

It was a sort of combination of a boarding-school and an agricultural community, and its plan was that the University should acquire tracts of land in central positions all over the country, and on these tracts should build and operate free schools. Each tract was to be under the joint management of a school master and a farmer, and the boys were to live at the school and spend half of each day in learning and the other half in cultivating the land. The University was to support these colonies for the first year, after which it was expected that the sale of the surplus produce would be sufficient to maintain them. The feeding and clothing of the boys themselves was to be a first charge upon the school funds in either case. Girls were to be admitted to the extent to which suitable work could be found for them. If after a school-colony had worked successfully for some years it was found that it had a sufficient surplus, it was to be allowed the honour of founding branches or offshoots, but all to be under the direct control of the University.

Boys who showed exceptional talent were to have facilities for entering other and higher schools, somewhat on the plan of the modern scholarship system, and if they could work their way up to the level required for the University itself, certain allowances were to be made to them and remunerative work of some sort was to be found for them when they had passed through its curriculum.

This scheme was first submitted to the ruler of the country who was graciously pleased to approve of it and to recommend his subjects to take advantage of it. Then Alcyone set vigorously to work, bought land in various places and got other tracts given to him and began to have schools built much on the general plan of the University itself - that is, not one large building, but a number of isolated rooms in a garden. The poor were at first a little shy in taking advantage of the establishments, mainly because the boys who went there were unable to earn any money for their parents; but soon the vast benefits of the scheme began to be generally understood, and they were all filled to overflowing.

Alcyone's plan for their management, was an economical one and as he was able to provide them with the right kind of seeds and cuttings from the vast estates of the University, they rapidly became financially independent, and a brisk competition arose among them for the honour of founding branches. Alcyone coupled with it his old idea of physical training, about which he was just as enthusiastic now as in the previous birth, so that the boys whom he turned out were not only far better educated but far healthier than the rest. To conclude this part of the subject here, Alcyone stayed altogether twentyseven years in North Africa and, before he left, a network of his schools had spread over the whole island, and the ruler had issued a decree making attendance at them compulsory upon all boys under a certain age until they had reached a certain level, with, however, discretion to local officials to make exceptions where they saw good cause to do so.

The plan on the whole worked exceedingly well, but it had one unexpected result. The care bestowed upon physical training and the direct affiliation with the University gave the pupils of these schools for the poor a considerable advantage over the sons of richer parents who attended private schools. A few merchants consequently began to send their sons to the school-colonies, and presently several of them joined together, bought some land, erected a school of the Alcyone type exclusively for children of their own class, and then offered it to the University. Alcyone accepted it and it proved a success, and soon there were many others like it. The natural result was that one after another of the old private schools closed for lack of pupils, and in a few years the whole education of the country was entirely under the management of the University, and Alcyone was practically Minister of Public Education.

All this kept him very busy and in such congenial toil the years slipped rapidly by. He and Sirius had agreed that their children should not be allowed to forget their native country and so they had sent each of them back once or twice on visits to their grandfather Mercury. During these visits the three boys had found themselves wives to accompany them back to the country of their adoption. Selene, a younger brother of Alcyone, had married Uranus, but died young, leaving a son Leo and a daughter Mira.

On his visit to Poseidonis, Vesta fell in love with and married Mira and when Selene's death occurred her brother Leo decided to return to Africa with his sister and brother-in-law. Alcyone at once found work for him in connection with the University, and he soon fell in love with and married Alcyone's eldest daughter Vega. Not long afterwards he met with a sad accident, being thrown from his horse, and receiving injuries which proved fatal; so Vega with her baby son Vajra returned once more to her father's house. After some years she married Pindar, a kind and capable man, and to them was born a daughter, Cygnus, who became a charming little girl, and was always a prime favourite with her grandfather Alcyone. They had also a son, Iris.

Alcyone worked on steadily for a number of years, and might have spent the whole of his life in guiding the University to which he was so closely linked, but that his father Mercury and his mother Brihat, finding themselves growing old and less active than of yore, wrote begging him to return and solace their last days with his presence. He felt it his duty to obey this call, though it was a great struggle for him to leave his African work. He discussed the matter with his wife, and she also agreed with him that they aught to sacrifice their own wishes, however strong they were, to the desire of the parents whom they so revered. So Alcyone made a journey to the capital and had an audience with the ruler, in which be told him exactly the facts of the case, and what he felt he aught to do.

At first the ruler flatly refused to give him permission to abandon the University; but after a night's sleep he sent for him again, and announced that if his son Bella (whom the ruler had seen and liked) would act as deputy manager of the University, Alcyone might go and obey his father's call. But he insisted that Alcyone should still remain the nominal Head of the University, and that all important questions connected with it should be submitted for his decision. Alcyone thankfully accepted his arrangement, subject of course to its endorsement by Bella; of which however he had little doubt. On his return home he summoned his sons to a family council, and told them the ruler's decision.

Bella was a business-like and capable man, and his wife Ulysses had also considerable administrative ability, so it seemed that the interests of the University would be safe in their hands; furthermore Vesta, who was psychic and impressionable, seemed in many ways better fitted for succession to the priestly office in Poseidonis than was his eldest son. After the first surprise of the proposal was over,

Fig. 4.8. <u>Neptune</u> = later St Paul, and later the Master Hilarion.

Drawn by Remote Viewing, from His appearance as it was in 1930 [4].

Note: The differences in appearance of ordinary people is said to be mainly due to their degree of advancement and karmic effects. But for the perfected Men shown in Figures 4.2 to 4.8 and in Fig. 4.10, these factors have dropped away and the Genesis (1:26) verse: "God said, Let us make man in our image, after our Likeness", applies without any constraints.

They all agreed that it was under the circumstances the best that could be done, and Bella in his turn journeyed to the capital to place his formal acceptance of the office in the hands of the ruler, and to receive from him a solemn charge with regard to the conduct of the University. On his return Alcyone set sail for Poseidonis, in the year 16,823 BC, taking with him Mizar, Vesta and Neptune.

On the voyage a great blow fell upon him in the death of his dearly loved wife Sirius by an accident. She was enceinte ((pregnant)) at the time, and in very bad weather she was thrown off a couch and fatally injured. Her husband was overpowered by grief, and declared that he could not live without her, and should not know in the least what to do. But she tried to cheer him, and begged him to grant her one last request. Of course he promised to do so, and she asked him to marry her sister Mizar at once, so that the home might go on just as before, and she might feel satisfied that everything was being made comfortable for him. She said that if she knew that this would be done she could die in peace, and she would also keep near them if it was permitted, and would even try to speak to them.

Alcyone and Mizar finally yielded to her request, and promised to marry as soon as they reached home; and when this was settled Sirius peacefully passed away, telling them with her last words not to grieve for her. She was buried at sea, and, true to his promise, Alcyone married Mizar as soon as possible after they reached Poseidonis.

Mercury, who mourned much over the death of Sirius, performed the ceremony for them, and they all felt the presence of the dead wife while the service was in progress. Indeed Brihat declared that she saw her standing smilingly beside them, and joining in some of the recitations. Brihat had had a dream or vision of the death of Sirius at the time when it occurred, and neither she nor Mercury was unprepared to hear the news of it on the arrival of the travellers. Mizar proved a true helpmate for Alcyone; she knew his ways so thoroughly that everything went on just as though Sirius had been still on the physical plane. She was also thoroughly in sympathy with all his interests and knew the whole of the University business, so that though he never forgot Sirius he soon settled down into the new condition of affairs, and his life ran smoothly along its grooves.

His old pleasure in the priestly work was soon revived, and he found that the manifold interests of the Temple left him little time for sorrowing over his loss. As soon as he was a little used to the management of affairs Mercury withdrew entirely into the background and lived the life of a recluse, coming forth only rarely and on special occasions.

Alcyone retained under these different conditions his strong interest in educational matters, and made an attempt to introduce into his native land a system similar to that which had been so successful in Africa. He founded a University on the lines of the old one, and opened a couple of his farm-colony schools for the poor. Both attempts may be said to have succeeded, but they were never taken up in the oligarchy with quite the same enthusiasm as in North Africa. Still, he worked hard at the arrangements, and his system slowly spread, and he was thanked by the council for introducing it; but as years passed on he was obliged more and more to delegate to others the business connected with it, for his priestly work became more and more engrossing.

He kept constantly in touch by correspondence with Bella and the University work in Africa, and frequent and earnest invitations reached him asking him to pay another visit to the scene of his earlier labours. He always promised that he would do this some time or other, but for years no opportunity presented itself. He was training his son Vesta to succeed him in the Temple work, but Vesta, though eager, zealous and psychic, was still somewhat too impulsive, and did not always distinguish impulses from intuitions, and so was sometimes hurried into unwise actions. His cousin and brother-in-law Auriga proved of the greatest assistance to him, and took up the educational work so enthusiastically that Alcyone soon turned over that department entirely to him. Auriga was a person of hard-headed common sense, and a good organiser, so under his management the schools soon began to flourish exceedingly.

Venus, the father of Auriga, had long before been called to the City of the Golden Gates to succeed Herakles, and he in his turn had summoned his eldest son Crux to support him in his old age, and to learn the way in which so cumbrous an Empire was managed, in preparation for the time when he himself should be called upon to hold the reins of power. In 16,811 BC Venus passed away and Crux came to the throne, and very shortly after that Mercury and Brihat died within few months of one another. Though this was not unexpected at so great an age, it came as a shock to Alcyone, all the more so as he had been overworking himself for a long time and was therefore not at his strongest. He felt the need of rest and change, and with considerable difficulty he was persuaded to pay the long promised visit to North Africa, the hope being that the sea voyage and the absence of responsibility might set him up again in health.

This anticipation was to a great extent fulfilled, for his passage was a pleasant one, and he received a most enthusiastic welcome at the University and was delighted to find that Bella had been managing everything with praiseworthy firmness and tact, so that both the University itself and the schools were in a most satisfactory state of efficiency. He declined to interfere in any way or to take any share in the management though he was of course feted everywhere, and expected to appear as a figure-head and make speeches on numerous occasions. He spent twelve months in Africa, and even then returned only because of an urgent request from Vesta. When he reached his native land he was already sixty-seven years old, and he yearned much for a life of meditation and repose, so he encouraged Vesta to continue as far as possible the work to which he had grown accustomed during his father's absence, and he himself remained rather in the background, coming forth only on great festivals or when special advice was needed.

He was regarded by all the people as a great saint and a person of marvellous wisdom, and those who could obtain his advice in their difficulties thought themselves highly favoured. On several occasions he mesmerically cured people suffering from various diseases, though he refused to make a regular practice of this, saying that he could help only those cases which he was specially inspired to help.

So he lived on for seventeen years, passing the evening of his life peacefully and contentedly, hale and vigorous, and keeping all his faculties to the last. Mizar remained inseparable from him (she had of course accompanied him to Africa) and their devotion to one another was touching. When Mizar died in the year 16,793 BC he seemed scarcely to mourn her, saying that it was not worthwhile to sorrow over so short a separation, as he knew he should follow her almost immediately.

His prediction was justified, for he passed quietly away the following year, leaving behind him a great reputation on two continents. Two exactly similar statues of him were made, and were set up in the central halls of his two Universities - in that in Africa beside that other statue of his earlier personality on the pedestal of which in his boyhood he had had his-present name engraved. The same sculptor produced the two statues and each University presented one to the other with a suitable inscription. The story of the founder who had so strangely returned and recognised his work was repeated in Africa for centuries, though later, when the statues had disappeared, it became confused, and ran that he was a great magician who had preserved the same body for fourteen hundred years, and so had revisited the scene of his former labours.

((End of verbatim text {p. 373-387 in [3]}.))

-=-=-=-=-=-=-=-=-=-=-=-=-=-=-

The 28th life of Alcyone was in Manoa (on the shore of the lake containing the island of Sacred City of Shamballa) in 15,995 BC. The Manu asked Surya to ask his sons Mars and Mercury to lead a vast host of emigrants to India, the largest ever, with Corona as one of the section leaders. Corona's path was through Kashmir to Punjab and to Bengal, while Mars went through Tibet to Nepal and thence to Bengal. A third group led by Vulcan went via Tibet to Bhutan and thence to Bengal. Al least 10 major figures who are now Masters were with this important emigration, along with their disciples, who followed them down the ages.

Below is a verbatim text {p. 393-405 in [3]}:

A ceremony in the Temple of the Sacred City on White Island was held, in the great Hall of Audience, with its massive chair carved from solid rock and covered with golden moldings, were gathered the most august of Figures. In the centre, in front of the chair but at the foot of its seven steps, towered the mighty form of Vaivasvata, the Manu, the typical man of the Fifth Root-Race. Clustering hair of dark brown shot with gold fell upon His shoulders, and the massive beard of like hue rolled, thick-curling, over His breast; eagle-eyed, with brows slightly arched and shadowing the eyes into darkness, save when the lids, normally somewhat drooped, were lifted suddenly and the eyes flashed out dazzlingly, compelling all who looked on Him to veil their gaze; the nose high and arched, the lips curved and set firmly. A King of men, truly; one whose word meant Law, whose lifted hand impelled or restrained at will.

Beside Him, on His right, stood the Mahaguru, His priestly Brother, the Head of the Religion of the community. Stately and mighty also was He, but while the Manu breathed resistless Will and every gesture spoke of Rule, this Blessed One breathed Love most compassionate, and a Wisdom as pure and deep as the Manu's Will was mighty. His hair dusky as ebony, His eyes of darkest violet, almost black, His mouth tender, easily curving into a gracious smile. Seeking His name we find many in the people's minds - as though reverence and love sought varied expression; often Pita, Deospita, Vyas, Sarvajnarshi, Sugata Ravidas, Ushadas, Mahamuni, Jnanaraj -such are some of the names by which the people love Him.

On the left side of the Manu stands Surya, with radiant hair and shining eyes -- eyes that dwell with deep affection on His noble sons, the chief figures in the crowd facing the altar, which stands between the Heads of the community and Their people.

They are clad with great magnificence; a long cloak of cloth of gold with heavy jewelled clasps falls around each, its folds sweeping the ground and lying in tumbled golden waves around Their feet; the Mahaguru and Surya have, beneath this, long white robes of finest material; the Manu wears a doublet-like garment of rich crimson, reaching below His knees, the legs and feet bare. They are waiting, expectant for the overshadowing presence of the Mighty Lords of the Flame, who are to appear to bless the departing hosts.

The Leaders of the army stand close to the ancient Altar, whereon each has placed his favourite weapon, mace, or axe, or sword, facing their Chiefs. Mars is in the centre, with his wife Brihat on his left, with Mercury upon his right. Mercury's wife, Saturn, is beside him, by her, again, stands Vulcan. On Brihat's left is Corona, who had once in a previous birth ruled as an Emperor of the City of the Golden Gates in far Atlantis. A noble quartet of warriors they looked, with their stately wives, full worthy of them.

A group of children sat beyond the Altar, a little awed by the great Figures on whom their eyes were fixed; they were the children of Mars and Mercury; Jupiter, a noble boy of ten years of age, the eldest Son of Mars, with his sisters Osiris, Uranus and Ulysses, his brother Siwa, a chubby boy of two, and in the arms of Osiris, the eight years old maiden, a baby boy, Viraj, who gazed with solemn eyes upon the Three. Mercury's eldest Son was Selene, a thoughtful-looking boy, about the age of Jupiter, his arm thrown round the youngest girl, Mizar, a restless babe scarce twelve months old; his brothers Leo and Vajra sat, with arms round each other's shoulders; the sisters Herakles and Alcyone sat nestled together, little maids of five and three, for Alcyone had been born in 15,995 BC, and Herakles was two years her elder and a little inclined to be protective of her junior.

There is a great hush, for a single note rings through the great Hall, clear as a silver clarion, and a brilliant Light blazes out above the rock-hewn chair. The assembly bows down, for in the chair is seated a marvellous Figure, dazzling, an embodied Power, and behind Him are three Others, only less great than He. They are the four Kumaras of Indian Scripture, the Lords of the Flame.

"Go forth, my children, and do my work; my strength is upon you. Having wrought, return".

The accents fall upon the hushed stillness; a hand is raised in blessing, and when the heads bent low in reverence are raised the chair is empty and the Light is gone.

Surya stepped out and blessed his sons, who bent the knee before him, and then, stooping, raised the little Alcyone, his favourite granddaughter, and drew close to him the sturdier form of Herakles:

"My little ones", he said, and his tender face grew gently solemn, "on a far rough way you go. Mothers of brave men you shall be, and fair women also shall call each of you 'mother'. Your race shall dwell long in the land and thither also you shall come again many times, to learn and teach. But this is the first of the lives of expiation, that old karma may be outworn, old wrongs made right. Death shall come to both of you together, in strange and violent way. In that hour, call on me and I will come to you, and the Light you have just now seen shall shine in the darkness then". Little Alcyone hid her face in his neck and laughed softly; she did not understand, but she loved her grandfather; and Herakles looked up boldly, unwitting the gravity of the prophecy: "I shall call loud, so that you will hear", and Jupiter, who always called Herakles his little wife, said proudly: "I will take care of you".

Long and arduous was the journey, and many years had passed ere the three commanders met again. Corona found his way south fairly easily, as the road through Kashmir was known, and the people of the settled portions were not unfriendly. But on reaching the Punjab he fell out with the inhabitants almost from the beginning, and presently he had to fight his way through a hostile country. He besieged the great Toltec city, now under Aryan rule, where Mars had been betrayed some fifteen hundred years before, and at length reduced it by starvation, and made its ruler swear fealty to himself; he next subdued Ravipur - near the site of the modern Delhi - and established there one of his own officers as a tributary King; he pressed southwards, ever fighting and reducing his enemies to submission, till he had carved himself out an empire, with half a hundred tributary chiefs. Forty years had rolled away ere he reached Bengal, an aged warrior of over seventy years of age, to find Mars settled in Central Bengal, having founded and established his kingdom.

Vulcan had found his way through Tibet and Bhutan a good sixteen years earlier, had joined his forces with those of Mars, and in 15,953 BC had invaded Assam, and had there established himself in fairly peaceful possession by the time Corona arrived, in 15,952 BC. Much, however, had happened ere that, and our hero, or rather heroine is with Mars, and to her fortunes we must turn.

The route of Mars on leaving Central Asia, took him in four years across the Great Range into Tibet, and he remained there for a full year, to rest the feebler of his army caravan, ere they began the toilsome road across the mountains to Nepal. During this time Castor was born, and much time was given daily to training the boys of the party in athletics of every sort.

Jupiter was the leader in all manly exercises, and among the boys whom he formed into a troop, which he trained in scouting and mimic warfare, we note his cousins Leo, Vajra and Selene, Vajra making up for his juvenility by his reckless daring and extreme activity - and their friends Albireo and Arcor. Alcyone, a girl between seven and eight, was a somewhat dreamy maiden, quiet and thoughtful, more apt to sit at home than to roam abroad. She would sing softly to herself the chants to the Angels of her people, and lose herself in visions as she sang.

At the end of the fifth year since leaving Manoa, the army started again on its way, and climbed slowly over the mountains which lay between Tibet and Nepal. It tried to follow the course of a mountain torrent, pouring eastwards and southwards, but was constantly forced to turn aside when the river plunged into impassable gorges and foamed through ravines where the cliffs almost closed above it. There were many skirmishes with hill-tribes, but no serious fighting until two years later they approached Nepal, where Mars found himself obliged to divide his army, leaving half under Mercury to guard the huge entrenched camp, and going out himself with the remainder of his troops to subdue the country sufficiently to make a safe road for his people. He took with him his eldest son, Jupiter, and his young troop, Mercury specially bidding his son Vajra learn the soldierly duty of obedience. One attempt was made to rush the camp during his absence, but Mercury repelled it without great difficulty and with little loss of life. It is a pretty scene to see Mercury seated with his wife and sister-in-law, with Alcyone nestling on his breast, and a girl-friend Capri, Herakles' special chum, leaning against his knee, as he told them stories of Surya and the Mahaguru, and sometimes, speaking softly and low, of the great Kumaras whom they had seen ere leaving Manoa. Herakles was a more restless child, and her eyes would rove eagerly over the camp - outside while her father was speaking, bringing on herself sometimes a solemn reproof from the more demure Capri. Osiris and Uranus also, with little Viraj, were interested auditors, while Ulysses was apt to sympathise with Herakles' wandering gaze.

Two years passed before the waiting camp again welcomed Mars, and joyous were the greetings which met the returning wanderers. He had secured a passage through Nepal, partly by fighting, partly by diplomacy, and the whole caravan set out, a couple of months later, in early summer. That winter they camped near the borders of Nepal, resuming their journey the following summer, and thus slowly they went forwards, marching during the summer, camping in the winter, and spending four weary years on the way ere they reached India itself.

Meanwhile the sisters had grown into stately and handsome maidens, inheriting something of the beauty and grace of their father and mother. Herakles was now eighteen, and Alcyone sixteen, and Mars sought his favourite niece as wife for his eldest son, while the sweet ways and gentle eyes of Alcyone had won the heart of Albireo, Jupiter's brother in arms. Demure Capri had become the ideal of Arcor, whose own somewhat stormy nature found rest and refreshment in her gentle household ways, and the three pairs were married ere the army left its winter camp in 15,979 BC.

Mars led his great host peacefully through the extreme north of Bengal that summer, and camped along a huge river when marching time was over. Here he determined to wait the arrival of Vulcan and Corona, in order that their united forces might take possession of the land, and that he might there build up his kingdom. Another two years, however, elapsed before the approach of Vulcan was reported to him. Nothing whatever was to be heard of Corona, and after waiting for a third year, Mars, Mercury and Vulcan decided to press on without him. They left the women and children in an entrenched camp in northern Bengal (15,975 BC) while they marched southwards, taking with them Jupiter, Albireo, Selene and Leo, through a fertile but only thinly settled land, and at intervals the army stopped and threw up strong embankments, protected by deep trenches which seem to have become easily filled up with water, the water being thus drained away from a considerable surrounding area, which was readily cultivable, and afforded splendid grazing grounds for cattle. Mars detached at each of these settlements a considerable body of troops, leaving them orders to make broad and firm roads between the camps; after five years of this marching and building, he placed Vulcan in authority over the whole of the conquered land, directing him to return to the northern camp taking with him all those who wished to settle down there with their wives and children, as well as a large force, sufficient to guard the great numbers that were to settle in the various camps established in Bengal. He himself determined to continue his march southwards and arranged to return to the place where they parted after another five years.

Vulcan accordingly started, visiting all the settlements on his way north; he found them prosperous and busy, the scattered inhabitants of the country having entered into friendly relations with them, often taking service as cowherds, labourers and so on. He pressed on northwards till he reached the original camp (15,967 BC) and was joyfully welcomed by its inhabitants. He found a few newcomers there; before they had parted Herakles had given birth to a son, Bee, and a daughter, Canopus; Alcyone to two sons, Neptune and Psyche, while Capri had borne Arcor a daughter, a pretty little girl, Pindar, and a son, Altair.

To these had been added Aletheia, son of Herakles, Rigel, daughter of Alcyone, and Adrona, son of Arcor. The three older children, Bee, Neptune and Pindar were of an age - eleven years old, having been born in the winter of 15,978 BC - and were as inseparable as their mothers, while the remaining trio, Canopus, Psyche and Altair were equally fond of each other. Each little maiden had her two knights, Pindar being everywhere accompanied by Bee and Neptune, Canopus by Psyche and Altair. A happy childhood was theirs, playing on foot and on ponyback, rough unkempt ponies, and gathering at eventide with their mothers, to tell of the day's delight, and to listen to stories of the land the mothers had left in childhood, above all to the story of the great Temple from the lips of Alcyone, and the august Figures their childish eyes had seen. Aletheia, Rigel and Adrona were but seven years of age, pretty healthy children, much petted by the uncles of the two first-named, Vajra and Castor, the younger sons of Mercury.

Vulcan gathered together all the families whose heads or elder members had followed Mars, and took them southwards, leaving each group with their long separated men relatives in the settlement where these were dwelling. Joyous were the meetings, saddened here and there by gaps in the family circles, when death had swept away by disease or violence those who were not to meet again their loved ones upon earth.

Meanwhile Mars had gone southwards, and soon found himself engaged in a long series of skirmishes and battles, for the country he invaded was thickly populated with people of Atlantean blood, and as he approached the sea-board these became more warlike, and offered more resistance to his advance. At last he had to fight a serious pitched battle, to which the King of the Orissa country had summoned all his hosts; his priests, followers of the Atlantean dark magic, had incited the troops to fury by fiery harangues, and had rendered them, as they believed, invincible by human sacrifices offered to their gloomy elemental deities in the huge temple near the sea which was the most sacred centre of their worship, a temple of unknown antiquity and cyclopean architecture of the Lemurian type, standing in what is now the town of Puri. In the dim recesses of that temple, on the night before the battle joined, the priests had gathered in unholy conclave, and with ghastly rites and furious invocations summoned their dark deities to give battle to the radiant angels of the Aryan invaders.

At daybreak the decisive battle began, and for five days it, raged; Mars and Mercury led their hosts with dauntless valour, well seconded by their sons and their faithful friends, among whom Arcor was conspicuous for his reckless courage. Great was the slaughter, but, as the fifth day darkened into evening, the hosts of Orissa were in headlong flight and the victorious Aryans chased them southwards, and encamped for the night in the camp that their enemies had left. Mars appeared to have carried a charmed life, but all the other leaders were wounded more or less, and very weary were the hosts that slept.

Rising ere daybreak, as was their wont, strange and new was the sight, before the eyes of those who, all unknowing had camped near the sea-shore. Never had they seen before the broad expanse of ocean, and loud cries of wonder and of awe burst from these children of the desert and the mountain as the huge plain of heaving waters burst upon their gaze in the dim twilight ere the dawn, and the waves rippled to their feet, making them start back in fear. Their leaders came out at the shouts of the soldiers, wondering if the enemy had returned in force. Transfixed they also stood, and, as they gazed, the eastern sky began to redden towards the dawn; they watched, breathless, and suddenly the crimson globe of the sun flung itself upwards from the waters, as though it leaped from the bosom of the deep, and Mars and Mercury threw themselves upon their faces and the red rays blazed across the ocean, and the cry: "Samudra ! Samudra ! " rang from a hundred throats. The Sun had been Pushan, the Nourisher, Pantha, the Path, as he guided them over the deserts ; now he was born of the sea, in the magical wonder of the dawning.

The neck of the resistance was broken, and Mars established the centre of his kingdom to the north of Orissa, in Central Bengal, leaving Jupiter, his eldest son, in charge of Orissa, with Albireo, Leo and Arcor as his lieutenants. He departed to keep his tryst with Vulcan, promising that Mercury should return and bring with him the families of all left to settle in that part of his realm. Immediately after this Vulcan parted from Mars and invaded and conquered Assam, setting up there his kingdom with little difficulty.

In due course Mercury returned, bringing with him his noble wife, Saturn, and his sons Vajra and Castor, and his three daughters, Herakles, Alcyone and Mizar. He brought with him also Uranus, to be the bride of Leo, and Aurora to wed Selene. Arcor joyfully welcomed his fondly loved Capri and his sons Altair and Adrona.

And now came many years of hard work, the building up of a kingdom, interspersed with occasional wars of defence - wars of aggression were forbidden by the Ruler, Mars; skirmishes with predatory bands, endeavours to conciliate the former owners of the country, and efforts to put down human sacrifices. Once during these years, Mars paid a visit to his children, bringing with him his sons Siwa and Viraj, and his daughter Ulysses. Osiris had married and could not leave her home. On this occasion Vajra and Ulysses were wed, and after much discussion, the parents decided to leave these two as rulers of Orissa, and to return themselves to the northern capital, taking with them Jupiter and his family; for Mars was very old, and wished to install his eldest son upon the throne and retire from the world with Mercury and their wives. This was done, and Vajra and Ulysses were left in charge.

For a time all went apparently well, but a storm was gathering below the surface. Vajra did not show the skill in conciliation characteristic of Jupiter, and his measures, aimed to bring about good results, were sometimes harsh. In 15,937 BC a great religious festival of the old religion was to be held, and Vajra had, the year before, forbidden its celebration, knowing the danger of such a concourse, excited by sacrifices and incantations. Herakles had come to spend some months with Alcyone, for the twain were not happy when apart, and she - having become learned in the deeper knowledge of the

Atlantean White Magic and having wedded it to the worship of the bright Gods worshipped in her ancient home, began teaching this mingled philosophy and religion to the younger men and women of her brother's kingdom, and she included in her classes some of the younger priests of the dark Atlantean faith. This was to strike a deadly blow at the still powerful priesthood, and ere long the muttering of hatred grew deep and angry.

As the months passed, the growlings grew louder, and a conspiracy was formed to attack the house of Albireo, where Herakles and Alcyone were living, while he was away on a projected journey with Vajra to a distant part of the country. The priesthood resolved that the forbidden celebration should take place, and with victims nobler than the common herd; and they diligently circulated rumours that a rising was to take place in the district whither Vajra and Albireo were going. The result of this skilfully planned deception was that Vajra took with him the main part of his army, leaving a comparatively small force under Arcor to preserve order and defend his household.

It was 15,937 BC and the high day, or rather night, of the forbidden festival was near. The early morning dawned clear and cool, but scattered groups might be seen slowly converging to a centre, and that centre the house of Albireo. The groups coalesced into a crowd; the crowd grew in number and denseness. Presently a deep clanging note clashed into the quiet: it was the note of the great bell of the temple, unheard for long, the bell that no longer might be sounded.

The roar of the crowd answered the brazen voice of the bell, and in a moment a riot had broken out. The house of Albireo was broken into, the guards slaughtered, and in front of the crowd, as it surged inwards, towered the tall gaunt form of the Atlantean High Priest, Scorpio, on whose head a price had long been set, and who had lain hidden in the underground vaults of the temple, known to none but the initiated priesthood.

"Ya-uli ! Ya-uli", shrieked the mob, half deeming him risen from the dead, and frenzied by religious excitement. A slow stern smile curved his iron lips as he heard his name re-echo, and turning, he waved back the yelling mob, and they stopped, silent.

"Wait, children of the Lords of the Dark Face; your day has come. I go to bring forth the accursed, the women of the barbarians of the North, who have crushed your worship and closed the temples of your Gods. Aiyo! Aiyo! the Lords have arisen; they cry for blood, and blood shall they have. Slay ! slay all but the two women who are theirs. They are mine, as the Priest of the Gods who drink human blood and devour human flesh. Tonight shall their thirst be slaked and their hunger appeased. Aiyo ! Aiyo! I have said!"

Into the house he stalked, grim as death and stern as an incarnate Hate. At the first alarm Arcor had sounded his conch to summon his men, and, as they flung themselves into the passages and held the stairways, a fierce but hopeless combat had ended in their extermination. Arcor himself had rushed to the private entrance into the ladies' apartments, had struck down the priests who led the crowd -- Ya-uli cautiously withdrawing till the way was clear - and had battled desperately, though alone, to bar the road. He fell, pierced by a score of wounds, and the Chief Priest stepped over his body to his prey.

Alcyone and Herakles were at their morning worship when the crash of breaking doors told them of danger, and as they rose, two tall and stately women - Herakles, now at the age of sixty, crowned with silver hair, and Alcyone with dark tresses, silver-streaked, falling below her waist - the door of their worship-chamber burst open, and the tall Priest stood on the threshold. The two women faced him, a proud interrogation as to such intrusion spoken by the uplift of the noble heads, the gaze of the steady eyes.

"Come, ye accursed ! the day of your oppression is over; the night of your doom is near. Come, for the Dark Lords call. I am their messenger of vengeance.
Herakles threw her arm round her sister's slighter form: "Priest you threaten those who know not fear. Begone ! Invite not death".

A harsh laugh grated on the air:
"Death, woman ! I give it, I do not accept it. Come forth: you are mine".

He made a gesture to some priests behind him; they came in and seized the women by the arms, drawing out cords to bind.
"Bind not!", said Herakles, "We shall not flee. Come, dearest, come. Our father's daughters know how to die". Alcyone glanced up at her sister, an angelic smile upon her face: "I am ready, sister beloved". And they moved slowly forward, surrounded by the priests, through the passages strewn with the bodies of the dead. Unblenching they went through the seething crowd, which yelled at them, shook clenched fists as they passed, and would have torn them in pieces had it not been for the priests they feared. Slowly they went onwards through the city to the place where yawned widely the mighty open gates of the temple, with long aisles of dark pillars glooming away into darkness. White-robed, fair-skinned, the two sisters looked like angels of light amid the tossing crowd of dark faces and dark bare arms flung high in air. At the gate the priests turned and Ya-uli spoke:
"Tonight, four hours after sunset, the gates will be opened; let all the children of the Lords of the Dark Face come to their festival." The gates clanged together, and Herakles and Alcyone were past all earthly help.

At first, no harm was wrought on them; they were offered rich food and wine, but would not eat. Only fruit would they take, and a drink of milk. Then commenced a long persuasive talk; Ya-uli strove to win their promise to take part in the worship of the Dark Gods that night, pledging himself that they should return home in safety if they would thus purchase life with dishonour. In his false heart he meant to slay them after they had worshipped, but he longed to proclaim them renegades to their faith and so win credit for his own.

Uselessly he strove against their steady will, and in wrath at last he bade the priests take them to the gloomy centre of the temple, and leave them there awhile.

A dread and awful place it was in which they were left. Dim shapes, some red, some black, some sickly grey, were half visible through the gloom. Low moans, as of something in pain, came, dully muffled, to their ears.

"Herakles", whispered Alcyone, "are these things alive or dead? they make me shudder". "Darling, I know not, but living or dead, they cannot hurt the soul". They whispered to each other in the gloomy cavern, spoke of home, of husbands, of children, and then of the days of happy childhood, and the glorious vision of the past.

"I think the time has come", said Alcyone, "and we shall see our grandfather again".
"And the Light!" breathed Herakles.

It was ten o'clock and a dense crowd filled the huge dark building, silent, expectant, awe-struck. At a sign the two women were seized, and lifted upon a high altar, in view of all, and a lurid light, blood-red, shone out, none could say whence, and threw the awful figures around into grim semblance of life. There was a sound of rending cloth, and the robes of the two women were torn from them, and the fair white bodies shone out nude and shrinking. A low cry of horror burst from them, and then Herakles threw up her proud head and flung her arms around her sister, striving to shield her from the gaze of the rough crowd: "You shame your mothers, men, in shaming us", she cried, and then stood silent.

"Look at them", called the Priest, "before the Dark Lord feasts upon them. When next ye see them, he shall have had his fill". And then the light faded, and the crowd filed out, to wait for the rites that none save priests might see and live.

How tell the horrors that ensued: flames rose from surrounding altars, and shrieking captives were led in, and the fire fed with fat skinned from their living bodies till the flames roared high; then their blood was set flowing and caught in iron vessels, and set to boil in huge iron pots, and poured upon the images set in the circle round; foul creatures of the slime, huge spiders, monstrous scorpions, fed on the remnants of the mutilated bodies; and presently one after another of the images woke into awful life, began to stir, to slip downwards from their pedestals, obscene shapes of unimaginable horror, and crawled and writhed towards the central altar where Alcyone and Herakles still stood, clasped in each other's arms.

"Fly! Fly!" yelled the priests, "the Dark Lord is coming, and his hosts are here !" and they tumbled over each other in a mad rush to escape from the terror they had invoked.

Out of the darkness loomed a gigantic face - a face of power majestic, of pain and wrath too deep for words: of intolerable weariness and despair.
A mighty hand was waved, just visible by its own dull glow, as of hot iron half-quenched and the fearful figures rolled up around the altar and reared up red gaping mouths and hairy tearing claws. Then rang out the voice of Herakles, loud and clear:
"Suryadeva, Suryadeva, Mahapita, come! Oh, come!"

And there, in the midst of all the horrors, there shone out the Light on which the children's eyes had rested, and beneath it the radiant form of the Surya they knew, with tender eyes and outstretched arms; and with a sob of joy Alcyone sprang forward, and her body dropped lifeless on the altar. And all the horrid shapes shrivelled into nothingness, and lay about like the cast-off skins of snakes, and the pillars broke, and the cavern walls fell in, and the bodies of the sisters had for tomb the mighty temple of the Lord of the Dark Face.

And that night in Puri, there was fear and trembling; for earthquakes rent the ground, and a huge tidal wave came rushing from the sea. But they who cowered in terror, and they who, remembering the two sisters, wept for their awful fate, they knew nothing of the outstretched arms that had carried them home, cradled on the Bosom that is to become the Refuge of the world; they knew nothing of the Light that had turned into heaven the darkness of that hell.

((End of verbatim text {p. 393-405 in [3]}.))

Fig. 4.9. Herakles & Alcyone in 15,937 BC [3].

-=-=-=-=-=-=-=-=-=-=-=-=-=-=-

The 29th life of Alcyone was in 15,402 BC in Rahana in the Oudh district of India.   In the 9th life of Alcyone, Surya had prophesied the tragic death which closed the 10th life and also foretold later trials and difficulties.   When an individual is approaching the entrance to the Path, there are likely to be some lives with considerable suffering and unpleasant conditions. There are two reasons:

(i) whatever evil karma remains to the individual must be cleared out of the way as speedily as possible, so that it is no impediment when the time arrives for the final effort, and,

(ii) any undesirable qualities in the individual must be quickly conquered, so that the necessary qualifications may be acquired, and the way may be clear.

(Author's (MGH's) note: there is said to be something called "the burning up of karma" when an individual has progressed beyond the point at which he/she would learn anything from suffering from past bad karmas which had not yet been discharged. These may be allowed to be suspended permanently in such cases. But the author (MGH) can recall no source for this information and it may not be correct. The previous (28th) life above, where Herakles and Alcyone were murdered, suggests that it is not correct.)

In the lives already described, Alcyone has had the privelege of frequent and close association with men and women who have since become Masters of Wisdom, and everything has been done to strengthen his/her character by example and precept. In the next 29th life, Alcyone is thrown, from birth, into evil surroundings and the help of the presence of the Great Ones is withdrawn. The purpose was to work off some bad karma and in so doing to give Alcyone an opportunity to show if he/she has sufficient strength and insight to break through an evil tradition, even though it has behind it all the weight of (evil) religious and parental authority, of immemorial custom and of personal passion.

Below is a verbatim text {p. 414-423 in [3]}:

Alcyone, then, was born this time in a female body in the year 15,402 BC, in Rahana in the Oudh district of India. Her father, Cetusa, was the priest of a religion about which there seemed to be much mystery. Although he himself was unquestionably of Aryan descent, the religion was certainly aboriginal for it was at the same time too elaborate and too barbarous for the joyous-hearted Aryans. It may well have been the seed from which Kali worship has since arisen, for it consisted mainly of gloomy rites to a bloodthirsty female deity. There was good deal of reckless gaiety about the outer side of this faith, but through it all there always rang a sombre note of gloom and fear.

Many secret services were held to which only the initiated were admitted; and at these the most horrible rites of the darker magic were freely practised. Many parts of some of these services were held in a language incomprehensible to the people, but at the same time some of the recitations were at least partially Sanskrit.

Alcyone's father was a fit priest for such a faith, a stern, reserved and gloomy man, but nevertheless a person of very great influence. He was supposed to have won many powers by sacrifices and austerities, and was further credited with readiness to use them for evil in a great many ways. Her mother, Cancer, was not unkind, but was always in a condition of anxiety and terror, which speedily communicated itself to the child. The latter lived a rather frightened and neglected life; she was not actually badly treated, and as she was not admitted to the inner services, she saw nothing definite of the more unnecessary horrors of her religion, but the gloom and the fear of the inner circles reacted upon her and made her childhood miserable with vague terrors.

She grew up without much education, and there was no event of special importance in her young life, until at the age of about sixteen she met Pollux, a bright handsome careless young fellow, whose appearance at once attracted her. The attraction seems to have been mutual so they fell in love in the ordinary way. Alcyone was too terrified to find it possible to propound the idea of love in the dark, uncertain atmosphere of the family life, so these young people met frequently in secret, and in course of time became too intimate. After a while Alcyone pressed her lover to make some arrangements as but when urged he declared this was an impossibility, as not only did he belong to quite a different religion, but there was also a hereditary feud between his family and that of Alcyone.

It took a long time to convince Alcyone that her lover was really heartless and did not intend to make any move in the matter; but when at last she realised the truth, she turned from him with disgust and told all to her mother, announcing her condition, and vowing to devote her life to being revenged upon the man who had brought her into it. Her mother was much shocked and upset, but when she learnt who the lover was she said at once that he came of a bad stock, and that his father before him had ruined a younger sister of hers in a similar manner. This story made Alcyone only the more fiercely indignant and, as has been said, she resolved to dedicate her whole life to a full and carefully-planned revenge. Her mother then unfolded to her the secret that revenge could be had through the secret rites of their religion, and she consequently become eager to be initiated into it.

The whole story had to be told to her father, who also was furiously angry, for by the customs of the time the birth of an illegitimate child doomed her to the life of a widow. He blamed her bitterly, but yet commended and encouraged her desire for revenge. He permitted her to learn the secret of the faith, by which she was deeply impressed, but also greatly terrified, for she had to pledge herself to a nightmare of horrors which she would have been glad to be able to forget. In order to cloak as far as possible the results of the undue intimacy, the father insisted upon her immediate marriage to a devil priest, Scorpio, a man much older than herself and of most undesirable type, one who was a medium for the most horrible influences.

Of course she shrank with loathing from all this, but yet accepted it as a necessary part of the revenge to which she had resolved to devote her life. The whole affair had become distorted by her long brooding over it, and her state of mind was such that she was open to a steady pressure from evil astral influences, a condition of practical obsession, which was considered a mark of great advancement in this peculiarly abominable religion. After extracting from her bloodcurdling oaths of secrecy, her mother unfolded to her a particularly ghoulish scheme of vengeance which she said had never been known to fail. Among other repulsive details it involved the crime of murdering her own child, and offering it to the deity invoked. In her rage against Pollux she agreed to this, because it would be his child; but when it was born her maternal instinct triumphed, and she refused to fulfil the agreement or to consummate the sacrifice.

Many of the ceremonies had already been commenced, for it was of the essence of the horrible pact that before the birth of the child she should already have dedicated both herself and it utterly to the service of this loathsome goddess. The culmination was to be the slaughter of the child upon the altar of the deity with certain tremendous invocations, in response to which the image was supposed to descend from its pedestal and to embrace the suppliant. In this embrace the goddess was to pass from the image into the body of the worshipper, who then, as the vehicle of the deity, was herself to devour the sacrifice. In the strength of that ghastly meal the obsessing entity was supposed to give to the body much the same powers which mediaeval superstition attributed to the Hand of Glory. At the approach of the avenger all doors flew open, and all living creatures became incapable of resisting his will, so that he could wreak his vengeance unopposed, and even unrecognised, for the goddess threw over him a mantle of invisibility.

Driven by mad rage and by the almost irresistible force of environment, Alcyone had begun the earlier stages of this appalling piece of witchcraft. But when the child was actually born she experienced a revulsion of feeling, and declined to continue the dedicatory ceremonies. Her father was exceedingly angry, and ridiculed her as weak and unworthy of the assistance and favour of the goddess. He even claimed that the child already belonged not to its mother but to the deity to whom it had been dedicated, and demanded that it should be delivered to him on her behalf. Alcyone firmly refused this, braving even the anger of her gloomy and terrible father. He insisted indignantly for a time, and then suddenly yielded with a sneer, saying that the goddess would obtain her rights in another way.

Soon afterwards the baby fell ill, and in spite of all that the mother could do its mysterious malady grew rapidly worse. She presently fell ill herself with watching and grieving over it, and when she recovered she was told that early in her illness the baby had died, and its body had been burnt in the usual way. But she always had certain lurking suspicions, and ever after this a dawning of hatred mingled with her fear of her father. The truth (which, however, she never actually knew, whatever she may have suspected) was that her father, really believing in his fanaticism that the child belonged to the goddess, and that her anger would descend upon him if he allowed her to be robbed of it, had contrived to administer repeated doses of slow poison, first to the child and then to the mother, and as soon as the latter was unconscious he had taken the child and sacrificed it himself to his bloodthirsty deity.

Human sacrifice formed a regular part of the secret rites of this horrible faith, and yet in the midst of all these abominations there were certain gleams of some original better influences – certain suggestions which may have been the reflection of a condition in which the faith was not so utterly degraded. The very phrase which was solemnly pronounced by the priest at the culminating point of a human sacrifice seemed to have in it some faint reflection of a better time, for the earlier part of it at least had a tone which reminds one of the Upanishads. It ran something like this:

"From the earth is the breath and the blood, but whence is the soul? Who is he who holds the unborn in his hand? The watchers of old are dead, and now we watch in turn. By the blood which we offer, hear us and save! The breath and the blood we give thee. Save thou the soul and give it to us in exchange".

These last words seem to point to the idea that the soul, or perhaps more exactly the astral body, of the sacrificed was to be given into their power to become one of their horrible band of obsessing entities, to be at once an instrument and yet in some strange way one of the objects of their degraded worship. As has been said, most of their incantations were entirely incomprehensible, and bore a considerable resemblance to those employed in Voodoo or Obeah ceremonies by Africans. Others, however, contained distinct Sanskrit words, usually buried in the midst of a series of uncouth exclamations delivered with a furious energy which certainly made them terribly powerful for evil.

One of their characteristics was the use of certain cacophonous combinations of consonants into which all the vowels were inserted in turn. The syllable "hrim" was used in this way, as also the interjection "kshrang". In the midst of these uncouth outbursts of spite, occurred what appears to be an evil wish in unmistakable Sanskrit: "Yushmabhih mohanam bhavatu", and the whole utterance concluded with some peculiarly explosive curses which it seems impossible to express in any ordinary system of letters.

Poor Alcyone led an exceedingly miserable life amidst all this chaos of obscene horrors. Her husband was an evil and crafty man, who preyed upon the credulity of the people, and was often in a condition of complete intoxication from the use of hemp and some form of opium. Soon Alcyone came bitterly to regret the fit of mad revenge which had led her into all this network of evil, but she was too firmly entangled in it to be able to make her escape, and indeed there still were times when the obsession dominated her and she felt that revenge would be right and sweet. Presently her father died, and the family fell back into a position of less influence.

This unnatural parent, however, was more terrible dead than alive, for he concentrated all his energies in the lowest part of the astral plane, and exercised a peculiarly malignant obsession over his daughter. She knew the influence well, and earnestly desired to resist it, but could find no method of doing so, though her suffering under it was indescribable and her whole soul was filled with uttermost loathing. Her mother and all the other female members of her family were under the same malign influence to a greater or less degree, but to them the whole thing was a matter of course, and they even supposed themselves to be specially favoured and to become in some way holy, when they were seized upon even for the most dreadful purposes.

Along with all this psychical influence there was a perfect labyrinth of the most complicated and ingenious plotting on the physical plane. Years were spent in the elaboration of a nefarious scheme to get Pollux into the power of the family, and at last the plan matured itself and he and his child Tiphys were in their hands - for he had married in the meantime and had with him a bright little boy.

Alcyone's mother and other female relatives were filled with fiendish exultation, and joined in a strange kind of orgy of hatred, the father impressing himself upon them all more strongly than ever. Alcyone felt the tremendous power of this combination, and was often carried away by it and unable to resist its action, although even then she was all the time in a condition of bitter protest and resentment. Pollux was to be poisoned in a peculiarly horrible way, and it was to Alcyone that the task was entrusted of the actual administration of the draught, under the guise of the most friendly hospitality. The man himself was bloated and broken down by years of debauchery and dissipation, and Alcyone felt nothing but repulsion for him; and, as at this critical moment the obsession by the father was almost perfect, there is little doubt that the crime would have been committed, but for a most fortunate shock which she received at the last moment. Just as she was handing the cup to her victim, she met the wide gaze of the child. His eyes were exactly those of his father, her joyous young lover of so many years ago, who had been the one bright spot in her dreary early life. In a flash those eyes brought back the past, and with it a realisation of what she was about to do now under the awful compelling power of this ghastly religion of hate. The instantaneous revulsion of feeling was complete; she dashed the cup to the ground and rushed from the house - from the house and from the city, dressed just as she happened to be at the moment, so overpowered by the horror of the thing that she never even paused for a thought as to what lay before her, or what would come of it, resolved only to have done for ever at any cost with all that evil life.

The violence of her feelings broke through the black pall of evil influences which had so long dominated her, and for the time she was entirely freed from the maleficent control of her father. She rushed out into the country, careless whither she went so long as she escaped for ever from that awful life. Unaccustomed to exercise and to the free air of heaven, she was soon sinking from fatigue, but still she pressed on, upheld somehow by a kind of frenzy of determination. She had of course no money, and only indoor clothing, but she thought nothing of these things until night began to fall. Then for the first time she looked about her and became conscious of her surroundings. She was already many miles from home, out in the open country, and, becoming conscious at last of severe fatigue and hunger, she turned her steps towards a country house of some size which she saw at a little distance.

She knew little what to say or do, but fortunately Achilles, the mistress of this house, was a kind motherly woman, who was touched by the exhausted condition of the wanderer and received her with open arms, and postponed her questions until she had eaten and rested. Then, little by little, the whole story came out, and many were the exclamations of wonder and pity on the part of the good old dame, as the horrors of the dark demon-worship were gradually revealed. The old lady made light of the fact that in leaving home Alcyone had lost her position in life and all her worldly possessions, telling her that all that mattered nothing now that she bad escaped from the other horrors, and that she just now devote herself to changing radically and entirely her whole attitude of and forget all about the past as though it bad been a mere hideous dream. She said wisely that life began afresh for her from that hour - indeed that she had not really lived until now, and she promised to do all in her power to help her and make the new life easy for her.

Alcyone feared that her husband, the devil-priest, might be incited to assert some kind of legal claim over her, for she knew that the worshippers of the dark cult would be fiercely angry that one of their initiates should escape from the fold. But the old woman, who was a brave and capable person, declared that she did not know exactly how the law might stand but that, law or no law, she was at any rate quite certain as to one fact - that she did not intend to give Alcyone up to her husband or any of her relations; and she felt quite confident that if the case were carried before the King of the country and all the nefarious proceedings of the dark demon-worship exposed, the authorities would be certain to take her side and decline to deliver her again into the slavery from which she had escaped.

Alcyone was most thankful to this kind protectress and in her condition of utter exhaustion of body and mind was glad to adopt the suggestion that at least they might leave all further discussion till the morrow, and to sink to rest in the comfortable quarters provided for her. The shock to her had been severe, and it would have been only natural if some serious illness had supervened; and indeed it seems as though that would have been the case but for a wonderful vision which came to her during the night. A man of commanding appearance and wonderful gentleness of mien (Mercury) appeared to her and spoke words of comfort and encouragement, telling her that the awful life which she had lived so far had two aspects of which she had been entirely unconscious. First, its terrible sufferings had paid off outstanding ((karmic)) debts from long-past lives and so had made the way clear for future advancement; and secondly. the whole life had been in the nature of a test, to see whether at its present stage her will was strong enough to break through an exceedingly powerful surrounding of evil. ((important points))

He congratulated her upon her success and determination in breaking away, and prophesied for her a future of rapid progress and usefulness. He said that the way was long before her, but drew for her also by his words a beautiful picture of two paths of progress, the slow and easy road that winds round and round the mountain, and the shorter but steeper and more rugged path that lies before those who, for love of God and man, are willing to devote themselves to the welfare of their brothers. She had, he said, the opportunity to take the latter line in the future if she chose, and if she took that path, though the work would be arduous, the reward would be glorious beyond all comprehension. This vision produced a profound impression upon her, and she never afterwards forgot the words or the face of the instructor, nor did she ever entirely lose the glow of enthusiasm with which she felt herself eagerly accepting the second of these alternatives which he placed before her.

Next morning she related her vision to her kind hostess, who was deeply impressed by it, and said that it quite confirmed the impressions which she herself had received. It had its effect even upon the physical plane, for it was largely owing to it, that Alcyone was better than might have been expected. Her dead father troubled her greatly by constant and determined attempts to reassert his old dominion over her.

She, however, called up all the latent reserves of her will and set them definitely against this influence, rejecting it with all the vigour which she possessed, without the slightest hesitation or compromise, with the strong resolution that she might die in resisting the obsession, but at least she would never again submit, to it. This struggle continued at frequent, intervals for many months, but whenever it came she always kept before her the face of the venerable messenger of her vision, and fortified herself by remembering his words.

All this time she stayed with her kind hostess, who would not, hear of her going anywhere else, or of her making any effort to support herself in any way. Apart from this constant astral pressure she had no trouble, for no attempt, to reclaim her was made on the physical plane on behalf of her husband. Indeed, it seems that the family somehow acquired the idea that she was dead, some rumours reaching them of the discovery of the body of a woman vaguely answering to her description. Her hostess always declared that the Gods had guided her footsteps to her, and that she accepted her as a charge from them. Alcyone was most grateful for all this kindness, and tried in every possible way to make herself of some use to her benefactress in return for it. She now began to learn something of the ordinary Aryan religion, which proved attractive to her after all the horrors of her early training. She devoted much time to its study, and soon knew much more about it than her hostess.

Little seems to have been at this time committed to writing, but she obtained much assistance and instruction from Vega, a Brahman, who made her acquaintance on the occasion of a visit which he paid to her hostess. He was much interested in her and profoundly touched by the story of her previous sufferings. He taught her a number of hymns, some of them of great beauty, and all of high moral tone and of beneficent intent. His advice was on the whole good and sensible, though in certain directions he was somewhat narrow and fettered. His wife Auriga was also of great help to Alcyone, for she was deeply interested in religious matters.

At the end of about a year the dead father ceased to make any effort to assert his influence, and Alcyone felt at last that all connection with the old evil life had been entirely severed. It seemed to her like looking back upon some past incarnation, when she tried for a moment to see anything of that earlier time, and soon she was able to cut herself off from it so far that some at least of its details began to fade from her memory.

After the influence of the father had entirely departed, she had the unspeakable pleasure and encouragement of seeing once more in dream the Hierophant who had shown himself to her on the first night of her escape. On this occasion he congratulated her upon her newly won freedom and gave her a promise of help and protection. She endeared herself much, not only to her hostess, hut also to other members of the family and to friends. She became practically a daughter of the house, or rather filled the place of one who had married and left the homestead. It seemed in fact, as though the family had forgotten that she was not one of themselves, for when the old benefactress died an equal share of what was left was offered to her as a matter of course, and when she protested against this it was pressed upon her with the utmost sincerity. She agreed at last to accept a certain small share, and continued for some years longer to live with this same family.

There came a time when the second generation was growing up and more room seemed desirable, so she transferred herself to a smaller house on the estate, to live there with one of the younger couples, Cygnus and Iris, to whom she acted as a kind of mother and adviser. Her interest in the religion never waned, and presently she had learned all that her Brahman friend was able to teach her, and was passionately desirous of still further information upon many points. The Brahman found himself unable to supply all this, but he told her of a holy man who, if he still lived, would be able to answer all her questions. He spoke of this man with the greatest reverence, saying that from him he had learnt all that he knew, and that he had always felt sadly conscious that he might have learnt much more if only he had had the power to grasp fully the words of wisdom which fell from this teacher's lips.

He spoke so earnestly and enthusiastically of this teacher, that after much consultation Alcyone resolved to make a journey in search of this man - a considerable undertaking for one who was now becoming an old woman. The distance was great, and as the Brahman had not heard of his teacher for a number of years, there was a good deal of uncertainty as to whether he would still be found in the same place, but there seems to have been no readily available method of making enquiries. However, Alcyone set off on this rather curious pilgrimage, and at the last moment the Brahman Vega resolved to throw up his position and his work and accompany her, and thus they journeyed together, taking with them only a couple of servants as attendants, one of whom was our old acquaintance Boreas.

After various adventures and more than a month's travelling, they reached the temple over which Vega's teacher presided, and heard to their great joy that he was still living. They asked for an audience, and Vega was overjoyed to fall once more at the feet of his ancient instructor. He then turned to introduce Alcyone, but saw with amazement that she was regarding the teacher with unspeakable wonder and reverence, and yet with an obvious recognition, while he in turn smiled upon her as upon some one with whom he was already familiar. A few words of incoherent explanation soon showed that this teacher was Mercury, the person who had twice appeared to her in vision, and of course this discovery put an entirely new complexion upon the affair, and linked them all together as already old friends.

Now began a happy time for Alcyone, for all her questions were answered and her most earnest desires fulfilled, and the teacher spoke often to her of a far distant future in which she should learn far more than she could at present know, and should hand on the knowledge to others for the helping of the world. But he told her that for this many qualities were needed which she did not yet possess, that there was much karma even yet to be worked out; that to this end she must be willing to forget self and to sacrifice herself utterly for the welfare of mankind, but that at the end of this effort would come triumph and peace at the last. Vega made up his mind to send for his wife and family and to stay for the rest of his life with this teacher, and Alcyone would gladly have done the same for a strong affection sprang up between them; but the teacher told her that this was not her destiny, and indeed that he himself would be but a little while longer upon the physical plane, while her duty lay with the family who had helped and rescued her.

So at the end of about a year she took leave of him with many regrets and travelled slowly back again to her old friends, who were heartily glad to welcome her. The rest of her life was spent quietly but happily in ministering to and helping the children and grand-children of those who had been so kind to her.

Alcyone acquired a wide reputation because of her remarkable knowledge on all religious points, and she became an authority to be consulted even by the priests and the Brahmans of the neighbourhood. So the life which had begun amidst such horrors of storm and strife, ended with the calm of a peaceful sunset, and she passed away deeply regretted by all those who knew and loved her so well.

((End of verbatim text {p. 413-423 in [3]}.))

-=-=-=-=-=-=-=-=-=-=-=-=-=-=-

In 13,524 BC Mars was born to Viraj (Ruler of the South Indian Empire) and Brihat was his queen. The Manu appeared astrally (i.e. in 4-dimensional space, the region just above what the catholics call "Purgatory") to Viraj and asked him to send his son (Mars) by sea to Ceylon (Sri Lanka) and thence to Egypt, with a chosen group of young men and women, to Aryanise Egypt, which at that time was an Atlantean colony (Toltec race). On arrival in Egypt, they were met by Jupiter, the then Pharaoh, who had one child only, his daughter, Saturn. His High Priest, Surya, had been directed in a vision, by Mahaguru, to ask Jupiter to give his daughter in marriage to Mars. Soon after this, all of the newcomers were also married to the indigenous Toltec Egyptians. The Pharaoh gave his approval to these marriages with the Aryans. This mixing of the two races produced a new and distinctive type, having the high Aryan features and the reddish Toltec colouring, seen in Egyptian statues and in a small percentage of the present day Egyptian population, the Copts, who are the descendants of the Ancient Egyptians. After the final sinking of Atlantis (Poseidonis), a vast tsunami flooded Egypt and presumably all archaeological evidence was swept away by it and anything left was deeply buried in sand and silt. Later the Manu himself was born there to unite the whole of Egypt.

Below are more details taken verbatim from {p. 341 in [2]}:

From the South Indian Aryan Kingdom an important mission to Egypt went out about 13,500 BC; the order came from the Head of the Hierarchy through the Manu, and the expedition travelled via Ceylon, by water up the Red Sea, then hardly more than an inlet.

It was not intended to colonise, since Egypt was already a mighty Empire, but rather to settle there under the Egyptian Government, a great and beneficent, as well as highly civilised, power.

Mars was at the head of the expedition, and Surya was a High Priest in Egypt, as he had been, in southern India nearly three thousand years before; as then, he smoothed the way for the coming Aryans, and he told the Pharaoh of their approach, and advised him to welcome them. His advice was taken, and a little later he counseled the Pharaoh to marry his daughter to Mars, and to name the latter his successor. This was duly done, and thus peaceably but effectively was an Aryan dynasty established in Egypt at the death of the ruling

CHAPTER 4   World History from 70,000 BC: 221
Buddha & Christ in their previous lives

Pharaoh. It reigned gloriously for many thousand years, until the sinking of Poseidonis, when it, with the Egyptian people, was driven to the hills by the flooding of Egypt. The flood, however, retreated comparatively soon, and the country recovered ere long. Manetho's history apparently deals with this Aryan dynasty; he makes Unas - whose date is given as 3,900 BC, while we make it 4,030 BC - the last King of the Fifth dynasty.

((End of verbatim text {p. 341 in [2]}.))

-=-=-=-=-=-=-=-=-=-=-=-=-=-=-=-

In his 34th life, in India, in 10,749 BC, Alcyone, in a trance state, apported (materialised) a ring talisman which he had previously owned in Peru where he had lived 800 years earlier, but most details of this fascinating life are omitted here as Surya was not involved. At this time (10,750 BC), Mars was the ruler of China and he married the daughter of Herakles, the high Priest. Surya was born as the grandson of Mars and great-grandson of the High Priest Herakles. Surya had 4 brothers, Viraj, Yajna, Naga and Sirius, who devoted their lives to Surya's service, for religious reform. See reference [14].

-=-=-=-=-=-=-=-=-=-=-=-=-=-=-=-

Alcyone's 47th life, in 630 BC, in India, was contemporaneous with the Mahaguru becoming the Buddha. The Buddha was born in 623 BC and was 35 years old shortly after obtaining "enlightenment" (Jnanodaya). Alcyone went to hear Him give many of His sermons. Alcyone describes one sermon to which he took Mizar, his cousin. The scene was of about 2000 people amongst trees, most sitting on the ground, some leaning against tree trunks, men and women together, and little children sitting with them or running about between the outlying groups of people. Buddha sat on a slightly raised platform – a grassy bank in the middle of the garden, surrounded by a group of His monks in their yellow robes, and with His musical glorious voice He made all that crowd hear without effort, and held them entranced day after day as they came to listen to Him.

Of Him it was indeed emphatically true, as was later said of another prophet, that, "never man spake like this man" (St John's Gospel).

Below is the verbatim text {p. 675-680 in [3]}:

The influence of His magnetism upon the people was incalculable. His aura filled the whole garden, so that all the vast crowd was directly under its influence - actually within Him, so to speak. The splendour of the aura attracted vast hosts of the higher devas of all kinds, and they also helped to influence the audience, so that we cannot wonder when we read in the sacred books that often at the close of a single sermon hundreds or even thousands attained the Arhat level ((pupil of a Master, see colour Fig. 4.12 near p. 63)). Many of the people then born in that part of India were those who had followed Him in previous incarnations in far-away lands, and were especially born in India in order that they might have this inestimable advantage of direct contact with Him after His Enlightenment had been gained.

Those whose vision was confined to the physical plane saw only a gracious Prince of commanding appearance and of winning manner, who spoke to them with a clearness a directness to which they were not at all accustomed from their Brahman teachers. The latter had for many years taught little but the necessity of frequent offerings to Brahmans and of constant sacrifice to the Gods, which of course always involved heavy fees to their priests. But now came this far mightier Teacher, who told them in the simplest and most direct language that the only sacrifice pleasing to the Gods was that of a pure and gentle life - that not animals but vices were to be destroyed and cast out, and that the great necessity was not gold for the temples, but purity and kindliness of life among the devotees.

On this occasion, when the two cousins went to hear Him, He took for His text the subject of fire. He pointed to a fire which was burning near, and told them it was no inapt symbol of delusion, in that the flame looked like what it was not; it seemed solid while it was not so, and it burned the man who touched it. Then He explained how all passion and all desire were like the burning flame - how with them, as with it, no half-measures were useful, since the fire was never safe until it was utterly stamped out – never certain not to reappear and cause devastation until there was no single spark of it left. So, He said, must anger, passion, desire, delusion, be stamped out of the human heart. Only then could peace be attained, only then could man enter upon the Path. ((important points))

The impression produced upon both the cousins was indescribable. At once, Alcyone announced his intention of giving up everything in the world, and devoting himself entirely to following the Lord. His wife Irene immediately agreed with him, and he proposed to turn over to Mizar his share in the temple, the headship of the family, and all his worldly wealth. Mizar, however, refused to receive this, and declared that if Alcyone devoted himself to religious life, he would do so too and even Thetis approved of this, though she said that she could not dare to offer herself for it, after all that had happened. Alcyone thought that the family should be perpetuated and the office of manager of the temple should be carried on, because of their promise to the father Jagannadha; and finally they went together to the Lord Buddha, told Him all that had happened from the beginning to the end, and put themselves unreservedly in His hands. The Blessed One heard their story and to Alcyone He said:

"Are you quite sure that there remains now no taint of hatred in your heart - that you forgive to the uttermost, even the death of your son, and that for all created beings you can feel nothing but love evermore, even for those who have injured you?"

And Alcyone replied: "Lord, this indeed is so; if my cousin's wife has injured me, I have forgotten it. I give him freely all my wealth, for l need it no longer. I have now in life only one desire, and though it take me a thousand lives, I vow here at Thy feet that I will never cease the effort until I shall have accomplished it. I vow to follow Thee, to give myself as Thou hast done to help the suffering world.
Thou hast freed me from my sorrow, and brought me to eternal peace. To that peace also will I bring the world, and to this l consecrate my future even until I shall be as Thou art, the Saviour of the World".

And the Lord Buddha bowed His head and answered:
"As thou sayest, so shall it be. I, the Buddha, accept that vow which can never be broken, and in the far distant ages it shall be fulfilled".

And so He stretched out His hand and blessed him, and Alcyone fell prostrate at His feet.

Then, turning to Mizar, He said: "You also shall follow me, but not yet. There is still much for you to do. Take up this charge which my new pupil has laid upon you. Take this which he gives you, for he needs it no longer, for the riches of the good Law excel all other wealth. Do justice and be merciful, and forget not that your time also shall soon come".

So He dismissed him with a blessing, but Alcyone remained with Him, and followed Him thereafter in all His wanderings up and down that fair northland of India.

Mizar after this returned home to fulfil his duties, as the Lord Buddha told him to do; but because the Buddha's teaching of mercy to all he steadfastly refused ever again to kill any animals for sacrifice, or to adopt any of the mean tricks by which Jagannadha had amassed so much wealth. Thus he lost much money, and made himself very unpopular with the other temple Brahmans, especially as he several times publicly announced his adhesion to the Buddha's saying that a Brahman, who does not live as a Brahman should, is not in reality a Brahman at all, no matter how high his birth may be, whereas even a Shudra who lives the life of a true Brahman is worthy of the respect accorded to a Brahman. The other Brahmans plotted against him, and reduced his revenues still further. Nevertheless, the King, being pronouncedly Buddhist, they could not procure his deposition, though they often lodged complaints against him.

He had a good reputation among the people for humanity and kindliness, in spite of all the stories which the Brahmans were constantly circulating against him, so as the years rolled on he grew richer in popularity, though poorer in pocket. It was a great triumph for him when King Bimbisara, moved by an eloquent sermon from the Lord Buddha, decreed that there should be no more slaughter for the sacrifice. The other Brahmans, though greatly incensed by this order, dared not disobey it, and because of the determined propagation of these ideas in earlier days Mizar stood well in the King's favour. Still, there were many who distrusted him because the hostile Brahmans had somehow come to hear a distorted version of the story of the poisoning of Alcyone's son, and of course they made the most of it.

Mizar still used some part of Jagannadha's organization to bring large bodies of pilgrims into his period of management of the temple, but not now in order to make money out of them, but in order to save them from the rapacity of his compeers – which naturally increased the hatred of the latter for him. His position was always a precarious one, for though he had the favour of the king and the gratitude of many people, he had to face ceaseless intrigues and scarcely-veiled malevolence in all sorts of small every-day matters. Still, for more than 20 years he contrived to carry on the work, and in that time introduced many useful reforms into the administration of the temple, in the teeth of much opposition. He was all the while quite openly and professedly a follower of the Buddha, and was living according to His teaching, though still remaining an orthodox Brahman; and in this he was by no means singular, for the Buddha did not take any people away from the older religion, and no-one except those who actually assumed the yellow robe attached themselves *exclusively* to Him.

The end of Mizar's life was from a worldly point of view unfortunate. In 566 BC Bimbisara was murdered by his unnatural son Ajatashatru, who thus seized upon the throne. His plot had been carried out by the aid of the Brahmans, and he therefore favoured them and their religion, and was openly opposed to Buddhism. So, when the Brahmans of the temple preferred a complaint against Mizar, the new King readily gave ear to them and deposed him, and confiscated most of his property. He still had a little land, and he retired and lived upon this in comparative poverty and obscurity until his death in 562 BC, at the age of sixty-six.

Meanwhile Alcyone had attached himself to the Lord Buddha, and never again left Him until death, but travelled with Him up and down the Ganges valley for many years, drinking ever more and more deeply at the fount of His wisdom, and participating in the private teaching which He gave only to His monks. He formed a close but reverential friendship for an older monk named Dharmajyoti, who was very kind to him, and helped him much along the road to perfect peace. This monk Dharmajyoti is known to us as Uranus; he was later Aryasanga, and is now the Master Djwal-kul. The name selected by Alcyone upon assuming the yellow robe was Maitribaldasa, which means "the servant of the power of kindness"; and the Lord said to him: "You have chosen well; that name is prophetic".

For Maitreya is the name of the Bodhisattva who succeeded the Lord Buddha in his office – the Christ who is to come; so the name may also be rendered "the servant of the power of Maitreya". Following thus in the train of the Lord Buddha, Alcyone naturally bore part in many interesting and historical scenes; for example, he was present when, in the year 580 BC, Chatta Manavaka (Selene) was called up by the Lord, and taught beautiful verses immortalized for us in the sacred books. Whenever the great Master's travels took Him to Rajagriha, Mizar invariably came to welcome Alcyone, and the affection between the cousins grew ever stronger as the years rolled on. Alcyone died in 559 BC at the age of 71, 16 years before the death of the Lord Buddha in 543 BC. The latter part of his life was passed in unalloyed peace and happiness.

A year after Alcyone's death came the great King Mars to hear the preaching of the Lord ((Buddha)). With him he brought his son Herakles, who listened to the Lord and followed Him thenceforth, and after His death became one of His great missionaries, carrying His Law into Burma and the East. Herakles in his turn had many enthusiastic disciples – his own son the disputatious Capri, and his nephews, the eager, earnest Polaris and Capella, the impulsive and blundering Gemini, and the ever smiling Adrona. The latter was, however, drawn away from him by the arguments of a wonder-working Brahman, Cetus, who had been acting as chief priest at the court of another Raja, Orpheus, whose daughter Herakles had married. An entire breaking up of the religious arrangements of that little state followed, for, after Adrona had pledged himself irrevocably to Cetus, Herakles succeeded in converting King Orpheus and his sons Siwa and Myna. Cetus was very angry about this, and eventually he and Adrona left the country with a small band of followers, and took up their abode in a neighbouring State, which they tried unsuccessfully to stir up into war against Orpheus. The first and closest follower of Herakles was his nephew, Ivy, with whom he had always a peculiarly string sympathy, born of an intimate relationship in the far-off past.

King Orpheus himself would have followed the Lord, but that the latter told him that he had a primary duty towards his kingdom, and that he must hold it on His behalf. The two kings, Mars and Orpheus, had an agreement between them that their children should intermarry, and they carried this out as far as possible, as will be seen from the accompanying chart. ((Geneological charts are given throughout the book [3].)) In this combined royal family it was not only Herakles who was so deeply affected; his brother Rama and his sister Naga were moved as profoundly, and both desired to offer to the Lord as followers not only themselves but their entire families – all their sons and daughters. Rama's wife Diana heartily agreed with him in this, but Naga's husband Myna hung back and was unwilling to make so great a sacrifice. Eventually the burning love of his wife overbore his scruples, and the two families were left entirely free to throw themselves at the feet of the Lord. Mars stipulated only that his grandson Theo should be left to succeed to his throne, and the Lord ordered that this should be so.

The effect of this life upon the characters of Alcyone and Mizar was enormous – as well it might be when they had earned so great a favour as to be born upon Earth at the same time as the Buddha, and to come under His benign influence. Every vestige of anger and revenge was wiped from the heart of Alcyone, and the qualities of compassion, forgiveness and true affection were developed in him to the utmost. How deep and essential in its nature was the result produced by this most fortunate of lives may be seen by the fact that the average interval between his lives has been entirely altered by it. Before this the average was about 700 years, and since then it has been 1200. Mizar too was powerfully affected, for in the beginning he had some scheming and selfishness in his character. Now most of that had disappeared for ever, and much of his earnestness and love had taken its place, while valuable links had been formed, the result of which lies yet in the future. In his case, however, the average interval was not changed and he therefore does not appear in the 48th life of Alcyone.
((End of verbatim text {p. 675-680 in [3]}.))

-=-=-=-=-=-=-=-=-=-=-=-

After the Lord Buddha resigned His physical body, His office as World Teacher passed to his successor, Surya, Lord Maitreya, the Christ. Taking advantage of the tremendous outpouring of power left by Buddha, Surya soon incarnated Himself as Krishna in India (not to be confused with the Krishna thousands of years earlier of the Baghavad Gita [3]). Almost simultaneously He sent Lyra to China as Lao Tsu (who produced the Tao Te Ching), Mercury to teach the Greeks as Pythagoras, and later He sent Pallas to Greece as Plato [3].

-=-=-=-=-=-=-=-=-=-=-=-=-=-

Fig. 4.10.   Brihat = now the Master Jesus.

Jesus was born in Palestine to prepare a body for Christ, who took it over 30 years later for his 3-year ministry as World Teacher.   See in Chapter 5.

Drawn by Remote Viewing, from His appearance as it was in 1930 [4].

Note: The differences in appearance of ordinary people is said to be mainly due to their degree of advancement and karmic effects. But for the perfected Men shown in Figures 4.2 to 4.8 & in Fig. 4.10, these factors have dropped away and the Genesis (1:26) verse: "God said, Let us make man in our image, after our Likeness", applies without any constraints.

Corona was born in 100 BC as Caius Julius Caesar (the well-known Julius Caesar, Roman Emperor). (The figure 100 BC differs from the conventional history figure due to small calendrical corrections.)

-=-=-=-=-=-=-=-=-=-=-=-=-=-=-

Brihat was born in Palestine as Jesus where he prepared and yielded up his body for Surya (Christ). Brihat was later reborn as Apollonius of Tyana who visited Rome during his travels.
Mercury appeared in India in about 180 **AD** as Nagarjuna.

-=-=-=-=-=-=-=-=-=-=-=-=-=-=-

The 48th life of Alcyone, in 624 **AD**, was in India. He fell in love with Ajax but decided to become a Buddhist monk. Ajax said she could never love another and she became a nun. In his previous life in which he had met the Buddha, the simple ceremony of joining the Sangha (being a monk) was that he bowed before the Buddha, answered questions and made some promises; he was then taken aside by Dharmajyoti and changed his ordinary clothes for the yellow monk's robe. Dressed thus, he had then prostrated himself at the feet of the Buddha, who blessed him solemnly as His new pupil, and told him to prove his life worthy of the robes he now wore. That was the Buddha's custom at that time, but about 1000 years later it had become an ornate ritual!

-=-=-=-=-=-=-=-=-=-=-=-=-=-=-

Author's (MGH's) Comments:

(1) Most people today are unaware that an important hidden "Path" exists: Masters take (selected) pupils who have reached an advanced stage of development, to train them to become Masters.

(2) Fig. 4.11 & 4.12 (between p. 63 & 64) show development from ordinary man to Master and beyond (as seen by Remote Viewing).

(3) It is seen from the above texts [2, 3] that individuals are reborn sometimes as men and other times as women.

The reason given is that if an individual is always reborn as (say) male, he gradually loses the valuable female parts of his psyche, and vice-versa, creating an unbalanced individual.

So if an individual is becoming (say) too female, after a series of lives as a woman, the next rebirth is likely to be as a man, but occasionally the consequence may be that he still feels female and is thus attracted by other men, leading to homosexuality.

This is a very simple explanation of homosexuality, and once realised, it is a great relief to the individual, who can then understand why it has occurred. It is a tragedy that the lack of this knowledge and explanation causes the many problems for homosexuals in today's society. These problems would vanish if they (and everyone else) understood this.

Chapter 4 Supplementary Material:
Comments on pre-Ice-Age-end civilizations

The last Ice Age ended 10,000 years ago. Sea levels were about 120 metres lower than they are now, due to frozen water then stored in the thick ice sheets extending far from the poles. Many ancient coastal settlements and cities are thus now lost, well below sea level.

In Alcyone's Life 32 (12,877 BC) near where Saharanpur, Punjab, India now is, the verbatim text from [3] is:

… an unfortunate accident befell Alcyone at this period, and caused him a great deal of suffering -- indeed, he never entirely recovered from it. He was always of an enquiring and experimental turn of mind, and when a rich Atlantean friend, Aletheia, imported one of the strange air-ships* from Atlantis, he willingly accepted an invitation to make a trial trip in it along with its owner. Some error in the management of the power caused one of the directing tubes to catch and become jammed at a critical moment so that the machine fell, and its passengers were thrown out with great violence.
((end of verbatim text from [3]))   ((**vimanas*, see below))

Besant & Leadbeater [2] also give the following verbatim account, describing the civilization of Atlantis:

… large towns, built on the plains, were protected by immense banks of earth, sloping towards the town, and sometimes terraced, while on the outward side, they were faced with thick plates of metal, clamped together … ((this was many millenia before the "conventional" start of the Iron Age, and are an improvement on our mediaeval stone castles which were no match against artillery))

But such a city lay open to assaults from above, and the Atlanteans carried the making of air-ships ... to a high state of excellence; and if such a city were to be attacked, these birds of war were sent to hover over it, and to drop bombs which burst in the air, and discharged a rain of heavy poisonous vapour, destructive of human life. Allusion to these may be found in the conflicts related in the great epics and Puranas of the Hindus. They also had weapons which projected sheaves of fire-tipped arrows, which scattered far and wide as they hurtled through the air like deadly rockets, and many others of similar kinds, all constructed by men well-versed in the higher branches of scientific knowledge. Many of these are described in the very ancient books above referred to, and they are mentioned as being given by some superior being. The knowledge required for their construction was never made common. ...

Forces, the knowledge of which has been lost, were known to the science of the day; one of these was used for the propulsion of both air- and water-ships; another for so changing the relation of heavy bodies to the earth, that the earth repelled instead of attracting them, so making the raising of gigantic stones to a lofty height a matter of the greatest ease. The subtler of these were not applied by machinery, but were controlled by will-power, using the thoroughly understood and well-developed mechanism of the human body, "the vina of a thousand strings". ((see section on levitation in Chapter 5)) ((also discussed further below))

Metals were much used and admirably wrought, gold, silver and aurichalcum (samarium, Sm) being those most employed in decoration and in domestic utensils. They were more often alchemically produced than sought for in the crust of the Earth ... ."
(("alchemically" produced – Star-Trek replication!? – as yet a method unknown to today's science, but it is reminiscent of the Gospels' accounts of the replication of loaves to feed 5000 by Christ – discussed in Chapter 5. Also see p. 323-324 on Sm produced "alchemically" in stars.))

((end of verbatim text from reference [2], published in 1913))

With this introduction, the ancient Indian books mentioned above are now discussed below. These are probably _far_ older than the "conventional 3000 years ago", as they were continually re-copied by scribes. Ancient Indian books, the Mahabharata and Ramayana have many accounts of large flying machines called vimanas. It may be thought that these are legends; but there is a possibility that they were real vehicles which have passed into legend. If they existed before the great divide at the end of the last Ice Age and the sinking of Atlantis, which caused vast Tsunamis destroying most archaeological evidence, they are now lost. The above relatively recent Remote Viewing [2] (done in about 1913) which describes a vimana in about 12,850 BC, indicates that they were factual and not legends.

It is of interest to read some of the accounts from ancient Indian literature, which seem rather too factual to be dismissed as legends. The author gives them without much comment and the reader must decide:

In the Vedic Brahmanas, the Samarangana Sutradhara which has 230 verses on vimanas and which is considered to be "manusa" or strictly factual, says:

"Manufacturing details of the vimanas are withheld for reasons of secrecy, not out of ignorance. Construction details are not mentioned, for ... were they publicly disclosed the machines would be wrongly used."
And:
"The body must be made strong and durable, like a great flying bird, of light material. Inside it one must place the mercury-engine with its iron heating apparatus beneath. By means of the power latent in the mercury which sets the driving whirlwind in motion, a man sitting inside may travel a great distance in the sky in a most marvellous manner." ((this reported use of mercury is discussed later))
And:
"Similarly by using the prescribed processes one can build a vimana as large as the Temple of the God-in-Motion. Four strong mercury containers must be built into the interior structure. When these have been heated by controlled fire from the iron containers, the vimana develops thunder-power through the mercury. And at once it becomes like a pearl in the sky."

The same words, "pearl in the sky" also appear in Tibetan books, the Tantjua and Kantjua, which also describe prehistoric flying machines by this name. The author (MGH) and others with him, have seen (and photographed) circles of "white dots" flying at extreme altitude, which could be described as tiny "pearls in the sky", over the Sierra Nevada mountains in California, which suddenly flew off (scattered) at very high speed (estimated at many thousand m.p.h.).

The mention of mercury is cryptic and there must be much more which is not written (discussed later below).

Samarangana Sutradhara continues:

"If this iron engine with properly welded joints be filled with mercury, and the fire be conducted to the upper part, it develops power with the roar of a lion."   This has been discussed by Leslie [11], who suggests that the "fire" may be electrical rather than thermal. Electric arc welding is mentioned in other ancient texts.   There is no mention of a conventional fuel.

Samarangana Sutradhara continues:

"The subdivisions of the vimana's movements are: slanting, vertical ascent, vertical descent, forwards, backwards, normal ascent, normal descent, progressing over long distances, through proper adjustment of the working parts which gives it perpetual motion." ((see later below))
"The strength and durability of these machines depend on the material used. Following here are some of the aerial car's main qualities: It can be invisible; it can carry passengers; it can also be made small and compact; it can move in silence; if <u>sound is to be used</u> there must be great flexibility of all the moving parts which must be made of faultless workmanship; it must last a long time; it must be <u>well covered in</u>; it must not become too hot, too stiff, nor too soft; it can be moved by tunes and rhythms."

If gravity shielding (see later) is used, passengers would become weightless and a top cover (also mentioned elsewhere) was essential. There is a suggestion here of sonic power.   Although beyond the scope of this book, in modern times, many UFO reports (q.v. on internet) mention UFOs making sounds, e.g. "a sound like an organ", "the sound of a tremendous chord of music", "a deep tuneful humming sound like huge bees or vacuum cleaners".   Cf the

Samarangana Sutradhara: "A vimana can be moved by tunes and rhythms." "Some were propelled by music alone" – Book of Oahspe.

In Ireland (Galway), folklore said: "In the old days everybody danced in the air, like leaves in the autumn wind ... people made a song to a plate." And in St Vincent, West Indies: "The wise people of old could fly quite easily. They clapped on gold plates, made music on them and flew." [9]. This has passed into legend. Montezuma gave Cortez two large flat gold discs about 10 to 12 inches diameter, as a gift to the King of Spain (who never received them!). The King's disc was about 6 mm thick and the Queen's was much thinner. The discs were said to be cut to a size and thickness corresponding to the size of the intended user, and the owner alone could use them to fly. The widely geographically separated locations of the above reports are surprising, but the modus operandi of such discs, if a fact, remains obscure.

On precision of control, Drona Parva (Mahabharata) says:

"Yudhishthira's vimana remained at the height of 4 fingers' breadth from the surface of the Earth."

This example could seem like a modern hovercraft, but many other accounts say they flew much higher than modern hovercraft are able to do. Also, modern VTOL (vertical take-off & landing) jets like the Harrier cannot be equated to vimanas, which would not have had modern aircraft fuel and engines, and some were so large that even modern engines could not possibly have lifted them. It is reported [2, 3] that Atlanteans had much better psychic ability than modern man and that this allowed them the insight to develop science to a higher level than we are yet able to achieve, especially in the power used to lift vimanas etc. (discussed further below).

Samarangana Sutradhara (Mahabharata) continues:

"By means of these machines, human beings can fly in the air and heavenly beings can come down to Earth." This suggests interplanetary visitors and is reminiscent of UFO reports of recent times, of which the authorities here do admit that a very few cannot at all be explained away as false reports, although most of course are false or mistaken.

There are several similarities, e.g. one similarity is that ancient descriptions of vimanas say they move in undulations rather than in a horizontal straight line; some modern UFO reports say UFOs move in undulations.

The Mahabharata (*Sanskrit*, Great India), written over 3000 years ago, probably from earlier sources, describes a weapon called the Agneya:

"A blazing missile possessed of the radiance of smokeless fire was discharged. A thick gloom suddenly encompassed the hosts. All points of the compass were suddenly enveloped in darkness. Evil-bearing winds began to blow. Clouds roared into the higher air, showering blood. The very elements seemed confused. The sun appeared to spin round. The world, scorched by the heat of that weapon, seemed to be in a fever. Elephants scorched by the energy of that weapon, ran in terror, seeking protection from its terrible force. The very water being heated, the creatures who live in the water seemed to burn. The enemy fell like trees that are burned down in a raging fire. Huge elephants, burned by that weapon, fell all around. Others, scorched, ran hither and thither, and roared around fearfully in the midst of the blazing forest. The steeds and the chariots, burned by the energy of that weapon, resembled the stumps of trees that have been consumed in a forest fire. Thousands of chariots fell down on all sides. Darkness then hid the entire army."
Then:
"Cool winds began to blow. All points of the compass became clear and bright. Then we saw a wonderful sight. Burned up by the terrible power of that weapon, the forms of the slain could not even be distinguished. We have never before heard of, nor seen, the like of that weapon."

This, if factual, was probably an event from over 10,000 years ago.

In the Mausala Parva (in Mahabharata), is written:

A weapon, "through which all members of the race of Vrishnis and the Andhakas were consumed to ashes. Indeed, for their destruction, Carna produced a fierce iron thunderbolt that looked like a gigantic messenger of death." "... in great distress of mind the King caused the bolt to be reduced to a fine powder ... and ... he employed men to cast that powder into the sea." Much damage was reported: hair and finger nails fell out overnight, gales blew, pottery cracked for no visible cause, birds turned white and their legs became red and blistered, and food went bad in hours.

Another weapon is the Brahmah weapon, also called (?) Indra's Dart, looking like a searchlight, burning the target. This description is many millenia before the modern invention of lasers, but it is not quite like a laser, as it was reported that winds blew and water boiled, and animals went amok, which suggests a sonic weapon; it was reported that two such weapons acting in opposition can neutralise each other, suggesting wave interference.

Drona Parva (in Mahabharata) reports:

"Drona's son touched water ((for earthing/grounding?)) and discharged the Narayana. Violent winds began to blow; showers of rain fell. Peals of thunder were heard, although the sky was cloudless. Earth shook. The seas swelled up … mountain summits split. Darkness came."

In the retreat after this, the text describes fleeing soldiers running to the nearest water where they strip off their clothes and wash themselves and their armour. This suggests a contamination. This is otherwise very strange behaviour if just a made-up story or legend. The text also mentions several times that only those touching or holding metal objects would be injured by the Brahma weapon, but if they dropped their weapons and left their chariots they were safe. There is more detail given.

Ghatotkacabhasma Parva (Mahabharata) reports vimanas used against armies:

"A huge and terrible vimana made of black iron, was 400 yojanas high and as many wide, equipped with engines … no steeds nor elephants propelled it. Instead it was driven by machines that looked the size of elephants."

A yojana is a Vedic unit of length, now lost in antiquity, but thought to be between 13 and 16 km. So 400 yojanas = 5200 to 6400 km, which is impossibly large and suggests that their "yojana" must have been much less than these modern estimates, or, a text error.

Karna Parva (Mahabharata) reports:

"We saw in the sky what appeared to us to be a mass of scarlet cloud resembling the fierce flames of a blazing fire. From this mass many blazing missiles flashed, and tremendous roars, like the noise of a thousand drums beaten at once. And from it fell many weapons winged with gold ((gold colour = rocket engine flames??)) and thousands of thunderbolts, with loud explosions, and many hundreds of fiery wheels. Loud became the uproar of falling horses, slain by these missiles, and of mighty elephants struck by the explosions. ... These terrible Rakshasas had the shape of large mounds stationed in the sky."

The above mixture of two cultures with totally different advancement, horse cavalry against stationary circular flying machines, is reminiscent of "Star-Trek", but where some visiting aliens did not follow the "Prime Directive" (non-interference with less-developed civilizations)!

It is then reported further in Karna Parva (Mahabharata), how these were defended against:

"Karna took up that terrible weapon, the tongue of the Destroyer, the Sister of Death, a terrible and effulgent weapon. When the Rakshasas saw that excellent and blazing weapon pointed up at them, they were afraid. ... The resplendent missile soared aloft into the night sky and entered the starlike formation ... and reduced to ashes the Rakshasa's vimana. The enemy craft fell from the sky with a terrible noise." The report continues, saying it was steered "Kamakasi-style" onto the soldiers below, so that "part of the army were crushed and pressed into the ground."

Ghatotkacabhasma Parva (Mahabharata) says:

"Gifted with great energy the Rakshasa once more came down to Earth in his golden vimana ... when it had landed it looked like a beautifully shaped mound of antimony on the surface of the ground." Another description is that Asura Maya travels in a "huge gorgeous golden circular construction, with four power plants and a circumference of twelve thousand cubits."

A cubit is about 20 inches or 50 cm, so 12,000 cubits is about 6 km! This circumference means about 3 km diameter. This is within Startrek proportions!

Some recent reports on UFOs (q.v. on internet) are reminiscent of the above texts, which suggest an extra-terrestrial origin, and technology. Ancient man-made lines in the Nazca Desert (Peru), which are only visible from an aircraft, are possible evidence of the existence of ancient aerial visitors.

Churchward [8] reports seeing an ancient manuscript in a temple in India about 80 years ago. The priests said it was a copy of ancient temple records from an earlier civilization preceding that of historically-known India:

"A drawing and instructions for the construction of the airship and her machinery, power, engines, etc. The power is taken from the atmosphere in a very simple and inexpensive manner. The engine is something like our present-day turbine in that it works from one chamber into another until finally exhausted. When the engine is once started it never stops until turned off. ((See next para below)) It will continue on if allowed to do so until the bearings are worn out. ... These ships could keep circling around the Earth forever without once coming down until the machinery wore out. The power is unlimited, or rather limited only by what metals will stand. I find flights spoken of which, according to our maps, would run from 1,000 to 3,000 miles." "All records relating to these airships distinctly state that they were self-moving, they propelled themselves; in other words, they generated their own power as they flew along ... independent of fuel. It seems to me ... we are about 15,000 to 20,000 years behind the times."

This agrees with Scott-Elliott's description of the vimanas [5]. Some suggestions of "perpetual motion" above need clarifying -- there is a class of motion called "apparent perpetual motion", e.g. the familiar dipping model bird which periodically dips its cloth beak in water, continuing indefinitely. "Apparent perpetual motion" heat engines, performing useful work have been patented and demonstrated but are difficult to keep within their narrow operating parameters [15].

The Mahabharata and Ramayana contain many descriptions of very large pre-historic aircraft. Ramayana reports:

"When morning dawned, Rama, took the Celestial Car (vimana) named 'Pushpaka', which Vibhisana had sent to him; it stood ready to depart. That car was self-propelled. It was large and finely painted. It had two storeys and many rooms with windows and was draped with flags and banners. It gave forth a melodious sound as it coursed along its airy way." [translated in ref. 8].
The Ramayana (written over 3,000 years ago, but probably from much earlier sources), also describes a flight to Sri Lanka.

Samsaptakabadha reports a vimana travelling from the Kailasa Mountains to the Malaya Mountains, saying it "proceeded through that region of the sky firmament which is above the region of the winds", i.e. the stratosphere. How could they have known that there is an airless stratosphere above the atmosphere?

Pre-historic stone forts on some hills in Ireland have their stones fused together in places, which must be due to great heat, suggesting a heat weapon, but some suggest it is due to lightening strikes, although lightening does not melt stone but fragments it.

Davenport [10] has shown that a ruined Indian city known as "the place of death" by archaeologists, was not destroyed by gradual decay. Originally, Mohenjo Daro, which is more than 5000 years old, lay on two islands in the Indus. Within a radius of 1.5 km, Davenport demonstrates three different degrees of devastation which spread from the centre outwards. Enormous heat unleashed total destruction at the centre. Thousands of lumps, christened 'black stones' by archaeologists, turned out to be fragments of clay vessels which had melted into each other in the extreme heat. The possibility of a volcanic eruption is excluded because there is no hardened lava or volcanic ash in or near Mohenjo Daro. Davenport assumed that the brief intensive heat reached 2000 degrees C, which melted the ceramic vessels. He suggested a nuclear explosion, but another type of weapon is more likely. (The author (MGH) has not seen this book [10] and so cannot report on if radioactive traces were found or not.)

Finally, a comment on Alcyone's vimana accident, mentioned earlier above. The vimana tubes which provide lift cannot be on the same principle as modern hovercraft or VTOL aircraft (discussed above) as the volume and velocity of air required would be far too high. Also, the tubes were fed by flexible tubes [5], very unlikely to be strong enough for a high hot air pressure. So, <u>to speculate</u>, the lift must be provided not by an airflow force but by the tubes discharging something which screens-off gravity. It is reported [2] that gravity was nullified by these tubes. As the only materials which could exit via such tubes is either air or a <u>cold</u> plasma, types of the latter need investigating to find any unusual lift-producing properties. (A plasma is atoms, ionised air and elementary particles.) In modern "Stealth"-type aircraft 50,000 volts is already in use to give extra lift (see Appendix 3B); more research is needed on this effect. The vimana maximum velocity was about 100 m.p.h [5], possibly limited to that by higher speeds sweeping away an anti-gravity plasma cloud.

Some high voltage induction coils used in the early 20th century (AD!) used a rotating toothed metal wheel with the teeth intersecting a liquid <u>mercury</u> jet, to act as the make-&-break, rather than the more usual spring switch which needs Pt contacts (probably not available in ancient times). This may (possibly) explain the strange mention of <u>mercury</u> in vimana engines (see above). A mercury stream being poured looks like milk, which may explain a description that the vimana engines "are fed with milk"! Vimanas are reported as glowing in the dark [5], whether made of wood or metal, presumably while flying. This suggests a high voltage discharge, related to plasma production. Modern UFO reports have similar mentions.

It is also reported [2] that "will-power" was used to drive some early vimanas, but in Alcyone's accident a physical tube malfunction caused the crash, so the lift there must have come from the tubes.

Looking into the future (from 1913 AD) with their Remote Viewing, Besant & Leadbeater [2] mention a new method of electric power distribution, which seems to the author (MGH) to be a possible extension of Tesla's ideas not yet extant. So a paper on this by the author is reprinted here in Appendix 7.

REFERENCES for Chapter 4

1. A. Besant & C.W. Leadbeater, "Oc Chemistry", published by TPS, Adyar, India (1908, 1919, 1951). Available as a free download from www.4-D.org.uk
2. A. Besant & C.W. Leadbeater, "Man: Whence, How and Whither", published by TPS, Adyar, India (1913)
3. A. Besant & C.W. Leadbeater, "The Lives of Alcyone", (2 volumes, 740 pp), published by TPS, Adyar, India (1924). (Available as a free download – but see important details in General References at end of this book).
4. D. Anrias, "Through the Eyes of the Masters", publ by Routledge & Kegan Paul (1932). (See copyright disclaimer, at end of Appendix 8)
5. W.J. Scott-Ellliot, "The Story of Atlantis", publ by Theosphical Publishing Co. (1893)
6. I. Donnelly, "Atantis, the Ante-Deluvian Continent", publ by Sidgwick & Jackson (1950).
7. Dr V.R. Ramachandra Dikshitar, "Warfare in Ancient India", publ by Macmillan (1952). The author was a professor at Oxford University.
See also: "Mahabharata", containing Drona Parva, Ghatotkacabhasma Parva, Karna Parva, Samsaptakabadha Parva, Vana Parva, Adi Parva.
Samarangana Sutradhara, etc.
8. J. Churchward, "The Childen of Mu", publ by Neville Spearman (1959).
9. Wilkins, "Secret Cities of South America".
10. D. W. Davenport, "2000 AC Diztruzione Atomica" in Italian ("Atomic Destruction 2000. BC"), (1979).
11. D. Leslie & G. Adamski, "Flying Saucers Have Landed", publ. by Werner Laurie (1953).
12. http://myths.e2bn.org/mythsandlegends/textonly24-the-green-children-of-woolpit.html summarises the following recorded event, but such events are rare:

In 12th Century, Suffolk, UK, villagers harvesting crops heard frightened cries coming from wolf-pits. They found two terrified children. The children spoke in a strange language and they were completely green! The children, a sister and her younger brother wore strange clothes of an unknown material.
The girl later learned English and said they came from a Christian country with churches. She said it was a place where the sun did not rise but where twilight settled on the land.
The girl could not explain how they were found in the wolf pits. She said that one day, they were feeding her father's flocks and heard a great sound of bells. They became entranced and then were in an underground cave, where they stayed, exploring, until they found a way out into our daylight and were in the wolf-pit and they then saw the villagers.

13. Nature, 209 (5018), 32-33 (1966) reports past exposure of the mid-Atlantic ridge to air -- geological scientific evidence for Atlantis.
14. M.G. Hocking (Ed.), "Remote Viewing of Ancient Times: A historical novel of actual events from 100,000 BC to 500 BC", www.4-D.org.uk/Books. Free download.
15. A.V. Serogodsky (Moscow), German Patent DE 42 44 016 A 1 (24-12-1992). (Contact M.G. Hocking for more details). (Search web for Serogodsky).

CHAPTER 5

Jesus : Christ

A Scientific Appraisal

Religion without Science is blind;
Science without Religion is lame.   Albert Einstein

(Note: Calendrical changes made in past centuries indicate that Jesus may have been born some years BC, but this is ignored for the present purposes.)

It is important to have read the previous chapters and the Foreword, before reading this chapter.

The similarity of some parables of the Buddha and the Christ has led some to suggest that the "missing years" (12 to 30 AD) of Jesus may have been spent in India where he studied the very large volume of Buddhist scriptures. But even otherwise, Christ would have been aware of such scriptures. There are many other similarities, e.g. Herod, and Krishna's uncle (500 years earlier), both killed many babies when Jesus, & Krishna, were born. Remote Viewing [3] finds that Krishna and Christ are the same individual, the World Teacher.

Jesus is reported as being virgin born. In today's terminology this is not biologically "impossible" (as some would have said 50 years ago) because it could now be compared to cloning, but with gender selection. In 0 AD, before our present science, this would have required divine intervention, but its achievement is not likely to be beyond the capability of future science; i.e. God acts within the laws of science (God is Science! Science is truth), but we do not have that degree of scientific knowledge yet.

In space with more than 3 dimensions, it is quite possible for very distant stars to be only a few millimetres away from us in a higher dimension [1]. Theoretically, only a small gravity leakage can occur from a higher dimensional space to ordinary 3-D space, but heat and light cannot be felt [1]. The dimensionality of space was discussed in Chapter 1 to avoid interrupting this text.

CHAPTER 5   Christ as World Teacher   243

So we should not ridicule Biblical astrology (because stars could be very close to us in a higher dimension); the Gospels (e.g. see Matthew 2) say Herod consulted astrologers. The chief priests told Herod of an Old Testament prophesy that the Messiah (the Hebrew word for Christ) will be born at Bethlehem in Judea. Herod asked the astrologers to find the child Jesus and report back, and they succeeded in finding Mary & Jesus but they were warned in a dream not to revisit Herod. The dream state is our awareness of the next higher (fourth) dimension but this was discussed in Chapter 1 to avoid interrupting this text. A series of such prophesies and dreams are recorded in the Gospels and are evidence of the existence of Jesus as an actual individual. In a dream, Joseph was told by an angel to leave and go to Egypt until Herod's death, at which time it had been prophesied (Old Testament) that God would declare, "*I called my son out of Egypt*" (Matthew 2: 15).

When Herod found that he had been tricked by the astrologers, he ordered all children around Bethlehem under 2 years old, to be killed. This fulfils a prophesy by Jeremiah (Old Testament), "*A voice was heard --- wailing and loud laments --- Rachel weeping for her children ---*" (Matthew 2: 18). (More prophesies: see p. 246-7)

Here we come to a point which many cannot understand -- how can God allow children to be killed like this? Some people then reject Christianity, concluding that God is unjust. This is the simplistic/naive reasoning of many atheists. But they do not see the whole situation: surely "God is just" must be the inviolable starting point (except for atheists), so logic dictates that some alternative must exist and so there must be something about the children which we do not know. This is explained as follows, taking relevant information from the preceding chapters above:

We have many lives, not just this one, and this was discussed in Chapter 2 to avoid interrupting this text. An absolute law of physics (and of everything else) is: *Action and Reaction are equal and opposite*. If I throw a ball directly at a wall, it will inevitably bounce back and hit me. I must not blame the wall -- it is a reaction caused solely by my own action.

This inevitable Law also applies to all our other actions, as Christ later says in 8 well-known different ways because it is so fundamental and so important, e.g. "*As a man sows, so shall he reap*", "*He who lives *by the sword, shall die by the sword"*, and six more such sayings, discussed in Chapter 2. This is identical to what is called the Law of Karma by Buddhism & Hinduism. So, logic dictates that the children who were killed must have killed others in a previous life (obviously, the children had killed no-one in their then-present life). God must not be blamed -- it was due to their own prior actions, exactly like throwing a ball at a wall. The only serious alternative**, that God is unjust, is quite unacceptable (except to atheists):

It was shown in Chapter 2 (to avoid interrupting this text) that there is strong recent evidence that God exists. Which is the more acceptable -- that God is unjust, or, that a child had murdered someone in a past life? If God were unjust, He could not be God, as it is a contradiction in terms for God to be unjust!

"*From zoophyte small to the Lords of All, the Law of God works for good* ", i.e. if a person has killed someone in a past life and is consequently killed in a subsequent life, that reduces his burden of "karma" (defined above) and so it can be described as a "good" event, although very sad to us. Every (passive) event is good, on this basis. This notion (reduction of our karmic burden) was expressed by Christ in his advice to "*turn the other cheek* " if someone strikes you (because he would not be able to strike you {either once, or twice} unless it was within your due karmic burden).

If one of the children mentioned above had not committed any murder in any past life, or had already (in recent previous lives) expiated any such bad karmas, then he would be impossible to kill, if God is just. If there is no action, there can be no reaction. The wall will not spontaneously throw a ball at us. If we don't owe anything, nothing will be taken from us. Only those children who owed a life, were drawn to birth in Bethlehem at that time.

There are almost no "innocent children" in the world -- all of us have done something which has harmed others in past lives (not necessarily including murder of course!). Tennyson's poem, "Intimations of Immortality" says:
Our birth is but a sleep and a forgetting:
The soul that rises with us,
Our life's star, hath elsewhere its setting, and cometh from afar:

** Contriving to avoid accepting "rebirth", some Christians say such children go to an eternal heaven, but this seems *ad hoc* and does not explain why they were murdered.

*Not in entire forgetfulness, and not in utter nakedness,
but trailing clouds of glory do we come from God, who is our home:
Heaven lies about us in our infancy!*

It could be added that some children may also come trailing dark clouds of horror from their possible past evil actions, but including this would spoil the poem!

This leads directly to the purpose of life: when we have exhausted (expiated) all our past bad karmas and have achieved an adequate level of wisdom, intellect, charity, compassion, etc, we will then be allowed to pass on to the next stage, which is to join the Communion of Saints (or the "Fellowship of the Holy Ghost" in theological terms) and need no more re-births in this ordinary 3-D world. From Revelations (New Testament, King James Version):
Him that overcometh I will make a pillar in the temple of my God, and he shall go <u>no more</u> out, ((my underlining))
and, in the Old Testament (Psalms) and also quoted in several New Testament Gospels:
The stone that the Builders rejected, I will make the Head of the Corner. (See in Psalms, Mark, Luke, Acts, and 1 Peter)

To believe that all this could be achieved in just one life, as modern Christianity does, is obviously incorrect and it is no wonder that many churches are empty! Furthermore, the failure of the modern church to explain suffering in any reasonable way, has also emptied many churches and allowed the rise of atheism.

A warning: *"An eye for an eye"* is an Old Testament statement of the Law of Karma; this statement is often dangerously mis-interpreted as a licence for <u>us</u> to perform an act of vengeance for an event that we may have witnessed locally, but it must be read along with another Old Testament statement, *"Vengeance is <u>mine</u>, saith the Lord".* So if one of us (person A) kills someone (person B) because that person (B) was seen to have killed someone else (C), this is entirely wrong and will create "bad karma" for person (A) who will be killed sometime in the future as a result! Only God can operate the Law of Karma.

Returning to the Gospels, when Herod died, Joseph had a dream in which he was told to go to Nazareth, which fulfilled an Old Testament prophesy, *"He shall be called a Nazarene"*. (See quoted in Matthew 2: 23). These Old Testament prophesies, written very long before

0 AD, are compelling evidence for the existence of Christ as a very significant important historical figure.

John the Baptist appeared as a preacher in the wilderness in Judea. John tried to dissuade Jesus from being baptised by him, but Jesus said that it was required by God. At the moment of baptism, Jesus said he saw the spirit of God descending like a dove to alight on him, and a voice was heard saying:
"This is my Son, my Beloved, on whom my favour rests". (*Matthew 3: 17*).

Jesus then went into the wilderness and fasted 40 days and nights. The 40 days may relate to the 40 days in esoteric Buddhism which is traditionally the time taken to awaken the "chakras" (spiritual centres), which are discussed in Chapter 1. Fasting reduces the ability of the brain to process sensory data but opens the mind to extra-sensory input, similar to a flotation tank (see Chapter 7).
After this 40 days of preparation, is probably the time when the Christ took over the body prepared for Him by His pupil, Jesus (= Brihat, from Remote Viewing – see Chapter 4): it is traditional in Buddhist texts for a pupil to be born and to develop a suitable new physical body to maturity which a Master can then take over. It is otherwise difficult to understand why the Christ would need 40 days to prepare himself!

Christ next went to Galilee and leaving Nazareth He went to Zebulun & Naphtali in Capernaum, as was prophesied in the Old Testament (by Isaiah), who had said, "the land of Zebalum, the land of Napthali, ---". (See quoted in Matthew 4: 14) (Other prophesies: on p. 253, 255).

Christ then went to the Sea of Galilee where He met disciples whom He had recruited (in previous lives: according to Remote Viewing [3]) into His group. He then cured many people in Galilee and His fame reached all of Syria.

Next Christ preached the Sermon on the Mount, which is well known and so this, and accounts of His healings, will not be repeated here.

He said, *"I tell you this: so long as heaven and earth endure, not a letter nor a dash"* (jot or title, in the King James version*) "will disappear from the Law ((of Karma)) before all that it stands for is achieved"*. (*Matthew 5: 18*).

This is another of his 8 statements of the Law of Karma. There is no hint in these 8 statements that one might escape from this Law, because:
(a) He also said, "... *deliver thee to the judge, and the judge deliver thee to the officer, and thou be cast into prison. Verily I say unto thee, thou shallt by no means come out thence, till thou hast paid the uttermost farthing."* (Matthew, King James Version of Bible),
(my underlining), and,
(b) Christ gave them The Lord's Prayer *(Matthew 6: 9-15)*, and added, "*For if you forgive others the wrongs they have done, your heavenly Father will also forgive you* " but this does not negate the Law (of Karma) because He also said, "*DO NOT SUPPOSE ((*sic*)) that I have come to abolish the Law ((of Karma))".* (Matthew 5: 17).
So " forgive" means "to cease to blame", but may not cancel or abolish the karmic consequences, which is why Christ stated the Law of Karma in 8 different ways, the others not using the word "forgive".

Distortion of this important point even led (in mediaeval times) some catholics to suppose that they could confess evil acts before doing them, in order to escape any (karmic) consequences!

Miracles

Christ performed many healings, and miracles like walking on water. In scientific terms, the healings can be described as an (undiscovered) ability to remove diseased cells and replace them with healthy ones. In modern medical science it is possible to grow specialised cells, even organs, and so His "miraculous" cures are an extension of medical research which is currently being developed. Of course, hidden powers are involved in His ability to apply them (with instant effect) to patients in the absence of any equipment. Buddha and Krishna also performed miracles. Miracles/magic result from the application of as-yet-undiscovered science.

Walking on water must be due to an ability to shield off the effect of gravity, i.e. levitation. This ability been famously demonstrated by Daniel Dunglas Home, a medium who levitated himself out of one third-floor window and into another window in view of witnesses. This demonstration avoided the accusation that wires were used. The scientist William Crookes claimed to know of more than 50 occasions in which Home levitated "in good light" (incandescent gas light) at least five to seven feet above the floor:

Fig. 5.1 The reported levitation at Ward Cheney's house interpreted in a lithograph from Louis Figuier, *Les Mystères de la science* 1887 (from Wikipedia: Daniel Dunglas Home).

Fig. 5.2. St Joseph of Copertino famously demonstrated his levitation to many observers.

There are very many recorded examples of levitation, well-known ones being as follows. St Teresa; Saint Francis of Assisi is recorded as having been "suspended above the earth, often to a height of three, and often to a height of four cubits." St. Alphonsus Liguori, when preaching at Foggia, was lifted before the eyes of the whole congregation several feet from the ground and Liguori is also said to have had the power of bilocation [4]. St Joseph of Copertino (17th century) levitated and carried heavy objects; and 200 other Saints are listed as being able to levitate, some in Egypt (Coptic) in recent times. See 'Saints and Levitation' in Wikipedia. More recently, Prof W.J. Crawford [5] described tests in laboratory conditions where heavy objects were levitated and moved around the room. Harry Edwards [6] made similar reports and there are many others. But these phenomena are confined to very few gifted individuals, which makes general acceptance difficult.

The author (MGH) has seen telekinesis done by Robert Monroe [7], and affirms there was no trickery (such as hypnotising – I was able to look out of the window etc quite normally). But public domain science has not yet reported an ability to make antigravity devices. As a speculation only, from Chapter 3, if the arnoo could all be oriented with their points facing either towards the Earth, or away from it, there could be an antigravity effect -- an experiment which the author is presently trying. It is known that about 60 kV DC is presently used to give extra lift to research aircraft;   see also Appendix 3 (3B) for a mention of the Kriegen Scheiben.

Poltergeist phenomena (objects being thrown without human contact with them) are very numerous and are well known worldwide, and also provide evidence of mechanical force being available to levitate those objects.

The feeding of a crowd of about 5000 with 5 loaves and later 4000 with 7 loaves means the possibility of somehow replicating atoms and molecules, or, apporting loaves from elsewhere.   At séances, demonstrations have been given of apportation, e.g. of fresh daffodils in the autumn (out of season, UK), but the sceptic will always disbelieve this type of report because it is of the anecdotal type (i.e. it cannot be repeated on demand by any experimenter, as is the normal requirement for an acceptable scientific experiment). An inability to repeat such a thing on demand does not, of course, mean that it is untrue. It just means that the event does not satisfy the requirements of a "scientific demonstration".

Other anecdotal reports (in recent times) include an experiment reported done with Sai Baba who produced a Swiss watch in India and a figure of Sai Baba was observed simultaneously present in India and in Switzerland where the watch was apported from. "Bi-location" has been reported in other cases. Uri Geller and others have demonstrated strange abilities. No scientific explanation for apportation is yet possible, but that does not mean that such events are impossible. Turning water into wine (John 2: 1-10) implies an ability to apport wine existing elsewhere, or to transmute elements, not impossible but belonging to a future science not yet known to us.

World Teacher

Christ appeared voluntarily, for our salvation. In John (14: 6) Christ says: "I am the way, the truth and the life; no-one comes to the Father except by Me". Christ is speaking here in his capacity as World Teacher {see Chapter 4} (*Bodhisattva* in Sanskrit). One of the Saints is appointed (by God Immanent) as World Teacher and Remote Viewing [2, 3] observation has found that Christ was Krishna in a previous life (He is the World Teacher at the present time); this Krishna is not to be confused with the Krishna of the Bhagavad Gita.

The present Buddha was the previous World Teacher. The present (5th) period was founded a million years ago but it effectively began only about 60,000 BC [2] and it considerably overlaps the 4th period. It contains 7 sub-periods (of unequal length). The 7 sub-periods also overlap considerably. Many human souls are expected to reach Sainthood at the end of this (5th) period in about 200,000 years from now [2]. Those who fail, go to the next upcoming batch, so there is no permanent failure. Over very many periods, animals will become humans and the plants will become animals. Thus progression is said to be a (fail-safe) batch process: see colour Fig. 4.1 (between p. 63 & 64), and ref. [2].

John (3: 16) says: *"God so loved the world that he gave His only Son, that everyone who has faith in Him should not die, but have eternal life".* This verse has led to much dispute and offence, as it seems to negate that the Buddha, Krishna, Thoth (or Hermes), Zoroaster etc, were also Sons of God. Remote Viewing observation [3, 2] shows that Krishna was a previous incarnation of Christ!

However, the Latin form of John (3: 16) is: *"Sic Deus dilexit mundum ut Filium suum unigenitum daret, ut omnes qui credit in eum non pereat, sed habeat vitam eternam."* The word *unigenitum* does not translate as *"only"* but literally is *"single genital",* or *"virgin born".* This is very different from God's "only" Son! The Latin word for *"only"* is *"solus".* John 3: 16 in Greek has ΜΟΝΟΓΕΝΗ, spoken as 'monogene', English translation: monogenetic, reproducing asexually. This removes a quite unnecessary and unfounded obstacle between Buddhism, Hinduism etc and Christianity.

There is an interesting example of the use of ESP (extra-sensory perception) by Christ given by John (4: 16-19) where Christ tells a woman, correctly, that she has had 5 husbands but the man with whom she was then living with was not her husband, to which she replied (with a dry humour ?), "I can see that you are a prophet" !

He continues (in John 4: 36-38): *"The reaper* ((= His disciples)) *is drawing his pay and gathering a crop* ((of souls)) *for eternal life, so that the sower and reaper may rejoice together. That is how the saying comes true: "One sows and another reaps". I sent you* ((= His disciples)) *to reap a crop for which you have not toiled. Others* ((= others like the disciples, working in past lives, who have now passed to Sainthood)) *toiled and you* ((= His disciples)) *have come in for the harvest of their toil".* Note: (( )) is the author's interpretation

Prophesies

The prophesy *(Matthew 24: 15): When you see "the abomination of desolation", of which the prophet Daniel spoke, standing in the holy place (let the reader understand), then those in Judea must take to the hills --- It will be a time of great distress, such has never been seen from the beginning of the world until now, and will never be again. If that time of troubles were not cut short, no living thing could survive; but for the sake of God's chosen it will be cut short"*, suggests an event like Hiroshima (an *abomination of desolation*) occurring in Jerusalem (*"the holy place"*), which can be seen inevitably approaching from the present political situation generating a massive accumulation of "bad karma". This is speculation. But the text in Matthew's and Mark's gospels continues: *"As soon as the distress of those days has passed, the sun will be darkened, the moon will not give her light, .."* which suggests a "nuclear winter",

but this could refer to a much later event than the current (2011) Middle East situation (cf page 256 below). In Matthew (24: 1-2), Christ says, "verily … there shall not be left here <u>one stone</u> upon <u>another</u> …", which would require a nuclear bomb (my underlining).
As a further speculation, it is curious that the gospels also warn that no-one should go back into the devastated area to retrieve possessions, which could suggest that the area may look safe but have a hidden lethal danger such as radiation contamination.

Matthew's and Mark's Gospels continue, *"I tell you this: the present generation will live to see it all. Heaven and Earth will pass away; my words will never pass away".* The term *"generation"* refers to the present <u>batch</u> of humanity, not of course just to those people alive at AD 33, i.e. "generation" refers to those souls who were in their early lives 70 thousand years ago and much earlier (see Chapter 4) and who will continue for about another 200 thousand years (the end of the present period ending with Judgement Day, according to Remote Viewing observations by Besant & Leadbeater [2]).   Luke (11: 49-51 and 21: 32-33) also refers to *"this generation"* in the same way; i.e. it includes all of <u>us</u>.

He continues: *"There will be two men working in the field; one will be taken, the other left; ---"* , meaning that those who have developed the required standard (of wisdom, intellect, compassion, etc), will pass away from the physical (3-D) level of existence and into the next levels (4th and higher dimensions), at (or before) the end of the present period. The others will be reborn in physical human form again in another period of human existence until they achieve the required standard.
The parable of the talents applies: they have to be developed, not left dormant.

This recalls a poignant verse from "The Light of Asia" (Edwin Arnold):

The Light of Truth's high noon
Is not for tender leaves,
Which shall blossom forth in later Suns
And in later aeons raise full crowned head of glory to the skies.

And expressed otherwise by Goethe:

Sees not the Gardener who prunes the trees in the morning,
The flowers and the fruit the future adorning?

In the Garden of Gethsemane there is some evidence that the disciples were able, while asleep, to observe Christ, or to have later seen it by Remote Viewing – evidence of Remote Viewing from within the Gospels! This is deduced because the Gospels describe what Christ was doing while they were asleep. He had asked them to remain awake. Peter's *"cock crowing = deny me thrice"* event was prophesied by Christ, one of many examples in the Bible showing the extent to which the future is already pre-ordained. But (from Remote Viewing observations) it is a mistake to conclude that everything is pre-ordained.

Judas' 30 pieces of silver were used to buy *"the potter's field"*, as prophesied centuries earlier by Jeremiah: *"They took the thirty silver pieces --- and gave the money for the potter's field"*.

Crucifixion

In John (19: 11) Christ told Pontius Pilate, *"You would have no power over me at all, if it had not been given to you from above"*, meaning that Christ has no bad karma remaining from his previous lives and so it was not possible for Pilate or anyone else to cause him any evil.

The crucifixion was thus of course voluntary and Christ was well able to apply the well-known (today) methods of self-hypnosis to avoid any pain if he wished. Christ did not need to suffer in any way, as he has passed beyond all karmas, which are the only cause of pain & suffering. Christ offered his own suffering to relieve the sins of others who repent. This has already been discussed in detail in Chapter 2 (see page 60). As a point of information, there is (presently) an annual crucifixion in the Philippines where a volunteer is actually crucified with nails (and then released!).

Christ told the man beside Him, who was being crucified with Him, that he would be in paradise that day. We are not told what He said to the other man on his other side, but He would have told him that He would come back for him "later" (in a future life), and not just consign him harshly (un-Christianly!) to eternal hell (unjust, if his life background chances had differed from that of the other man). Some may interpret the first man as negating karma + reincarnation, as he was a thief but was offered paradise. But that man could be reborn centuries later from paradise (the normal inter-incarnations state), for a later life.

Matthew (27: 46) cautiously/dubiously gives the words of Christ in their original (untranslated) language, as, *"Eli, Eli, lema sabachthani"* which Matthew then says (incorrectly, by wrongly assuming it was spoken in Aramaic) to mean:
"My God, My God, why hast Thou forsaken me?"
But it is found from Remote Viewing observations that it was actually spoken in an ancient Atlantean language and means:
"My God, My God, how Thou dost glorify me".

There are many prior instances of the word "glorified" in the Gospels, e.g. *"after Jesus had been glorified"* in John (12: 16) and Christ said, *"The hour has come for the Son of Man to be glorified"* in John (12: 23) and in John (12: 28) is said, *"A voice sounded from Heaven, 'I have glorified it and will glorify it again' "*. Also see John 13: 31 & 17: 1 & 5. Christ says in John (16: 14): *"He will glorify me".*

Because He did not irreversibly "die" (see below), Christ then passed into a state of clinical death or suspended animation, as has (recently) been demonstrated by yogis. Mark (15: 44) says, *"Pilate marvelled if he were already dead"* (as it takes much longer for death). He was taken down much earlier than usual and put into a tomb, in which he recovered and was later seen by many. Luke (24: 39) clearly says that He was a physical form, reporting that Christ said: *"Look at my hands and feet. It is myself. Touch me and see; no ghost has flesh and bones as you can see that I have".* Luke (24: 41) reports that He then said: *"Have you anything here to eat?' They offered him a piece of fish that they had cooked, which he took and ate before their eyes".* Clearly, he had converted his death into life, somewhat like an NDE (see pages 29 & 52) but in a <u>fully controlled</u> way; clearly it was His <u>intention</u> to <u>show</u> them that He was not dead. The "Gospel of Peter" also supports this view.

John (20: 6-7) says that he (*"the other disciple",* who arrived at the tomb first) saw the linen wrappings lying there, and the napkin which had originally been placed over His head, not lying with the wrappings but rolled together in a place by itself.

Of His life, Christ said (in John 9:18): *"I have the right to lay it down, and I have the right to receive it back again; this charge I have received from my Father".* (So He could die & resurrect himself at will.)
"I am the resurrection and the Life" (John 11: 25, & see I Cor 15: 26).

"I am with you alway, even unto the end of the World" (Matthew 28:20).

This concluded His 3-year project in Judea as World Teacher. But as World Teacher or Messiah, He has other communities to care for: in John (9: 16) Christ says, *"There are other sheep of mine, not belonging to this fold, whom I must bring in; and they too will listen to my voice. Then there will be one flock, one shepherd".*

500 years earlier (in an earlier incarnation) He had been in India as Krishna [3] and it is said (from other sources) that He went again to India just after His ministry in Palestine, to continue His duties there. He must have gone somewhere, as he was not dead (see above).

Later He is reported [8] as being (voluntarily) re-born as St Patrick to continue the essential development of His new religion into Europe.

The Book of Revelation

In Rev (4: 1) John says that a door opened in Heaven and he describes what he saw by his "Remote Viewing" ability. The "door opened in heaven" can be interpreted as an ability to view into the next higher dimensions beyond our 3-D world (see Chapter 1).

Revelations chapter 8 lists major events in World history, some of which have occurred and others are still to occur:

The second event mentions many ships sinking when a *"blazing mountain was hurled into the sea",* which could refer to the 120 metre rise in sea levels at the end of the last Ice Age, 10 thousand years ago, but as there was no *"blazing mountain",* it could better refer to the volcanic sinking of Atlantis 12 thousand years ago (lost in legend, although there are about 1000 (scholarly) books and publications written about Atlantis; see Chapter 4).

The third event is clearer: *"a great star shot from the sky, flaming like a torch; and it fell on a third of the rivers and springs. The name of the star was Wormwood; and a third of the water turned to wormwood, and men in great numbers died of the water because it had been poisoned".* Although it is a bit garbled, an interpretation is:

In the Ukrainian language translation of the New Testament, the Greek word apsinthion (English: Wormwood) is "Chernobl", and the nuclear power plant there is surrounded by a forest which was killed in 1986 by high radiation levels when the nuclear plant went into meltdown. Dead wood is eventually converted into worm-eaten wood [9]. The lake there has radioactive contamination. Many people were killed in that event and the region will be radioactive for tens of thousands of years -- clearly a major World Event. John's account in Revelation is a bit garbled, saying the star was wormwood (the accident involved red hot uranium, hardly a star, but perhaps seems so if Remote Viewing magnified U atoms: see Chap. 3); he also says the water turned to wormwood, which makes no sense. There is a big lake there -- maybe the translation is poor and should have said the lake or water is in the forest (which turned to worm-eaten wood), not that the water *"turned to"* wormwood? It is difficult for to write down from memory all the details of lengthy and complex visions.

If the above is correct for the third event, then the fourth event has yet to occur and so is very important for us to understand. It is fairly clear what it will be, saying, *"The third part of the sun was struck, a third of the moon, and a third of the stars, so that the third part went dark and a third of the day failed, and of the night".* The significance of the word *"third"* is obscure -- it is also used in the previous events (e.g. a "third" of the water, in the wormwood one)! Ignoring this obviously incorrect (or mis-translated?) word, this fourth event looks like a nuclear winter and suggests a future nuclear war or a major volcanic eruption (e.g. the Yellowstone super-volcano event -- now due and measurably bulging -- which would destroy the North American continent and create a nuclear winter type of effect -- dust clouds). A major volcanic eruption in the 6th century AD created such an effect worldwide for 2 years (no sunlight) (tree ring data etc), which led to famines and the "Dark Ages" before the mediaeval period. The conversion of North America into a series of islands is predicted to occur within the next few hundred years [2].

The overall impression of Revelations is mixed-up observations made by a rather untrained Remote Viewing observer who had personal prejudices which he has allowed to colour his text. There can be no (reasonable) doubt that Revelations does refer to the Chernobl disaster, which caused a large area to become uninhabitable for tens of thousands of years, which thus classes as a major World Event. Now that we know the Chernobl details (as it has happened), we can see the errors in the Revelations 8 account of it and we can conclude that there are likely to be similar errors in other parts of Revelations.

But finally, the author of Revelations (Rev 22: 18-19) writes a threat or curse, which reveals his prejudice (but these verses could be a later addition by someone else): "--- *should anyone add to the words of prophesy in this book, God will add to him the plagues described in this book; should anyone take away from the words in this book of prophecy, God will take away from him his share in the tree of life and the Holy City, described in this book".* In saying this the writer not only takes the name of God in vain, but he also shows that he had an exaggerated idea of his prophetic ability.

Modern scholarship considers the author of Revelations (written towards the end of the first century) is not the same as the author of the Gospel of St John, because of their Greek grammar differences etc, although traditionally they have been believed to be by the same author. It is possible that the Apostle John wrote Revelations and someone else wrote the Gospel of St John. But lack of genuine authorship does not mean that a tract is not inspired, and it was written in the last half of the first century AD and so is likely to contain the Theology of early Christianity.

Modern scholarship also suggests that the strong anti-women verses in St Paul's Epistles may have been added later by someone else. Remote Viewing also supports this possibility, as St Paul was seen as a woman in several of his previous lives.

REFERENCES for Chapter 5

1. N. Arkani-Hamed, S. Dimopoulos, G. Dvali, Scientific American, page 48 (August 2000). (article on multidimensional space).
2. A. Besant & C.W. Leadbeater, "Man: Whence, How and Whither", published by TPS, Adyar, India (1913). (See Note in reference 2).
3. A. Besant & C.W. Leadbeater, "The Lives of Alcyone", (2 volumes, 740 pp), published by TPS, Adyar, India (1924). **Note**: This is available as a free download from: www.AnandGholap.net and many other books by Besant & Leadbeater are available there. Or type "The Lives of Alcyone" into Google search engine to find other sources. Caution: With any text obtained by scanning a book into a computer, there is always the chance of (usually minor) errors, unless the scanned result was checked very carefully. See important note on this point in the General References section at end of book.
4. Search Google or other internet search facility for searchwords: Saint Levitation.
5. Prof W.J. Crawford, "The Reality of Psychic Phenomena".
6. H. Edwards, "The Mediumship, of Jack J. Webber".
7. The author's private visit to see Robert Monroe at his Institute in the Blue Ridge Mountains. He is known for his "Hemi-Sync" binaural beat tapes/CDs. Books by R. A. Monroe: "Journeys out of the Body", publ. by Anchor/Doubleday (1977). "Far Journeys", Doubleday (1985). "Ultimate Journey", Doubleday (1994).
8. D. Anrias, "Through the Eyes of the Masters", Routlege & Kegan Paul (1932).
9. The Greek apsinthion means the plant Wormwood, so 'worm-eaten wood' could be dubious. In C15 Old English *Wormod* changed to Wormwood. Did the prophet know this?!

CHAPTER 6: Conclusions

◆ **Religion in the present day.**
◆ **Possible integration with other religions.**

My son, when you were born, you were crying,
and all around were laughing.
Go through life likewise, so that when you are dying,
and all around are crying,
You will be laughing!
                                    A Sufi Poem

Religion in the present day

This book has been written mainly to counteract increasing atheism and books on atheism. In recent decades there has been a rise in agnosticism and atheism, which is related to:

(**a**) the occurrence of random events like earthquakes, crashes etc, sometimes called "acts of God", which appear to strike at apparently "innocent people". But if God is just, these must be the result of some action done in their past life, if there is no apparent reason for them in their present life. This is very simple, but is studiously ignored and people then wrongly conclude that God is unjust, or does not exist! An unjust God is a contradiction in terms (a God cannot be unjust).

(**b**) an apparent absence of the effects of actions -- it seems that one can do anything without any consequences (beyond those of the country's Law, if discovered). E.g. 200,000 abortions are done annually in the UK (in each of the past few years) and the people involved seem to think that if a doctor says it is alright (on medical grounds), then they can just proceed without the same horrific karmic consequence to themselves (in a future life). (The UK welfare state would give them minimal (but adequate) financial help if they did not proceed, so lack of finance is no "excuse", and even otherwise it would be a very bad karmic action.)

The same incorrect "mind-set" applies to many other serious actions. Chapters 2 & 5 set out the unseen consequences of these beliefs, which are mainly due to actions done in one life being carried forward <u>unobserved</u>, to one's next life, like a financial balance-sheet.

Possible integration with other religions

It is noticeable that there are many common beliefs in different religions, and the points of difference can be explained. E.g. the Christian belief that Christ is the "only" Son of God is due to a mistranslation {*unigenitum* (Latin) & *monogene* (Greek) do not mean "*only*", explained in Chapter 5, p. 251}.   John 10: 16 says, "*Other sheep I have which are not of this fold: them also I must bring*" ; this is in accord with the Remote Viewing observation [page 227] that He had previously incarnated as Krishna** in the Hindu religion. This is how His statement, "*I am the Way, the Truth, and the Life*" should be understood, and, "*None cometh to the Father except through Me*", because <u>He</u> is the appointed World Teacher. Hinduism is a monotheistic religion (as mentioned earlier, its multiplicity are <u>aspects</u> of God but there is only <u>one</u> God). Nothing in Hinduism is against Christianity -- how <u>could</u> there be, if they were both promulgated by the same perfected Man (the World Teacher)!

The last 2000 years is less than 0.1% of the total time that humans have lived on Earth. Yet many proponents of "modern" Christianity claim that <u>only</u> <u>Christians</u> can obtain salvation, which leads to the absurd consequence that none of the billions of people living before the year 0 AD could have achieved salvation and were deprived of it! Krishna lived well before 0 AD. Christ's predecessor, the Buddha, {as Vyasa, Thoth (Hermes), Zoroaster (Zarathustra), and Orpheus}, cannot be simply brushed aside, nor studiously ignored!

For Christianity to spread to Europe and not to remain in Palestine as a minor religion, it was essential for it to be promulgated within Europe itself, and Remote Viewing showed that Christ was (voluntarily) reborn a few hundred years after His appearance in Palestine, in Ireland as St Patrick, who organised many others to successfully spread His new religion throughout Europe.   He would have visited England, and Patterdale in the Lake District is said to be named from St Patrick's Dale. Hence William Blake's poem: "*And did Those feet, in Ancient time, walk upon England's mountains green*". See also Chapter 4 page 117 (Surya's life in Ireland, 1000 generations earlier, when Ireland was still joined to England).

**Not to be confused with the Krishna in the Bhagavad Gita ([3] in Chapter 4]).

CHAPTER 7

SENSORY DEPRIVATION - FLOTATION TANKS

Silence is the gateway to the Soul;
The Soul is the gateway to God.

This is a chapter (Chapter 10) reprinted from the author's book**:**
"Exploring the Sub-Conscious using New Technology",
by M.G. Hocking, published by CMC Ltd. (1993).
Available from: www.4-D.org.uk/Books

This is a simple effortless method for obtaining first-hand visionary experiences and insights. It is a great improvement on the traditional monastic method of simple "silence".   See also Chapter 1, section 2a, for another method; see especially: T. Yuschak, "Advanced Lucid Dreaming", ISBN 978-1-4303-0542-2 -- **Highly recommended reading.** It covers use of herbal supplements (which, unlike narcotic drugs, do not alter one's waking behaviour & so are morally acceptable) to produce immediate lucid dreams and OBEs, including seamless transition from waking to lucid dreamstate consciousness. See page 36 for present author's (MGH) comments on suppliers. See ref. 14 on p. 241 for important details not included in this present book.

Flotation tanks

This section tells of the remarkable effects obtained from a very simple procedure: merely floating on 10 inch deep pool of water in a dark enclosed chamber less than 8 feet by 4 feet in size (Fig. 7.1). Of all the methods described in the above book, this and CES (cranial electrical stimulation) are the most universal in their effect.
A flotation tank is perhaps comparable to the ancient "mysteries", like the Elusinian Greek Mysteries, etc, but is not secret.

Sensory Deprivation:

At first, the mind will do anything to prevent the removal of the sensory input to which it has got used. Research has been done on

the effect of sensory deprivation on volunteers in a black room [25]. People stay in a totally black room in total silence. After an initial long period of sleep their brainwaves drop to lower frequencies spontaneously. They see bright visions. A faster method is to float in a bath of very dense solution of Epsom salt in water controlled at normal skin temperature, in the dark and silence. This removes all the senses (touch, sound and vision) and one has no sensation of even having a body because its boundary (the skin) cannot be sensed. Remarkable results have been reported [6,2,41].
It might be thought likely to be boring, but it is not!

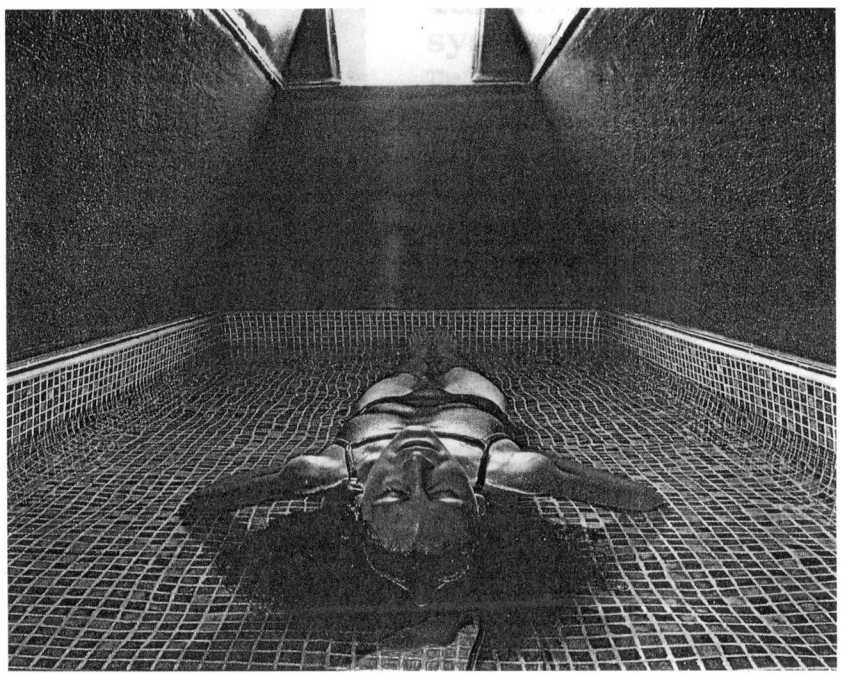

Fig. 7.1. Flotation Tank (picture: courtesy of Visionary Designs, Brighton, UK)

Contains 700 litres of water and 500 kg of $MgSO_4 \cdot 7H_2O$ B.P. (Epsom Salts) to give buoyant floating. (B.P. = British Pharmacopoeia grade, possible supplier: William Blythe Ltd, Church, BB5 4PD).
Water temperature = 35ºC (= skin temperature).

DIY Caution: floor loading!

Total darkness and silence are essential for complete 5-sensory deprivation.

The flotation tank has water only 10 inches deep but it contains 800 pounds weight of Epsom salts to give extreme buoyancy. An air pump provides ventilation. The optimum water temperature of 93.5°F is maintained by a thermostat. At this temperature it is impossible to feel the water (by its temperature) and so you cannot distinguish where your skin ends and the water begins! Also, in a fluid you have no sensation of any pressure (unlike lying on a bed). Such floating completely removes the sense of touch. It is comparatively easy to remove the other senses (vision - just keep out light; sound - just keep out noises; taste - just don't eat!; smell - just don't have any scented materials around). Clearly, of the 5 senses, touch is the most difficult to avoid. It requires a flotation tank.

Flotation feels like being weightless in black silent space and it is difficult to know if you are on your back or front or vertical. But there is no vertigo or discomfort nor any kind of claustrophobic sensation. As each part of one's body becomes relaxed, it seems to vanish from awareness until there is nothing left but consciousness. It is an almost indescribable experience, unlike anything else whatever. Soon after this, visualised events brightly appear, often scenes from childhood.

On emerging from an hour in the tank, the world seems to have changed in your absence! Things are seen anew - the world is fresh, illuminated, glowing bright, luminous, intensified. William Blake described such an awareness as "cleansing the doors of perception"; see Aldous Huxley's book of this title [1]. Tests on subjects of sensory deprivation experiments show beneficial results such as increased visual acuity, tactile sense, auditory sensitivity and taste sensitivity, lasting up to 2 weeks [41]. Improved learning, memory, IQ scores, perceptual motor tasks, enhanced visual concentration and increased short term visual storage also resulted.

Some physiologists estimate about 85% of our brain's activity is spent in dealing with balance and counteracting gravity. The effect of gravity is annulled in flotation and this releases a large proportion of brain activity. This quasi-release from gravity also allows blood circulation more freely, reaching parts not well-supplied due to cardiovascular constriction (due to smoking, cholesterol clogging, tension, etc.) and this reduces the effort needed by the heart. The blood pressure falls beneficially and the pulse rate slows. Relief of gravity-caused pressure on joints etc. will alleviate temporarily chronic pain due to arthritis, sprains etc.

Lilly & Shurley were the pioneers of flotation [6] and laid the foundations, but following Ptolemy's 1800 year-old account of the mental effects of a stroboscope (his spinning spoked wheel between closed eye and sun), one is tempted to speculate on the use of the deep sarcophagus in the Great Pyramid which can still be seen there, and in which one can lie [100, 111]. There is nothing new under the sun...[42]!

The granite coffer in the Great Pyramid is over 6 ft 6 inches long and over 2 ft wide (internal measurements) and is too wide to go through the door. Clearly it would not have been removed by those who put it there, so the inference must be that they did not want 'those who came after' to remove it either. One may then speculate that they 'built it in' to be an 'indelible message' for those who came after. As a speculation, could the message be a recommendation to use a flotation tank to gain very easily the enormous benefits of deep meditative states?

To speculate further, the Coptic Christians in Egypt are the lineal descendants of the Pharaonic or Osirian religion of ancient Egypt, founded by Thoth: Isis (or Asset), the mother of Horus, was later replaced by Mary. Coptic churches contain large fonts. Churches across the world contain fonts; are these a 'folk-memory' that a religious building should contain a flotation tank? It is very likely that flotation tanks were used anciently as part of initiation procedures, but were kept secret and have since been forgotten. The Egyptian Mysteries were transmitted onwards to become the Greek (Orphic) Mysteries and were later passed on to Rome before almost disappearing in the Dark Ages.

Lilly [6] reports that he could control various states of consciousness while floating, such as waking dreams (vivid daydreams); events could occur with such brightness that they seemed real and could possibly be mistaken for events in the outside physical world. Truly an 'astral doorway' [see references 34, 37-40, 12, 7-11].

Meditation conventionally (with no equipment) takes many years to achieve. To 'achieve' means, in measurable terms, the ability to go into a state in which the EEG trace is symmetrically bilateral and includes a small beta peak accompanied by larger alpha and theta peaks; see Fig. 1.11 in Chapter 1 (in the present book, page 33). But in a flotation tank these years of preparation (which overcome the external stimuli of touch and the other senses) are completely removed! It is 'instant mysticism'. People can go almost immediately to deep levels of meditation and contemplation [41, 72], often on the first float but several may be required.

Psychological problems have also been rapidly solved in the tank: an acutely shy person became able to perform confident public speaking, to his amazement [41]. Depression is linked to raised levels of pituitary and adrenal activity; floating is associated with decreases in these two activities. Hence it relieves depression.

Pain is substantially relieved by flotation and rates of recuperation increased. This includes cases of psychological damage as well as physical. Lawyers report greatly enhanced ability to marshal facts for cases during flotation, resulting in impressive results; this clearly can apply to any kind of job requiring mental organisation [41]. In some cases even non-religious people have described flotation experiences as 'religious experiences'; it is clear that something very unusual indeed is occurring while floating in the tank.

Green [4] points out that to produce theta consciously, it is necessary to have a quiet body, emotions and thoughts, simultaneously. It takes Zen monks some 20 years, but the flotation tank does it all for you in minutes! Studies have shown that after only 1 hour the theta level is significantly raised (without sleep intervening) [41]. The tank is ideal for maintaining wakefulness while in theta [74]. This is a very important remark. The tank is thus ideal for promoting creativity, as an example for the outer world; or for deep meditation, as an example for those interested in the inner world.

Maxwell Cade [5] discovered from studies of over 4000 people that unusual abilities like self-control of pain, healing, telepathy etc. are associated with changes in EEG pattern to a bilaterally symmetrical form [5] (see Fig. 1.11 in Chapter 1). Other workers (and Maxwell Cade) [76-78] found the same thing for those in deep meditation. The same happens during floating; the tank does not block left brain activity but makes it a partner with the right brain, whose activity floating brings forward [41].

Before trying floating, many people think it must be boring. In fact the opposite of boredom occurs. One floater said that every time she floated she felt like an explorer or adventurer and was proud of herself [41]: "Consciousness - the ultimate frontier".

Some suggest the tank experience has obvious parallels with a return to the womb. Babies have theta brainwaves. Tanks cause increased endorphin levels [79]; for comparison pregnant women have up to 8 times the usual blood endorphin level. Endorphins, like heroin and morphine, relieve pain and create euphoria.

An alternative to the "return to the womb" parallel may be replaced by a return to the paleontological environment of our distant ancestors, living in a warm sea [41].

Relaxation in the Tank

A common experience on entering the tank is to try hard to relax and not succeeding; but then, on just giving up trying, relaxation comes immediately! Relaxation comes without any effort being needed, very unlike in the world outside. Any efforts in the tank will cause hindrance. Biofeedback workers know this as the 'Law of Reversed Effort': whatever you "try" to do, the opposite result will be obtained. See the example given under BIOFEEDBACK (Chapter 6 in reference [0]) - making one's finger hotter/colder. Like users of biofeedback machines, floaters soon learn the knack of 'letting go'. This knack becomes a 'body memory' and a familiar deep relaxation occurs in minutes after entering the tank [41]. This state, once entered, is then ideal for beginning various techniques, such as: self-hypnosis, visualisation, healing, meditation, etc. and these work much more strongly in the tank. But the best advice is to set no such goals for the first few floats [41].

Precaution: Do not touch your eyes, as salt will cause great discomfort and cause you to end the session.

Deepening relaxation in the tank: Imagery

While avoiding the pitfall of over-trying to relax, just mentioned, focusing the mind on breathing helps relaxation. Relax the abdominal stomach area muscles. Try concentrating on the coolness at the tip of the nose on inhaling, and then the warmth there as you exhale. Try counting these from 1 to 10, repeatedly. Do not resist random thoughts but allow them to dissipate away.

Try concentrating your being in the centre of your forehead on a count of 1 breath. On breath 2, concentrate your being in your throat; 3 - right shoulder; 4 etc. - right elbow, wrist, each finger and back up arm to throat; similarly left arm; chest; abdomen; pelvis; right leg - toes - back up to pelvis; left leg; return up abdomen, to forehead. Each part will probably become warm and glowing as tension is released. Deep relaxation is achieved [41, 4].

Especially if you are a 'non-visualiser' type of person, practice visualising things while in the tank - e.g. a daffodil. You will find this remarkable and profitable acquisition with many varied applications.

The above nose-breathing exercise can be combined with a visualisation to increase its effectiveness:

Visual imagery in the tank can be facilitated as follows [41]: As you breathe in, visualise a pure white light entering your nostrils and going into your lungs and abdomen, thence radiating out to all parts of the body. On breathing out, imagine the reverse of this process. Exclude all thoughts from your mind except breathing.

On the in-breath, imagine the life-giving power of the oxygen and on breathing out imagine that as dark blue or grey, containing toxins, fatigue etc. which are being expelled. With this process, your body will slowly increase in brightness with time until the tank is filled with bright light.

A variant is to focus on a particular area of he body and visualise that as warmly glowing. This could be, for example, one of the chakras [113].

A further exercise in control of imagery is to imagine, for example, a blackboard and to 'see' on it various coloured shapes, such as a sphere, cube, pyramid, etc. Try a red triangle against a blue 'blackboard'.

You can also 'see' familiar cartoon figures from TV and imagine them dancing about.

An advanced visualisation is to imagine yourself getting smaller and smaller (like Alice in wonderland). Such visualisations have been described many centuries ago by Patanjali [114], who indicates that a trained observer may be able to observe atoms. A century ago, trained meditators reported using this (claimed) ability and drew string-like pictures of what may be recognised today as strings of quarks. The author has discussed this extensively in Chapter 3 and elsewhere [115]. If such an ability can be developed, it would have profound scientific consequences. Similar visualisations stopping at the cellular level are the basis for the healing described below and in Chapter 4 in reference [0]. Visualise entering your body and visiting various organs.

Further revealing visualisations are [41]:

(1) Visualise someone well-known to you, face, skin texture, hair colour, eyes; see effect of a smile, of him/her moving, speaking (hear his/her voice) - note what is being said: probably important.

(ii) Visualise your own face; smile; notice all details. Anything you don't like? Visualise in some activity - like climbing. Let the image speak to you - what do you hear?

(iii) Visualise yourself in a scene from childhood.

The results of such visualisations will often surprise you and will reveal directly what is hidden in your mind.

Self-Hypnosis in the tank

Under self-hypnosis in the tank, Hutchison describes a successful self-diagnosis which proved correct ([41] page 138). The tank is ideal for self-hypnosis, which requires deep relaxation and focused attention. People highly resistant to hypnosis and self-hypnosis can be hypnotised in the tank [135,136].

Children are right-hemisphere active, and if admonished by a parent saying "You're no good" etc, then this effectively powerfully negatively programmes the child and he/she needs to be deprogrammed later in life to remove the hidden blockage caused. This can be done under hypnosis or self-hypnosis. For self-hypnosis, try the standard method of counting backwards, but in the tank you will not need to count from 100 to 1 (conventional) but only from 10 to 1 due to the tank-conferred relaxation effect. Consult the many books on self-hypnosis for further details of procedure. Use the present tense in all hypnotic suggestions; try to image beneficial suggestions as being already true and make them as positive not negative ones (i.e. avoid using the word "not" in a suggestion) [41]. Link them with an image where possible, for added power.

Suggestions are more effective if accompanied by stately, flowing melodious and light gentle background music.

For medically permissible pain relief, e.g. arthritis, imagine a bright light focused on the area, while in the tank [41]. Floating has surprising and durable analgesic effects, due to endorphin release.

Superlearning

Many studies have conclusively demonstrated [41] that in a state of deep relaxation and focused attention, a state of hyper-suggestibility is reached where very fast learning of large amounts of information is possible. The tank is ideal for producing this state, due to the RAS 'turning up the volume' and to unusual access to the right-brain [41]. ("**RAS**" is explained below.) Light & sound machines can also perform this function, by placing the brain in the theta state.

In a survey [41] with a control group and a float tank group, taped chemistry lessons were heard, containing: (a) basic information, (b) application, (c) problem solving. Results showed floaters were better than the control group for (a), much better for (b) and very much better for (c). EEG traces showed greater theta waves during flotation and this was shown to be related to the depth of understanding involved. It was also found that visualisation helps learning [41].

Lozanov [83] has invented a (non-tank) superlearning system in which a deeply relaxed state is invoked and while rhythmically breathing, the students listen to lessons spoken against slow background music [82]. The book, reference [82] is recommended as applicable to tank learning also. The deeper the state of relaxation, the more the student can learn.

There are Flotation Tank Associations -- see on the Internet.

The relation of human brain evolution to Flotation:

The human brain evolved in 3 stages:

(1) the brain stem, from the reptilian period, (2) the limbic system (mammalian) and the cerebellum added to the stem, (3) the higher cortex. The last two of these are divided into left and right hemispheres. For more details, see reference [44].

A psychologist [70, 71] has remarked that the brain is hindered by a 'design error' - not enough communication between the higher cortex and the two older levels of the brain. This is a quite separate effect from the left/right hemispheres of the cortex; this lateral split is bridged by the corpus callosum and although industrial culture has caused left brain dominance, this can be corrected (there are ways to integrate and synchronise the left and right brains, as explained in this book). But the vertical division has few and slow connections.

In the (primitive) brain stem from the reptilian period, there is the 'reticular activating system' (**RAS**). This sets the arousal or awareness level. In the (advanced) cortex, all our advanced mental functioning occurs but the RAS is essential to keep the cortex awake. Consciousness is impossible if the RAS is destroyed.

Input from the senses to the cortex are passed to the RAS. If these signals are too intense, the RAS 'turns down' the brain's level of awareness (arousal state); if this fails, anxiety states occur. Conversely, if inputs from the senses are very low, as in sensory deprivation, then the RAS turns up the awareness of the cortex. This is the principle of the ganzfeld effect and of flotation tanks.

A ganzfeld device is an optical system which presents a uniform featureless light field to the eyes. As the RAS turns up the gain, visions may present themselves which otherwise would be swamped out. This effect is enhanced by simultaneously listening to "pink noise" on earphones. A primitive ganzfeld device is to place a half Ping-Pong ball over each eye to give a uniform featureless light field. Commercially produced more advanced ganzfeld devices are available. The RAS also 'switches on' the cortex during the sensory deprivation of sleep; visions occur (dreams), but as the brain's awareness level in sleep is very low (no beta waves), these dreams cannot be manipulated. Electronic units described in Chapter 9 in reference [0], are designed to overcome this and convert the dream to lucid. Meditation while we are awake is the manipulation of the RAS to create reverie and higher states of consciousness, with awareness. Floating in an isolation tank has a strong effect on the RAS, causing deep relaxation combined with great alertness - a very unusual state. The sensory deprivation caused by the tank causes the RAS to 'turn up the volume' on all the senses, bringing even the involuntary functions like heartbeat under conscious control, thereby achieving a unity between the reptile brain (RAS), limbic system (autonomic system) and neocortex (conscious awareness and voluntary control) [41]. After some time, subliminal visions may begin to present themselves.

Hutchison [41] points out that paradoxically, the RAS also interprets the experience of floating as a type of sensory overload on some channels and so responds by causing deep relaxation. Hence the unique result of floating is deep relaxation with intense conscious awareness.

Floating also causes the limbic system to inhibit stress hormones such as epinephrine and adrenaline, while increasing the production of endorphins which are beneficial neurochemicals. This results in a reduction of anxiety and induces euphoria, and helps the cortex to synchronise its hemispheres, to generate theta waves and visualise, etc. Habitual, chronic stress has replaced earlier primitive threats like loss of life, territorial combat and starvation, and modern man is in a perpetual state of non-specific arousal [128]. Many experts consider about 85% of all illness is stress-related. In the flotation tank there is sanctuary from stress situations: there are no other people, no noise, no light, and nothing that needs doing; in this absence of threats, body chemicals return to (better than) normal and one has the chance to examine one's life calmly and objectively.

Tank treatment is not passive but has an equal and opposite effect to what any stress could have. Hutchison [41] points out that our Judeo-Christian ethos imbues a tendency to regard relaxation as something opposed to productive activity: if you're relaxed this means you're not performing any worthwhile activity! Its apparent wasting time and laziness aspects makes it appear to be a luxury.

The limbic system atop the brain stem is responsible for some effects of altered states of consciousness such as euphoria, feelings of divided consciousness, loss of awareness of body boundaries, feelings of floating or flying and strange visual experiences like white or golden light [5].

Most people have been taught to avoid solitude, isolation and confinement. TV sets are anti-isolation devices! So most people have a negative attitude to solitude and isolation [6]. It is thus necessary to make a (small) mental effort to overcome this feeling.

In Summary, the benefits of flotation are:

- Floating stimulates endorphin production, described under CES, in Chapter 11 in [0]. It quickly and considerably reduces stress and anxiety and reduces any tendency for heart disease and other stress-related illnesses by lowering the levels of stress-related biochemicals.

- Floating immediately brings forward the right brain hemisphere, giving unusual access to imagination, creativity, visualisation and problem solving.

- It allows remarkable "superlearning", verified and used by many universities and schools mainly in the USA. Tape recorded and video information is accurately assimilated.

- Two hours of floating are more restorative than a full night of sound sleep and the deepest rest ever experienced is attained.
- Athletes improve performance using flotation.

- Floating quickly reduces smoking and drug use and counteracts withdrawal symptoms (by raising endorphin production).

- Floating is effective if weight loss is desired and its effect lasts for months after floating.

- Dogmatic attitudes and beliefs are realised for what they are, while floating. Unreasonable resistance to new ideas is thus weakened - a very beneficial effect.

Remote Viewing should become possible for most people in the flotation tank but much practice is essential to avoid interference from one's own preconceived ideas about an event. Although not requiring a flotation tank, the gifted CIA Remote Viewers were able to go back in time to repeatedly view ("replay") the Pan-Am aircraft which crashed at Lockerbie, just before the bomb exploded, to discover exactly where the explosion originated in the aircraft's baggage hold. Similarly, Besant & Leadbeater used Remote Viewing to view the events given in Chapter 4.

-=-=-=-=-=-=-=-=-=-=-

There are many websites on flotation tanks, including those in which time can be hired, but be aware that the quality may vary greatly for those tanks. Some tanks do not have the total blackness facility. It is essential to have **all** five senses receiving zero inputs, if the normally subliminal higher senses are to be perceived. This can take years by normal meditation techniques. Normal (everyday life) inputs to the 5 senses drown out the much smaller inputs to the higher senses. Depriving the normal 5 senses of all input allows the very faint subliminal 6th sense to be perceived.

Additional:

To attain higher states of consciousness also requires perseverance and single-mindedness and rejection of any feelings of selfishness. Absence of anger, hatred, desire, and delusion are essential. Many others have achieved this. Concentration is a valuable faculty and the exercises in ref [a] are highly recommended. The subject of a meditation must then be chosen, the simplest example being just God in the abstract. Using the practice of concentration achieved by these exercises, a determined effort must be made to raise one's consciousness and merge with the subject chosen. Persevere -- others have succeeded in doing it. See references [b] & [c] below.

a. Mouni Sadhu, "Concentration", publ by Unwin Paperbacks, and others.
b. Swami Vivekananda, "Raja Yoga", publ Bharatiya Kala Prakashan, or others (same text) (first publ about 1880). Note: does not warn against excessive use of pranayama (breath control).
c. Monks of the RamaKrishna Order, "Meditation", publ RamaKrishna Vedanta Centre, London (1972). Recommended.

A Final Comment:

As everyone eventually has to leave the 3-D physical world, it seems common sense to explore the next 4-D world, just as we might look at travel guides for any destination before going there for a holiday. It seems very strange to the author that very few people consider doing this! Two important quick methods are: the flotation tank, and, lucid dreaming herbal supplements. On the latter, it is **essential** to see the highly recommended reference cited on page 260 (T. Yuschak). Important: See also: http://dreamstudies.org and: http://dreamstudies.org/galantamine-review-lucid-dreaming-pill/
http://www.dreamviews.com
See ref. 14 on p. 241 for **essential** details not in this present book.

REFERENCES for Chapter 7

As this may be useful for readers, the list of references below is from the whole book (all chapters in it), called: "Exploring the Sub-Conscious using New Technology" by M.G. Hocking, published by CMC Ltd., London, 2nd edn (1993).

0. "Exploring the Sub-Conscious using New Technology" by M.G. Hocking, published by CMC Ltd., London, 2nd edn (1993).
1. A. Huxley, "The Doors of Perception" , Penguin, (1960). .
2. M. Hutchison, "Megabrain", Morrow, NY (1986)
3. W. Grey Walter, "The Living Brain", Pelican (1961).
4. E & A Green, "Beyond Biofeedback", Delacorte Press, USA, (1977). Also see Research Centre J. (T.S.), <u>16</u>, no 4, p 87 (1972).
5. C. Maxwell Cade & N. Coxhead, "The Awakened Mind", Element Books, UK, (1989).   & G.G. Blundell, C. Maxwell Cade, "EEG Measurement", Published by Audio Ltd, London.
6. J. C. Lilley, "In the centre of the Cyclone",   Paladin (1973).
7. R. A. Monroe, "Journeys out of the Body", Anchor (1977).
13. W. Sargant, "Battle for the Mind",   Great Pan (1957).
14. M. Sadhu, "Concentration",   Unwin Paperbacks (1977).
15. H. Hewitt, "Meditation", Teach Yourself Books, Hodder & Stoughton (1978).
16. A. E., "The Candle of Vision" , University Books Inc., NY.
17. R. M. Bucke, "Cosmic Consciousness",   Dutton, USA, (1969).
18. J. Silva & P. Miele, "The Silva mind control method", Souvenir Press (1978).
19. G. I. Gurdjieff, "Views from the Real World", Routledge & Kegan Paul (1976).

20. C. Castaneda, "The Teachings of Don Juan", (& other titles), Penguin (1977).
21. Paramhansa Yogananda, "Autobiography of a Yogi", Rider, London (1961).
22. Maharishi Mahesh Yogi, "Commentary on the Bhagavad Gita", Penguin (1969).
23. "Highways of the Mind", "The Shining Paths", "Inner Landscapes" and other books by D Ashcroft-Nowicki, Aquarian Press (1980s).
24. Translations of books by F. Lefebure
    Some titles are:
    "Brain development by alternative audition" by Dr F.Lefebure   [English translation of book below]
    "L'activation du cerveau par l'audition alternative (alternophonie)" by Dr F. Lefebure (Editions Jacques Bersez, Paris)
    "Phosphenism - the art of visualization" by Dr F.Lefebure   [English translation]
    "Developing clairvoyance by phosphenism" by Dr F.Lefebure
    "L'exploration du cerveau par les oscillations des phosphenes doubles (phosphenisme)" by Dr F.Lefebure (Editions Jacques Bersez, Paris, 1982)
    "Du moulin a priere a la dynamo spirituelle (Kundalini)" by Dr F.Lefebure (Editions Jacques Bersez, Paris, 1982)
    "Le pneumophene (phosphenisme)" by Dr F.Lefebure (Editions Jacques Bersez, Paris, 1982)
    [Translations of some of these may be available from The SEED Institute, 10 Magnolia Way, Fleet, GU19 9JZ, UK]
25. J. Vernon, "Inside the Black Room", C. N. Potter Inc., New York, (1963).
26. VIDEOS: "The Lion, Witch & Wardrobe" (cartoon version, Vestron CAD 14194); "The Neverending Story"; "A Flight of Dragons" (Channel 5, CFV05732); these are just a few highly recommended examples.
27. L.E. Walkup, Perceptual Motor Skills 21, 35-41 (1965).
28. R. May, "Creativity & its cultivation", ed. H.B.Anderson, Harper & Bros., NY, (1959).
29. A. Kasamatsu & T. Hirai, Psychologia, 6, 86-91 (1963).
30. VIDEO: Koyaanisqatsi (Polygram Video Ltd, 083 448 3).
31. "Auditory beats in the brain", Sci. Amer., 94-102, (Oct. 1973)
32. C. Green, "Lucid Dreams", (Inst of Psychphysical Research, Oxford, 1968)
33. K. Harary & P. Weintraub, "Lucid Dreams in 30 Days", Aquarian Press (1989)
35. H.de Saint-Denys, "Dreams & How to guide them", Duckworth, London, (1982)
36. M. Watkins, "Waking Dreams", Spring Publ.Inc, Dallas (1984)
37. S.J. Blackmore, "Beyond the Body", Paladin (1986)
41. M. Hutchison, "The Book of Floating", Quill - Morrow, New York (1984)
42. The Book of Ecclesiastes, The Bible.
43. Michael S Gazzaniga "The Split Brain in Man", Scientific American, 24-29, Oct. 1967.
44. S. Rose, "The making of memory", Bantam (1992)
45. F.C. Happold, "Mysticism" (1963)
46. C. Tart, "Altered States of Consciousness" (Wiley, NY, 1969)
47. E.C. Steinbecher, "Inner Guide Meditation", Aquarian Press, UK (1988)
48. M. Zdenek, "The Right-brain Experience", Corgi (1983)
49. Megabrain Report, 1990

50. O. Simonton et al, "Getting well again", Tarcher, Los Angeles (1978)
51. M. Westcott & J. Ranzoni, "Correlates of Intuitive Thinking", Psychological Reports, 12, 595-613 (1963).
52. R.L. Walford, "Maximum Life Span", Norton, NY (1983)
53. G. Leonard, "The Silent Pulse", Dutton, NY (1978)
54. P.H.C. Mutke, "Selective Awareness", Celestial Arts, Millbrae, CA (1976)
55. R.D. Willard, Am J Clin Hypnos 19, 195 (1977)
57. J. Hooper, "Interview with C. Pert", Omni, (Feb. 1982)
58. G Rattray Taylor, "The Natural History of the Mind", Dutton, NY (1979)
59. P.B. Applewhite, "Molecular Gods: how molecules determine our behaviour", Prentice-Hall (1981)
60. J. Olds, Scientific American 195, 105-116 (1956)
61. J. Olds, Am Psychologist 24, 707-719 (1969)
62. J.E. Adam, Pain 2, 161-6 (1976)
63. J.B. Levine et al, Nature 272, 826-7 (1978)
64. D.J. Mayer et al, Brain Res 121, 360-73 (1977)
65. H. Benson & R.K. Wallace, "Decreased drug abuse with Transcendental Meditation: A study of 1862 subjects", Congressional Record, 92nd Congress, 1st Session, June 1971.
66. M. Shafii et al, Am J Psychiatry 131, 60-3 (1975)
67. Idem, ibid 132, 942-5 (1975)
68. Time, 19 Jan 1981
69. M.Csikszentmihalyi, "Beyond boredom & anxiety", Jossey-Bass, San Francisco & London (1975)
70. P.D. MacLean, "A triune concept of the brain & behaviour", Univ of Toronto Press (1973)
71. "How the brain works", Newsweek, 7 Feb 1983
72. J.C. Lilly, "The deep self", Simon & Schuster, NY (1977)
73. H. Benson, "The Relaxation Response", Morrow, NY (1975).
74. T. Budzynski, Proc First Intl Conf on R.E.S.T. & Self-regulation, Denver, CO, (18 March, 1983)
75. Alan Richardson, "Mental Imagery", Springer-Verlag, NY (1969).
76. BC Glueck & C.F. Stroebel, Comprehensive Psychiatry 16, 303-21 (1975)
77. J.P. Banquet, J Electroencephalography & Clin Neurophysiol 33, 449-58 (1972)
78. Idem, ibid 35, 143-51 (1973)
79. J.W. Turner, Proc First Intl Conf on R.E.S.T. & Self-regulation (Denver, CO, 18 March, 1983)
80. Diana Deutsch, "Musical Illusions", Scientific American, 92-104, Oct (1975).
81. Nathaniel Kleitman, "Patterns of Dreaming" Scientific American, 82-88, Nov. (1960).
82. S.Ostrander & L.Schroeder, "Superlearning", Delacorte Press, NY (1979)
83. G.Lozanov, "Suggestology & outlines of suggestopedy", Gordon & Breach, (1982)
84. N. Drury, "Music for Inner Space", Prism Press, UK (1985)
85. A. Puharich, "The sacred mushroom", Doubleday (1959)
86. "Beyond Telepathy" idem, Doubleday (1962)
87. J. Dunlap, "Exploring Inner Space", Harcourt Brace & World (1961)
88. D. Solomon, "LSD, the consciousness-expanding drug", Putnam, London (1964)
89. "Drug Experience" ed. by D. Ebin (Orion Press)

90. Lama Anagarika Govida, "The Way of the White Clouds" Rider (1966)
91. I.Tweedie "The Chasm of Fire", Element Books, UK (1979)
92. Gopi Krishna, "Kundalini", Robinson & Watkins (1971)
93. "The Secret of Yoga" idem, Turnstone Books, London (1972)
94. Mouni Sadhu, "Samadhi", Allen & Unwin (1962)
95. Dr R.A. Moody, "Life After Life", Bantam Books, (1977) (a best seller)
96. Dr R.A. Moody, "Reflections on Life After Life", Bantam Books.
97. H.Wambach, "Life Before Life", Bantam (1981)
98. Dr M.B.Sabom , "Recollections of Death", Corgi, UK (1982)
99. D.Scott Rogo, "The Return from Silence", Aquarian Press, (1981) and: "Life after Death", idem, Harper-Collins, Glasgow, (1992)
100. P. Lemesurier, "The Great Pyramid decoded", Element Books (1977)
101. Suzuki, "Zen mind & beginner's mind", (& other titles), Weatherhill, NY, (1970)
102. E. Herrigel, "Zen in the Art of Archery", Routlege & Kegan Paul (1959)
103. Z'ev ben Shimon Halevi (Warren Kenton), "Kabbalah", Thames & Hudson, London (1979)
104. S. Court, "The meditator's manual", Aquarian Press, UK (1984)
105. P. Russell, "Meditation", published by BBC, London (1979)
106. D. & J. Beck, "The pleasure connexion: how endorphins affect our health & happiness".
107. "Altered States of Consciousness" ed. by C.P. Tart (1990)
108. M. Csikszentmihalyi, "Flow: the psychology of optimal experience".
109. P. Kelder, "Tibetan secrets of youth & vitality", Harper Collins, Glasgow (1991)
110. Graham Greene, "A World of my own: a Dream Diary", Reinhardt (1992).
111. P. Brunton, "In search of secret Egypt" (& other books).
112. R.O. Becker & G. Selden, "The body electric" and R.O.Becker, "Cross currents".
113. C.W. Leadbeater, "The Chakras", Theos Publ House, London.
114. Patanjali, "Yoga Aphorisms", published about 400 BC.
115. M.G. Hocking, "ESP observation of atoms & molecules", Bull Th Sci Stdy Gp $\underline{21}$, 53 (1983) & $\underline{22}$, 5 (1984). Free downloads available from: www.4-D.org.uk
116. Dr Joel Funk, Professor of Psychology, Plymouth State College, New Hampshire.
117. Nature $\underline{345}$, 463 (1990)
118. Science News $\underline{137}$, 229 (1990)
119. The Lancet, Jan 1983, p.246
120. New England J Medicine $\underline{307}$, 249 (1982)
121. New Yorker, 12 June 1989, p.69
122. Cancer Research Aug 1988, p.4222
123. E. Swedenborg, "Heaven & its wonders and hell", (Swedenborg Foundation, New York; distributed by Popular Library, New York, (1960)
124. "Neurochemical responses to CES & photo-stimulation via brainwave synchronization" study by Dr R.K. Cady & Dr N. Shealy at Shealy Institute of Comprehensive health care, Springfield, Missouri (1990), 11 pages.
125. A. Koestler, "The act of creation", Pan Books, London (1984)
126. B. Edwards, "Drawing on the Right Side of the Brain".
127. Doreen Kimura, "The Asymmetry of the Human Brain", Scientific American, 70-78 (1973).
128. K. R. Pelletier, "Mind as Healer, Mind as Slayer", Derlacorte Press, NY (1977)
129. G. Oster, 'Auditory Beats in the Brain', Scientific American $\underline{229}$, 94-102 (1973).
130. A.R. Luria, article in "Recent Progress in Perception" publ. Freeman, San Francisco (1970).

131. F.H. Atwater, Hemi-Sync Journal, <u>11</u>, No 1 (1993)
132. F.J. Boersma & C. Gagnon, Medical Hypnoanalysis Journal <u>7</u> (3), 80-97 (1992).
133. T.H. Budzynski, "Cranial Electrical Stimulation (CES) and the Practitioner" (publ. by CES Labs., 14770 N.E. 95th, Redmond, WA 98052, USA.)
134. R. Beck, "Bibliography of Cranial Electro-Stimulation" (publ. by Allied Forces Inc., PO Box 1530, Stanwood, WA 98292, USA.)
135. I. Wickramsekera, "Sensory Restriction and Self-Hypnosis as Potentiators of Self Refulation." Paper delivered at First International Conference on REST and Self-Refulation, Denver, Colorado, March 18, (1983).
136. "Restricted Environmental Stimulation & the Enhancement of Hypnotizability: Pain, EEG Alpha, Skin Conductance and Temperature Responses", The International Journal of Clinical and Experimental Hypnosis, Vol. 2, 147-166 (1982).

Note: Light & Sound machines, flotation tanks, etc can be obtained from: www.toolsforwellness.com (See page 36 for other items.)
This is given for information only. The author has no connection with this firm, but has found it reliable.
Many other suppliers exist; a low-cost "DIY" flotation tank construction booklet is available from: Design Mobility Inc, 280 Nevins Street, Brooklyn, NY 11217, USA. The author (MGH) can give advice on request: to contact, see www.4-D.org.uk/Books

NOTICE & DISCLAIMER: For legal reasons we make the following statements. The items described in the book are experimental consciousness-enhancing products, for which no medical or safety claims are made or implied. The descriptions given for each product are reports of the effects produced as given by the manufacturers and by professional users. None of these statements by professionals should be construed as medical claims. The claims made centre around relaxation, meditation, hypnosis and learning.

Warning: Light & sound and CES units should not be used by persons with a history of epilepsy or other neurological disorders. But the following finding is quoted, for information, from Maxwell Cade & Coxhead [5]: "After 4 years use of the lights, with more than 4000 pupils including 25 known epileptics, there have been no mishaps, and most of the epileptics have reported a marked improvement in their condition. ...subjects were only exposed to the lights after they had become very relaxed..." (see page 49 in reference 5 for more details). This is quoted for information only and our advice is that medical advice should be obtained before proceeding in such cases.

Additionally, it is essential to allow half an hour after using light and sound machines and other mind-enhancing units and tapes before operating machinery or driving a car.

No responsibility can be accepted for any adverse effects.

LIST OF APPENDICES

<u>NOTE</u>: References for each appendix are at the end of that appendix.

1. Heart Transplants transferring donor characteristics to recipients
    279

2. Some remote viewing personnel   291

3. General Relativity & Gravitation   292

4. Mass increase with velocity   302

5. Remote Viewing of atoms discussion (1983)   306

6. Linking String and Membrane theory to Quantum Mechanics and Special Relativity equations, avoiding Special Relativity assumptions: M.G. Hocking, Journal of Scientific Exploration $\underline{21}$ (1), 13-26 (2007)
    332

7. The high-voltage research of Dr N. Tesla: predictions   354

8. Remote Viewing Mechanism   361

GENERAL REFERENCES   373

APPENDIX 1

Heart Transplants which transferred the donor characteristics to the recipients

This article was originally published under the title "Changes in Heart Transplant Recipients that Parallel the Personalities of their Donors" in the *Journal of Near-Death Studies*, vol. 20, no. 3, Spring 2002.
For further information in connection with this article, contact Dr Gary E. Schwartz, Professor of Psychology, Department of Psychology, University of Arizona, Box 210068, Tucson, AZ 85721-0068, USA, telephone (520) 318 0286, email: gschwart@u.arizona.edu   Also see his websites:

http://veritas.arizona.edu
http://www.openmindsciences.com

FULL TEXT (verbatim):

Note: In the author's (MGH's) opinion, the paper below is not evidence for a so-called "cell memory" for the reasons given in Chapter 2, but is good evidence for soul transfer.

Acknowledgement: Below, is the full text, originally published 2002 in volume 20 of the Journal of Near-Death Studies, the quarterly scholarly publication of the International Association for Near-Death Studies (IANDS), reprinted here with their kind permission:

Pearsall, P, Schwartz, G E R, and Russek, L G S (2002), Changes in Heart Transplant Recipients That Parallel the Personalities of Their Donors.
Journal of Near-Death Studies 20 (3), © 2002 Human Sciences Press, Inc.
Ten cases reported by P. Pearsall, G.E. Schwartz, and L.G. Russek:

APPENDIX 1   Heart transplants transferring donor character to recipient

Case 1
The donor was an 18-year-old boy killed in an automobile accident. The recipient was an 18-year-old-girl diagnosed with endocarditis and subsequent heart failure.

<u>The donor's father, a psychiatrist, reported:</u>
My son always wrote poetry. We had waited more than a year to clean out his room after he died. We found a book of poems he had never shown us, and we've never told anyone about them. One of them has left us shaken emotionally and spiritually. It spoke of his seeing his own sudden death. He was a musician too, and we found a song he titled, "Danny, My Heart is Yours." The words are about how my son felt he was destined to die and give his heart to someone. He had decided to donate his organs when he was 12 years old. We thought it was quite strong, but we thought they were talking about it in school. When we met his recipient, we were so . . . we didn't know like what it was. We don't know now. We just don't know.

<u>The recipient reported:</u>
When they showed me pictures of their son, I knew him directly. I would have picked him out anywhere. He's in me. I know he is in me and he is in love with me. He was always my lover, maybe in another time somewhere. How could he know years before he died that he would die and give his heart to me? How would he know my name is Danielle? And then, when they played me some of his music, I could finish the phrases of his songs. I could never play before, but after my transplant, I began to love music. I felt it in my heart. My heart had to play it. I told my mom I wanted to take guitar lessons, the same instrument Paul [the donor] had played. His song is in me. I feel it a lot at night and it's like Paul is serenading me.

<u>The recipient's father reported:</u>
My daughter, she was what you say . . . a hell raiser. Until she got sick, they say from a dentist they think, she was the wild one. Then, she became quite quiet. I think it was her illness, but she said she felt more energy, not less. She said she wanted to play an instrument and she wanted to sing. When she wrote her first song, she sang about her new heart as her lover's heart. She said her lover had come to save her life.

Case 2

The donor was a 16-month-old boy who drowned in a bathtub. The recipient was a 7-month-old boy diagnosed with tetralogy of Fallot, a syndrome involving a hole in the ventricular septum, displacement of the aorta, pulmonary stenosis, and thickening of the right ventricle.

<u>The donor's mother, a physician, reported:</u>

The first thing is that I could more than hear Jerry's [her son, the donor] heart. I could feel it in me. When Carter [the recipient] first saw me, he ran to me and pushed his nose against me and rubbed and rubbed it. It was just exactly what we did with Jerry. Jerry and Carter's heart is 5 years old now, but Carter's eyes were Jerry's eyes. When he hugged me, I could feel my son. I mean I could feel him, not just symbolically. He was there. I felt his energy. I'm a doctor. I'm trained to be a keen observer and have always been a natural born skeptic. But this was real. I know people will say that I need to believe my son's spirit is alive, and perhaps I do. But I felt it. My husband and my father felt it. And I swear to you, and you can ask my mother, Carter said the same baby-talk words that Jerry said. Carter is 6, but he was talking Jerry's baby talk and playing with my nose just like Jerry did. We stayed with the [recipient family] that night. In the middle of the night, Carter came in and asked to sleep with my husband and me. He cuddled up between us exactly like Jerry did, and we began to cry. Carter told us not to cry because Jerry said everything was okay. My husband and I, our parents, and those who really knew Jerry have no doubt. Our son's heart contains much of our son and beats in Carter's chest. On some level, our son is still alive.

<u>The recipient's mother reported:</u>

I saw Carter go to her [the donor's mother]. He never does that. He is very, very shy, but he went to her just like he used to run to me when he was a baby. When he whispered "It's okay, Mama," I broke down. He called her Mother, or maybe it was Jerry's heart talking. And one more thing that got to us. We found out talking to Jerry's mom that Jerry had mild cerebral palsy, mostly on his left side. Carter has stiffness and some shaking on that same side. He never did as a baby and it only showed up after the transplant. The doctors say it's probably something to do with his medical condition, but I really think there's more to it.

One more thing I'd like you to know about. When we went to church together, Carter had never met Jerry's father. We came late and Jerry's dad was sitting with a group of people in the middle of the congregation. Carter let go of my hand and ran right to that man. He climbed on his lap, hugged him, and said "Daddy." We were flabbergasted. How could he have known him? Why did he call him Dad? He never did things like that. He would never let go of my hand in church and never run to a stranger. When I asked him why he did it, he said he didn't. He said Jerry did and he went with him.

Case 3
The donor was a 24-year-old female automobile accident victim. The recipient was a 25-year-old male graduate student suffering from cystic fibrosis who received a heart–lung transplant.
The donor's sister reported:
My sister was a very sensual person. Her one love was painting. She was on her way to her first solo showing at a tiny art shop when a drunk plowed into her. It's a lesbian art store that supports gay artists. My sister was not really very "out" about it, but she was gay. She said her landscape paintings were really representations of the mother or woman figure. She would look at a naked woman model and paint a landscape from that! Can you imagine? She was gifted.
The recipient reported:
I never told anyone at first, but I thought having a woman's heart would make me gay. Since my surgery, I've been hornier than ever, and women just seem to look even more erotic and sensual, so I thought I might have gotten internal transsexual surgery. My doctor told me it was just my new energy and lease on life that made me feel that way, but I'm different. I know I'm different. I make love like I know exactly how the woman's body feels and responds — almost as if it is my body. I have the same body, but I still think I've got a woman's way of thinking about sex now.
The recipient's girlfriend reported:
He's a much better lover now. Of course, he was weaker before, but it's not that. He's like, I mean he just knows my body as well as I do. He wants to cuddle, hold, and take a lot of time. Before he was a good lover, but not like this. It's just different. He wants to hug all the time and go shopping. My God, he never wanted to shop. And you know what? He carries a purse now. His purse! He

slings it over his shoulder and calls it his bag, but it's a purse. He hates it when I say that, but going to the mall with him is like going with one of the girls. And one more thing: he loves to go to museums. He would never, absolutely never, do that. Now he would go every week. Sometimes he stands for minutes and looks at a painting without talking. He loves landscapes and just stares. Sometimes I just leave him there and come back later.

Case 4
The donor was a 17-year-old African-American male student, a victim of a drive-by shooting. The recipient was a 47-year-old Caucasian male foundry worker diagnosed with aortic stenosis.
<u>The donor's mother reported:</u>
Our son was walking to violin class when he was hit. Nobody knows where the bullet came from, but it just hit him and he fell. He died right there on the street hugging his violin case. He loved music and his teachers said he had a real thing for it. He would listen to music and play along with it. I think he would have been at Carnegie Hall someday, but the other kids always made fun of the music he liked.
<u>The recipient reported:</u>
I'm real sad and all for the guy who died and gave me his heart, but I really have trouble with the fact that he was black. I'm not a racist, mind you, not at all. Most of [my] friends at the plant are black guys. But the idea that there is a black heart in a white body seems really . . . well, I don't know. I told my wife that I thought my penis might grow to a black man's size. They say black men have larger penises, but I don't know for sure. After we have sex, I sometimes feel guilty because a black man made love to my wife, but I don't really think that seriously. I can tell you one thing, though. I used to hate classical music, but now I love it. So I know it's not my new heart, because a black guy from the 'hood wouldn't be into that. {See Note 1} Now it calms my heart. I play it all the time. I more than like it. I play it all the time. I didn't tell any of the guys on the line that I have a black heart, but I think about it a lot.
<u>The recipient's wife reported:</u>
He was more than concerned about the idea when he heard it was a black man's heart. He actually asked me if he could ask the doctor for a white heart when one came up. He's no Archie Bunker, but he's close to it. And he would kill me if he knew I told

you this, but for the first time, he's invited his black friends over from work. It's like he doesn't see their color anymore, even though he still talks about it sometimes. He seems more comfortable and at ease with these black guys, but he's not aware of it. And one more thing I should say. He's driving me nuts with the classical music. He doesn't know the name of one song and never, never listened to it before. Now, he sits for hours and listens to it. He even whistles classical music songs that he could never know. How does he know them? You'd think he'd like rap music or something because of his black heart. {see Note 1}

Note 1: This case illustrates why many cases go unreported. When the Caucasian foundry worker received the heart of a 17-year-old African-American student, he presumed that the donor would have preferred rap music. Hence, he dismissed the idea that his new radical change in preference for classical music could have come from the heart of the donor. However, unbeknownst to the recipient, the donor actually loved classical music, and died "hugging his violin case" on the way to his violin class.

Case 5
The donor was a 19-year-old woman killed in an automobile accident. The recipient was a 29-year-old woman diagnosed with cardiomyopathy secondary to endocarditis.
The donor's mother reported:
My Sara was the most loving girl. She owned and operated her own health food restaurant and scolded me constantly about not being a vegetarian. She was a great kid — wild, but great. She was into the free love thing and had a different man in her life every few months. She was "man crazy" when she was a little girl, and it never stopped. She was able to write some notes to me when she was dying. She was so out of it, but she kept saying how she could feel the impact of the car hitting them. She said she could feel it going through her body.
The recipient reported:
You can tell people about this if you want to, but it will make you sound crazy. When I got my new heart, two things happened to me. First, almost every night and still sometimes now, I actually feel the accident my donor had. I can feel the impact in my chest. It slams into me, but my doctor said everything looks fine. Also, I

hate meat now. I can't stand it. I was McDonald's biggest moneymaker, and now meat makes me throw up. Actually, when I even smell it, my heart starts to race. But that's not a big deal. My doctor said that's just due to my medicines. I couldn't tell him, but what really bothers me is that I'm engaged to be married now. He's a great guy and we love each other. The sex is terrific. The problem is, I'm gay. At least, I thought I was. After my transplant, I'm not. . . I don't think anyway.

. . I'm sort of semi- or confused-gay. Women still seem attractive to me, but my boyfriend turns me on. Women don't. I have absolutely no desire to be with a woman. I think I got a gender transplant.

<u>The recipient's brother reported:</u>
Susie's straight now. I mean it seriously. She was gay and now her new heart made her straight. She threw out all her books and stuff about gay politics and never talks about it any more. She was really militant about it before. She holds hands and cuddles with Steven just like my girlfriend does with me. She talks girl talk with my girl friend, where before she would be lecturing about the evils of sexist men. And my sister, the Queen of the Big Mac, hates meat. She won't even have it in the house.

Case 6
The donor was a 14-year-old girl injured in a gymnastics accident. The recipient was a 47-year-old man diagnosed with benign myxoma and cardiomyopathy.

<u>The donor's mother reported:</u>
Look at her [showing picture]. My daughter was the picture of health. There wasn't an ounce of fat on her. She was a gymnast and her coach could lift her above his head with one hand. She was so excited about life that she would just hop and jump all the time like a kitten. She had some trouble with food, though. She would skip meals and, for a while, she was purging. I think they would call her a little anorexic. We took her to therapy about it, but she just wasn't much into food. And she had this silly little giggle when she got embarrassed. It sounded like a little bird.

<u>The recipient reported:</u>
I feel new again. I feel like a teenager. I actually feel giddy. I know it's just the energy of the new heart, but I really feel younger in every way, not just physically. I see the world that way. I'm really young at heart. I have this annoying tendency to

giggle that drives my wife nuts. And there's something about food. I don't know what it is. I get hungry, but after I eat, I often feel nauseated and that it would help if I could throw up.

The recipient's brother reported:

Gus is a teenager. No doubt about that. He's a kid — or at least he thinks he's a kid. Even when we're bowling, he yells and jumps around like a fool. He's got this weird laugh now. It's a girl's laugh and we tell him that. He doesn't care. His appetite never did bounce back after the surgery. He's pretty much nauseated almost all the time. After Thanksgiving dinner — and he loved it — he went upstairs and vomited. We took him to the emergency room, but it wasn't anything to do with his new heart. They said it was probably a reaction to something in the meal. None of the rest of the family got sick, though. He's going to have to watch it. His doctor is concerned about his weight.

Case 7

The donor was a 3-year-old girl who drowned in the family pool. The recipient was a 9-year-old boy diagnosed with myocarditis and septal defect. The recipient's mother, who knew who the donor had been, reported:

He [the recipient] doesn't know who his donor was or how she died. We do. She drowned at her mother's boyfriend's house. Her mother and her boyfriend left her with a teenage babysitter who was on the phone when it happened. I never met her father, but the mother said they had a very ugly divorce and that the father never saw his daughter. She said she worked a lot of hours and wished she had spent more time with her. I think she feels pretty guilty about it all . . . you know, the both of them sort of not appreciating their daughter until it was too late.

The recipient, who claimed not to know who the donor was, reported:

I talk to her sometimes. I can feel her in there. She seems very sad. She is very afraid. I tell her it's okay, but she is very afraid. She says she wishes that parents wouldn't throw away their children. I don't know why she would say that.

The recipient's mother further reported about the recipient:

Well, the one thing I notice most is that Jimmy is now deathly afraid of the water. He loved it before. We live on a lake and he won't go out in the backyard. He keeps closing and locking the back door. He says he's afraid of the water and doesn't know why. He won't talk about it.

Case 8

The donor was a 19-year-old woman who suffered a broken neck in dance class. The recipient was a 19-year-old woman diagnosed with cardiomyopathy.

The donor's mother reported:

We've met Angela [the recipient], and she is the image of our daughter. They could almost be twins. They're both bright girls; I mean, my daughter was bright, too. She wanted to be an actress, but we thought she had too much academic potential for that. Her father is a doctor and really wanted her to follow in his footsteps.

The donor's father reported:

Stacy [his daughter, the donor] was extremely bright. It's so tragic. She would have made an outstanding physician, but she wanted to dance and sing. That's how she died. She fell in dance class. We always argued good-naturedly about how disappointed I would be if she went to Hollywood instead of Harvard. I hope she knew I just wanted her to be happy.

The recipient reported:

I think of her as my sister. I think we must have been sisters in a former life. I only know my donor was a girl my age, but it's more than that. I talk to her at night or when I'm sad. I feel her answering me. I can feel it in my chest. I put my left hand there and press it with my right. It's like I can connect with her. Sometimes she seems sad. I think she wanted to be a nurse or something, but other times it's like she wanted to be on Broadway. I think she wanted to be on Broadway more. I want to be a nurse, but I could be a doctor too. I hope she will be happy, because she will always be my angel, my sister in my chest. I carry my angel with me everywhere.

The recipient's mother reported:

We can sometimes hear her talking to her heart. It's like a "Dear Diary" thing. She puts her hand on her chest and talks to who she thinks her donor is. Once, we found her holding a stethoscope to her chest to try to hear her new heart. I think she still does that sometimes. And the only other thing is that she really wants to go to medical school now.

She never wanted to do that before, but that's because I don't think she thought she would live. She's already changed her college classes.

Case 9

The donor was a 3-year-old boy who fell from an apartment window. The recipient was a 5-year-old boy with septal defect and cardiomyopathy.

<u>The donor's mother reported:</u>

It was uncanny. When I met the family and Daryl [the recipient] at the transplant meeting, I broke into tears. We went up to the giving tree where you hand a token symbolizing your donor. I was already crying when my husband told me to look at the table we were passing. It was the [recipient's family] with Daryl sitting there. I knew it right away. Daryl smiled at me exactly like Timmy [her son, the donor] did. After we talked for hours with Daryl's parents, we were comforted. It somehow just didn't seem strange at all after a while. When we heard that Daryl had made up the name Timmy and got his age right, we began to cry. But they were tears of relief because we knew that Timmy's spirit was alive.

<u>The recipient reported:</u>

I gave the boy a name. He's younger than me and I call him Timmy. He's just a little kid. He's a little brother like about half my age. He got hurt bad when he fell down. He likes Power Rangers a lot, I think, just like I used to. I don't like them anymore, though. I like Tim Allen on "Tool Time," so I called him Tim. I wonder where my old heart went, too. I sort of miss it. It was broken, but it took care of me for a while.

<u>The recipient's father reported:</u>

Daryl never knew the name of his donor or his age. We didn't know, either, until recently. We just learned that the boy who died had fallen from a window. We didn't even know his age until now. Daryl had it about right. Probably just a lucky guess or something, but he got it right. What is spooky, though, is that he not only got the age right and some idea of how he died, he got the name right. The boy's name was Thomas, but for some reason his immediate family called him "Tim."

<u>The recipient's mother added:</u>

Are you going to tell him the real Twilight Zone thing? Timmy fell trying to reach a Power Ranger toy that had fallen on the ledge of the window. Daryl won't even touch his Power Rangers any more.

APPENDIX 1   Heart transplants transferring donor character to recipient

Case 10

The donor was a 34-year-old police officer shot attempting to arrest a drug dealer. The recipient was a 56-year-old college professor diagnosed with atherosclerosis and ischemic heart disease. <u>The donor's wife reported:</u>

When I met Ben and Casey [the recipient and his wife], I almost collapsed. First, it was a remarkable feeling seeing the man with my husband's heart in his chest. I think I could almost see Carl [her husband, the donor] in Ben's eyes. When I asked how Ben felt, I think I was really trying to ask Carl how he was. I wouldn't say that to them, but I wish I could have touched Ben's chest and talked to my husband's heart. What really bothers me, though, is when Casey said offhandedly that the only real side effect of Ben's surgery was flashes of light in his face. That's exactly how Carl died. The bastard shot him right in the face. The last thing he must have seen is a terrible flash.

They never caught the guy, but they think they know who it is. I've seen the drawing of his face. The guy has long hair, deep eyes, a beard, and this real calm look. He looks sort of like some of the pictures of Jesus*.

<u>The recipient reported:</u>

If you promise you won't tell anyone my name, I'll tell you what I've not told any of my doctors. Only my wife knows. I only knew that my donor was a 34-year-old, very healthy guy. A few weeks after I got my heart, I began to have dreams. I would see a flash of light right in my face and my face gets real, real hot. It actually burns. Just before that time, I would get a glimpse of Jesus*. I've had these dreams and now daydreams ever since: Jesus and then a flash. That's the only thing I can say is something different, other than feeling really good for the first time in my life.

<u>The recipient's wife reported:</u>

I'm very, very glad you asked him about his transplant. He is more bothered than he'll tell you about these flashes. He says he sees Jesus* and then a blind flash. He told the doctors about the flashes, but not Jesus*. They said it's probably a side effect of the medications, but, God, we wish they would stop.

-=-=-=-=-=-=-=(end of paper)=-=-=-=-=-=-=-

* "Jesus" is unfortunately used here infelicitously, to refer to a criminal seen.

Miscellaneous: An interesting Biblical mention of transfer is found in Luke 8-31, where Christ was about to exorcise "devils" from a man, and they asked Him not to cast them into the abyss, so He allowed them to enter a herd of swine instead! The man was then reported as cured.

Remote Viewing (Chapter 5) reports Christ as taking over the body prepared for him by his pupil Jesus, for his 3-year ministry.

In considering explanations for the above ten cases, bear in mind Occam's Razor Principle:
The simplest explanation is probably the correct one.

A recent paper reports that thinking about our ancestors enhances intellectual performance, which is difficult to explain unless one subconsciously links to one's own past lives, which increase one's intellectual ability! See Fischer, P., Sauer, A., Vogrincic, C., & Weisweiler, S. (in press, 2010 or 2011). "The ancestor effect": *European Journal of Social Psychology* (Fast Track Report).

For information:

The 10 cases paper by Pearsall, Schwartz & Russek was also reprinted in *Antimatters*, an open access journal addressing issues in science and the humanities from non-materialistic perspectives   ISSN: 0973-8606   Volume 1, Issue 1, August 2007

From the journal's Focus and Scope: *AntiMatters* is for those who are uncomfortable with (or unconvinced of) materialism, or who favor a non-materialistic world view. Such persons are oftentimes unaware of how much of what is claimed to have been scientifically established is actually spurious. For their benefit, the journal aims to critically examine the alleged scientific evidence for materialism. While authors are expected to respect and take account of all relevant empirical data, they should bear in mind that empirical data are inevitably theory laden and paradigm-dependent, and that theories and paradigms, being to a considerable extent social constructions, are always relative.
Science operates within an interpretative framework that formulates questions and interprets answers. This framework is itself not testable. *AntiMatters* wants to serve as a platform for the study of alternative interpretative frameworks.
AntiMatters is published quarterly by the Sri Aurobindo International Centre of Education in Puducherry (formerly Pondicherry), India. • MANAGING EDITOR: Ulrich J Mohrhoff, Sri Aurobindo International Centre of Education, India.

Appendix 2

Some military personnel of the CIA Remote Viewing programme

Fig. A2.1 (**see colour picture** between p. 63 & 64):
Official US Navy photograph of Admiral Stansfield Turner,
Director of CIA from 1977 – 1981,
Founder of the CIA Remote Viewing Group.

Some other senior staff are shown in the videos [1], but video copyright prevents them from being shown here:

Major General Albert Stubblebine (who was responsible for 30,000 soldiers of the US Army).

Major General Edward Thompson (Assistant Chief of Staff, US Army Intelligence, 1977-1981).

Colonel Alexander (US Army).

References
1. Three DVDs of remote viewing: (1) "The Real X-Files", Channel 4 TV (UK), Presenter: J. Schnabel, Director: B. Eagles, Producer: A. Graham, "Wall-to-Wall TV Productions for Channel 4" (interviews with CIA) (1995); (2) "Strange but True", LWTV programme for ITV (UK), Presenter: M. Aspel, Producer & Director: N. Miller (interviews with Dr D. Morehouse – see also refs. 14 & 16 below) (1997); (3) Ingo Swann, Remote Viewing Conference 2006 (Intl Remote Viewing Assoc): DVD available from www.irva.org

See also reference list of Chapter 1 for many books on Remote Viewing.

APPENDIX 3

General Relativity and Gravitation

"A surprising fact was noted, that the first variety of ozone .... always rose in the air. It cannot be lighter, because the number of Arnoo in both varieties of ozone are the same, i.e. 435."
Besant & Leadbeater [1]

The author (MGH) is investigating this (see page 301) and any more results will be reported on the website: www.4-D.org.uk

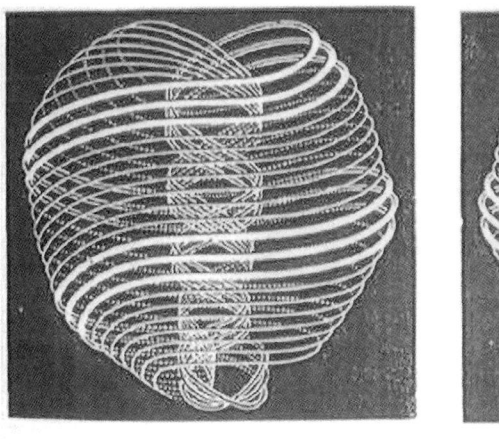 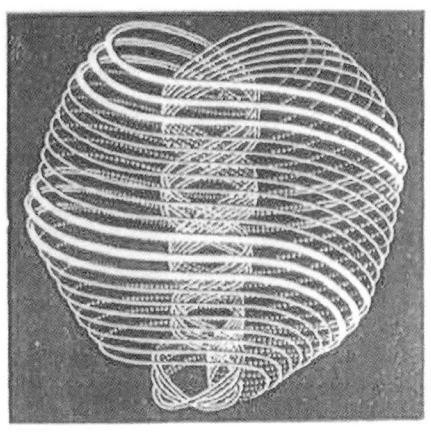

POSITIVE   NEGATIVE

Fig. 1. Arnoo, showing chirality (described as "positive" and "negative").

The arnoo is the ultimate smallest elementary particle observed (see Chapter 3) [1]. In a proton, there are 3 arnoo in a quark and 3 quarks in a proton. (The plural of the Sanskrit word arnoo is also arnoo.)

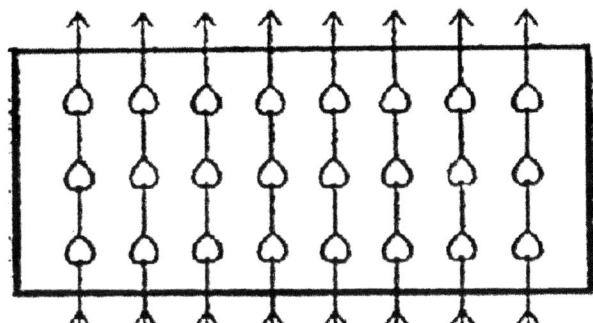

Fig. 2 Arnoo lined up by electric current.

Introduction

The above quotation, plus Besant & Leadbeater's [1] observation of arnoo lining-up (see Fig. 2 above) are the basis for an experimental prediction given below in the Gravity section.

The high accuracy of the equality of inertial mass and gravitational mass for all materials, is a basis either for General Relativity (via Mach's Principle) or for supposing that all forms of matter (including associated binding energies) are composed of identical ultimate particles. The latter view is suggested here, with the universal ultimate particle (mass quantum) as the arnoo (not the arnoo of 3-D but the ultimate Arnoo in 10-D). Besant & Leadbeater report that each dimension (3 to 10) have their ultimate smallest particle of the same appearance as the other dimensions, but with different orders of coiled-coiling (explained in Table 1 in Chapter 3).

A test for Einstein's General Theory of Relativity is that the deflection of a starlight beam passing close to the sun is about twice that calculated from just assuming the photon's mass and velocity (c) passing near the sun. But if the "hidden matter" of the Universe is that in higher dimensions [4], which still are subject to gravity, the sun may be more massive than expected (having extra "hidden" mass) and may thus deflect the light beam more than predicted by Newton's gravity equation. It was found experimentally that the deflection of a light beam is about double the Newton prediction. This is explained in the Gravity section (later below). About 95% of mass in the Universe is hidden mass possibly suggesting an equipartition of mass to each useable dimension, 2-D to 10-D, which is 9 dimensions (discounting 1-D space).

General Relativity depends partly on the Principle of Equivalence, which is explained in many textbooks. It asserts that there is no experimental way of distinguishing any difference between acceleration due to gravity and acceleration due to being inside an accelerating vehicle. But in a vehicle it is surely a matter of common observation that fast acceleration causes one to be compressed into one's seat, but free-falling in a helicopter from which the blades have broken off would be quite different and there would be no sensation of being pressed or compressed by any part of the cabin (one could float about freely and this effect is used to train astronauts by diving in an aircraft).

A simple quantitative experiment would be to place a cube made of soft spongy plastic in the cabin of a rocket-ship in free space (away from any gravity field) and it would be pressed flat against the floor if the rocket motors were started. But if the rocket-ship were falling towards a planet (e.g. Jupiter) without any motors firing, the sponge cube would be free to drift about in the cabin and would not be pressed flat against the floor or ceiling by the gravitational acceleration. So gravitational acceleration is fundamentally different from (and distinguishable from) being in an accelerating vehicle: gravity acts on every atom throughout the body independently, but vehicular acceleration acts only on those molecules in direct contact with the vehicle and the force is then transferred purely mechanically to other parts of the body and the result is compression of the body.

If the above paragraph is not convincing, please see Appendix 3A for an alternative model.

So, with these simple and obvious examples, it is very surprising that so many textbooks assert that there is no observable difference between these two types of acceleration! The difference has been studiously ignored and the Principle of Equivalence of acceleration is clearly wrong.

The equivalence of accelerations is required for General Relativity because it can easily be shown that a clock runs slower if accelerated and the proof is trivially easy for an accelerating vehicle, but is not obvious for acceleration by a gravitational field.

If a clock runs slower in a gravity field, this leads to the notion of "space-time" distortions required for general relativity. But as with Special Relativity, the results derived from the theory of General Relativity can be obtained without assuming the Principle of Equivalence and General Relativity is not required. (See p. 296.)

Gravity

The arnoo are described as being cardioid in shape (Fig. 1 above), with an inrush of a force from 4-D space into the top of the "positive" type of arnoo (source), and an outrush of force from 3-D space back to 4-D space, from the "negative" type of Arnoo, which acts like a hole in space (sink). If smaller arnoo from the higher dimensions (call these gravitons) are pulled into the top and ejected from the point, this would propel the arnoo towards the source of the gravitons. The mass of the Earth will have large numbers of arnoo in it and in its vicinity, so that there would be a tendency for physical (3-D) level Arnoo in the Earth's atmosphere to move in the direction of the centre of the Earth. All arnoo are described as spinning and moving chaotically and so an arnoo will sometimes be pointing towards and sometimes away from the Earth. When its top (Fig. 1) is pointing towards the Earth, the arnoo will move towards the Earth; when its top is pointing away from the Earth it will move away, but because more gravitons are nearer the Earth than further away from it, there will be a (small) net movement of arnoo towards the Earth (more graviton throughput per second), which explains gravity.

As mentioned earlier, one observer [2] reports that the arnoo is actually the top of a carrot-like structure and the point of the "carrot" is attached to the centre of the Earth. Since B&L do not mention this "carrot", and since they always very precisely confine their observation to the particular level of interest (which is the 3-D world, if looking at physical 3-D arnoo), they do not digress to describe any associated 4-D structure, but they do mention [1] that the physical 3-D arnoo is a point of entry (positive type) or exit (negative type) (see Fig. 1) of an energy from or to the next higher dimension (4-D). This suggests that the arnoo is a vortex. Using the Hermetic adage, "As above, so below", it may be comparable with a waterspout or tornado. It could draw in the much smaller arnoo of higher dimensions. If higher-dimensional arnoo (gravitons) are drawn into the top of a 3-D arnoo while it is facing downwards towards the Earth, this could propel that arnoo towards the Earth; these gravitons could then be passed down the "carrot" structure which exists in 4-D, back to the centre of the Earth, and thus re-cycled ; the return process, being in 4-D, is not in any direction in the 3-D world. This may give an insight into gravity. A relevant paper on vortexes is available [3].

Reducing Gravity

The above discussion suggests that a way to reduce the gravity acting on a material could be to pass a vertically oriented electric current through it, or place it in a vertical electric field, to line-up its arnoo as in Fig. 2 above, which would orient the Arnoo's point toward the Earth. Experiments to study this are in progress. See also Appendix 3B below.

Deflection of a starlight beam by gravity

This is the classic experiment, which is cited as proof of Einstein's General Relativity theory because it gives the correct deflection.

Gravity in 3 dimensions, like the electrostatic force, follows an inverse square law with distance. Unlike electrostatic and electromagnetic forces, gravity propagates through the higher dimensions. If gravity propagates outwards from a sphere like the Earth, the amount of propagation passing through unit area of the sphere's surface falls off with the square of the sphere's radius, because the area of a sphere is proportional to its radius squared. This explains why gravity falls off with the square of the radial distance. See Chapter 1, page 22.

But for the 4 dimensional analogue of a sphere, the fall-off is proportional to the cube of the radial distance. For higher dimensional sphere analogues, the fall off is proportional to $(\text{dimension})^{(n-1)}$ where n is the dimension. Thus, on the above model the total gravitational effect of the sun will be more than that calculated from assuming it to be only a 3-D object, and will include extra gravitational terms up to the 10th dimension. Thus a light beam will be deflected more than calculated from assuming the sun is just a 3-D object. This is explained further, below.

To keep the figures simple, define an arbitrary unit of distance such that the radius of the sun is 2 such distance units. Then using Newton's Law of Gravity: (NOTE: $\propto$ means "proportional to")

APPENDIX 3   General Relativity and Gravitation

At the sun's periphery : Force $\propto 1/d^2 \propto 1/2^2 \propto 1/4$ (for ordinary 3-D matter).

And the gravity force from the "hidden matter" (4-D matter) in the sun, on the ordinary 3-D matter in the photon beam:

Force $\propto 1/d^3 \propto 1/2^3 \propto 1/8$ at the sun's periphery. [This assumes equipartition of mass into each dimension (see above), which may not be correct.]

An additional gravity force on the 4-D hidden matter in the photon beam from the hidden 4-D matter in the sun is also:

Force $\propto 1/d^3 \propto 1/2^3 \propto 1/8$ at the sun's periphery. 5-D and higher terms are small.

But at the Earth's distance from the sun, which is about 200 times the sun's radius:

Force $\propto 1/(2 \times 200)^2 \propto 1/160000$.

And the gravity force from the hidden matter (4-D matter) in the sun :

Force $\propto 1/(2 \times 200)^3 \propto 1/64000000$ etc.

The gravity force from hidden matter in the 5th and higher dimensions adds further but very small attractive forces, as will "cross-dimension" gravity forces (between 4-D mass of sun and 5-D mass of the moving object, etc). Neglecting these, the gravity force on a starlight photon tangential to the periphery of the sun $\propto$ 1/4 + 1/8 + 1/8 $\propto$ 1/2, which is double 1/4, whereas the gravity force on it if it is tangential to the Earth's orbit $\propto$ 1/160000 + 1/64000000, in which the second term is negligible (only 0.24% of the first term).

So the hidden matter's gravitational force causing deflection of a starlight beam at the sun's periphery is as great as the ordinary (3-D) matter's gravitational force, whereas at Earth orbit distance from the sun, the sun's hidden matter gravitational force is only 0.24% greater than its ordinary (3-D) matter's gravitational force. So the gravitational force of the sun falls off rapidly with distance, reminiscent of the commonly seen net or mesh diagrams of curved "space-time" at a star. Long ago Tesla made an interesting objection to curved "space time", saying that something with no structure (a noumenon) cannot be curved.

This means that the mass of the sun (as is normally calculated from the Earth's orbit parameters) will only be about 0.24% in error from its normal 3-D mass, which means only a small error due to ignoring its hidden matter content.

Also, the deflection of a starlight beam tangential to the sun's periphery will be double that calculated classically using Newton's Law, i.e. the same prediction as that from General Relativity, but without assuming General Relativity.

The value for the sun's mass is calculated from the parameters of the Earth's orbit, which is so far away from it that the 3rd & higher order terms (for its hidden matter), in the gravitational force equations, will be quite negligible. But if a beam of starlight photons grazes the solar periphery, this is near enough to incur these higher order terms and so the deflection of its trajectory is greater than conventionally expected, by a factor of 2. But as mentioned above the simple calculation above assumes equipartition of mass into each dimension, which may not be correct.

Notes: The deflection for a photon passing tangentially through the sun's periphery is the same for one photon or for a group of any number of photons, as it is independent of mass [just as the Moon and Earth, having different masses, remain in the same solar orbit and do not diverge, and just as a feather and a cannonball hit the ground together if dropped from a tower (down an evacuated tube)]. The sun's mass (M_s) is calculated from the Earth's orbit as follows:

Using Newton's Laws, $F = GM_s M_e / d^2$ and $F = M_e a$, so that M_e cancels out when equating these, i.e. the Earth's mass (M_e) is not required to calculate the mass of the sun -- so it does not matter if it is not accurate (for this purpose).

Gravitons, if they are arnoo, would need to be identified as the arnoo of the 10th dimension. If they were the arnoo of (say) the 8th dimension, this would mean that the 9th and 10th dimension arnoo would drift away from the location of the lower dimensions!

APPENDIX 3 General Relativity and Gravitation 299

B&L report that the 4th dimensional (so-called "astral") level is located concentric with the physical world; if there were no gravitational force on 4-D matter, the entire 4-D region would drift away into outer space!

A problem with this is, what prevents the Arnoo of the 10th dimension from drifting away? The answer may be that motion is impossible for 10-D particles if the model of motion given earlier [4] is correct - that motion requires a jump via the next higher dimension (the "air bubbles can't move in ice" model, explained in reference 4), but there is no dimension beyond the tenth.

APPENDIX 3A: Further comment on the Equivalence Principle.

Imagine two cases :

Case g : Your room and contents as it is, and,

Case a : The same room but located in deep space (away from any planet or other source of gravitational field) within a large lift which is accelerating upwards at an acceleration with the same value as g on Earth, i.e. 9.8 m or 32 ft per second per second.

In this room, imagine a large cube which has a small cube with a side-length of (say) a million atoms, stuck to its side at the top. Now imagine the adhesive comes loose and the small cube drops off. The small cube of Case g will have a different size to the cube in Case a:

The small cube in Case a, while falling, is in no gravitational field, unlike the cube in Case g. While falling, the cube in Case g has a slightly larger gravitational force acting on its lower atoms than on its upper atoms, due to the inverse square law of the Earth's gravitational field (as the lower atoms are nearer to the Earth). Thus the cube of Case g will be slightly stretched (in the vertical direction) compared to the cube of Case a. So its atom spacing in the vertical direction will be slightly greater than that of the Case a cube.

This experiment would distinguish between the effects of the two types of gravitational fields, which contravenes the Principle of Equivalence which is the main principle of General Relativity. The

Principle of Equivalence requires that there is no way to distinguish Case g from Case a.

APPENDIX 3B: Anti-gravity devices.

In 1942 German scientists wrote to Hitler that they are developing a flight device that would be decisive in the war. They worked at a secret location in the Sudeten Mountains on the Polish/Czech border, excavated underground several km into a mountain, called the Wenceslas Mine. A high voltage bell-like structure about 3 metres diameter, was later developed above ground inside a restraining circular concrete structure about 60 metres diameter, which is still there. It was named the Kriegen Scheiben ("war disc").

(The above sizes are a guess from the film, shown on More4 TV (UK) about 2006, as "UFOs, The Secret Evidence", a 2 hour documentary made in about 1993 by Oxford Film & Television Productions for Channel 4 TV (UK) and presented by Nick Cook, a competent defence journalist). The vast underground tunnel system was blown up and sealed off in 1945 by the retreating Nazis, but later some items were removed by the Americans. Many tunnels were never re-opened and their secret contents remain sealed off.

Also in the documentary is a present-day small model made of light balsa wood with an extra-high voltage (EHT) wire to it, which levitates off the ground, in the workshop of Tim Ventura, an inventor, in Seattle. It is a triangle of about 50 cm side held by multiple balsa wood spars radiating from its centre, each spar covered in Al foil and a thin wire on insulating spacers is about 3 to 4 cm above it which is connected to a terminal atop a roughly 150 cm high plastic tube about 10 cm diameter, which is probably a typical diode ladder giving EHT. They suggested the lift may be due to an ion wind effect, but the discussion here above (page 296) is the first (speculative) explanation involving Arnoo being lined up electrically, to orient them against gravity.

Caution: " Don't try this at home folks ", unless you use a Wimshurst Machine, or a Van de Graaff Generator with dome less than 25 cm, or similar, which cannot (check this!) generate lethal currents at EHT voltages. Do not use capacitors with it in case they charge up to deliver lethal currents. Also, capacitors may explode if their working voltage is exceeded.

References for Appendix 3:

1. A. Besant & C.W. Leadbeater, "Occult (Hidden) Chemistry", available as a download on www.4-D.org.uk

2. Babbit, "Principles of Light & Colour" (details unknown, but published in UK over a century ago)

3. K. Wakelam, "Vortexes" (request copy from: www.4-D.org.uk/Books)

4. M.G. Hocking, "Linking String Theory and Membrane Theory to Quantum Mechanics and Special Relativity Equations, avoiding any Special Relativity assumptions", Journal of Scientific Exploration 21 (1), 13-26 (2007). This is downloadable (full text, free) from Page 1 of this website: www.4-D.org.uk
It is also printed in this book as Appendix 6, page 332.

-=-=-=-=-=-=-=-=-=-=-=-

Experimental Results on Ozone {by the author (MGH)}:

Besant & Leadbeater observed ozone at 3000 ft altitude, which probably originated from the solar UV effect on oxygen. They called this Type 1 ozone which rose in the air, unlike the normal (Type 2) ozone. The author (MGH) has tested ozone from 9 volt DC electrolysis of 3-M sulphuric acid in a glass U-tube, and ozone from a Tesla coil high voltage electrical discharge, but neither was lighter than air. It is presumed that these are "Type 2" ozone. Ozone was verified by turning starch iodide paper brown immediately. Ozone from both these methods was warm and so it was led via a water cooler at room temperature into the open end of the inverted plastic beaker for the following weight tests:

Polypropylene has no double bonds which ozone may attack. A 200 mL polypropylene beaker (weight 3 grams), was suspended inverted from a thin cotton thread hanging from a precision pyrex spring. Spring specification: 10 grams maximum weight and sensitivity 0.02 mm per milligram. The lower end of the spring was focussed on with a telescope held in a clamp. No visible weight change occurred (detection limit 1 mg).
Type 1 ozone observed at 3000 ft is presumably produced by ultraviolet light, but ozone-producing uv light bulbs cost over £100 and this type of ozone has not been tested yet. Any results will be reported later on the author's website: www.4-D.org.uk
Test for gravitational repulsion: A Tesla induction coil was connected to a 15 cm long Cu wire sealed into a polyethylene tetraphthalate (PET) bottle. The bottle was placed inside an earthed (grounded) steel can.   Ozone was produced when the coil was switched on. The bottle (with cap and glass tube) was weighed before, and immediately after using the induction coil, using an external chemical balance accurate to 0.1 mg. As expected, there was no weight change, because the total number of <u>atoms</u> remained constant, because the bottle was <u>sealed</u>. This is quite different from the other above experiments, which used an inverted plastic beaker with its end <u>open</u> at ambient atmospheric pressure.

APPENDIX 4

Evidence for mass increase with velocity (i.e. with kinetic energy):

Besant & Leadbeater [1] report that for every <u>700</u> bubbles on each of the "highest" 7 whorls of the physical world (3D) Arnoo (see Fig. 1 below), there are <u>704</u> on each of the "lowest" 3 whorls. This is seen in Fig. 1 as a <u>thickening</u> of the 3 of the 10 whorls (or strings).   See detail in Fig. 2 below.

If the Arnoo were at (absolute) rest, there would be no such thickening, if the following assumption is correct:

<u>Only</u> if there is <u>absolute motion</u> (see Appendix 6) and if rest mass is <u>observed</u> as a certain number of bubbles per whorl, then at speed there should actually <u>be</u> an <u>absolute</u> number of <u>more bubbles</u> per whorl.   This is supported by the above observation of an increase from 700 to 704. (But see note in Fig.1 caption below)

With this <u>speculative</u> assumption, a simple but speculative calculation can find the speed producing this amount of enhancement:

$m_o^2 / m^2 = (c^2 - v^2) / c^2$ is the formula for mass dilation (derived in Appendix 6).

Speculating that the ratio of rest mass to moving mass is 700x10 / (700x7 + 704x3), (see above), then:

$7000^2/7012^2 = (186000^2 - v^2) / 186000^2$, where c = 186000 miles per second, using the above mass dilation formula.

Hence v = 10860 miles per second = 17400 km/second, which is about 6% of the velocity of light.

<u>Note</u>: The above calculation allows for only 3 of the 10 strings in the Arnoo are thickened, but if it is assumed instead that (for some unknown reason) Einstein's mass dilation equation does not

APPENDIX 4 Mass increase with Velocity

apply to the 7 non-thickened strings, then the above calculation is simpler:

Taking the ratio of rest mass to moving mass be 700 / 704, $700^2/704^2 = (186000^2 - v^2) / 186000^2$, where c = 186000 miles per second,

whence v = 20000 miles per second = 36000 km/second, which is about 11% of the velocity of light.

The velocity of the solar system relative to the local group of 35 galaxies is about 600 km/second (Hubble), but the author cannot find a value for the velocity of the solar system relative to the centre of the Universe (assuming the Universe is expanding from this point of origin, as the point of origin of the Big Bang). But a figure of 17,400 km/second is of the right order.

<u>Later note</u>: After writing the above, a paper found in Scientific American [2] gives about 180,000 km/s as the relative velocity of recession of a very distant super-nova, so very roughly half this may be the order of our velocity relative to the centre of the Universe, if the distant super-nova and ourselves are about equidistant from the centre of the Universe (on the opposite sides of the Universe). But if it is much further from the centre than we are, then our velocity relative to the centre may be much less than 90,000 km/s. There may be other data that the author has not found yet.

<u>Note</u>: The Pythagorean triangle equation for m/m_o as derived in the author's paper [3] could mean that space is flat to a very high accuracy -- so there is no curved space or space-time distortion as postulated by Relativity theory. Tesla raised an objection to curved "space-time", in that something structureless cannot be said to be curved.

Curved space does not obey the plane geometry Pythagoras theorem:-

304   APPENDIX 4   Mass increase with Velocity

E.g. a triangle drawn on a sphere surface can have 3 right angles in it, totalling 270° instead of the 180° of a flat triangle -- e.g. the equator and the two lines of longitude of 0° and 90°.)

References for Appendix 4 are below the figures.

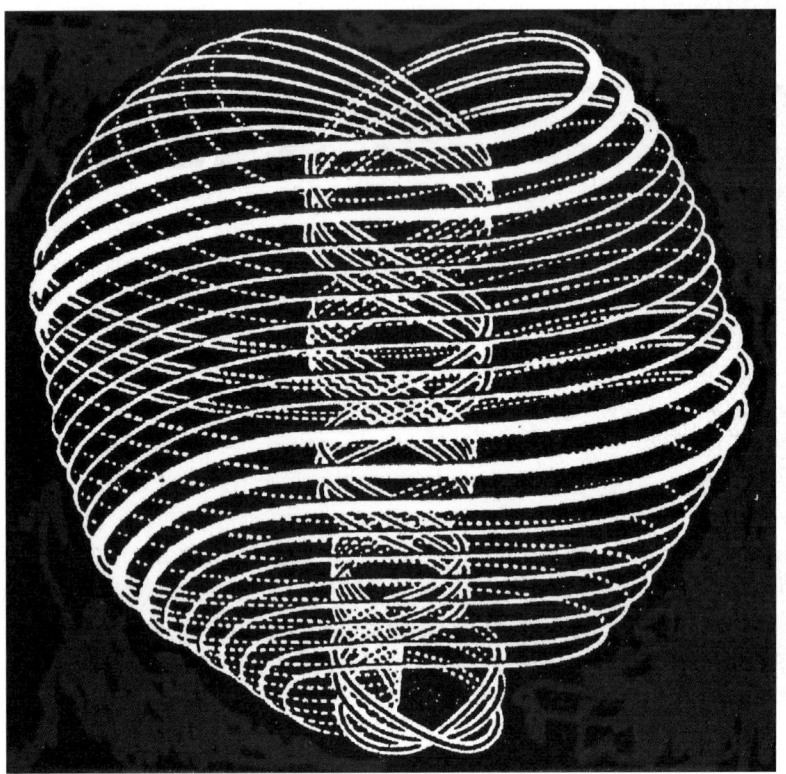

Fig. 1. Arnoo. The 3 "thicker" strings have 704 instead of 700 bubbles (per a certain arbitrary length), which gives the thicker appearance.

NOTE: The number of bubbles in the whole arnoo is given as 49^6 [1], which is about 13,841,200,000. This could avoid a possible problem that velocities would be quantised in single-bubble steps which would be noticeable, but would require some speculative assumptions.

Fig. 2. Detail of the strings in the arnoo, showing bubbles which give an appearance of "thickness" of the string (see Besant & Leadbeater's book [1]).

References for Appendix 4
1. A. Besant & C.W. Leadbeater, "Oc. Chemistry" T.P.H., London (1908), Second edn. (1919), Third enlarged edn. TPH Adyar (1951), reprinted (1994).
2. A.G. Riess & M.S. Turner, Scientific American page 50-55 (2004).
3. M.G. Hocking, "Linking String Theory and Membrane Theory to Quantum Mechanics and Special Relativity Equations, avoiding any Special Relativity assumptions", J. Scientific Exploration 21 (1), 13-26 (2007). This is printed in this book as Appendix 6, page 332.

APPENDIX 5

Paper published in Theo. Sci., Vol 21, (5 & 6), 61 (1983)

A model for ESP observation of atoms & molecules-Part I
M.G. Hocking

Introduction Due to resolving power limitations of all microscopes, any alternative method of direct observation of atomic structure is worthy of investigation. Detailed publications of claimed ESP observations of atoms and molecules have appeared [1] but until recently these have been set aside as they apparently disagree with scientific observations.

The method of ESP observation [1] is said to involve a claimed ability, long ago described by Patanjali [2], of being able to indefinitely reduce the size of an ESP sensor in the mind of the observer so that objects appear increasingly large. But descriptions given of atoms and molecules showed always an unexpected doubling of the structure and the structures were also 'unexpectedly' outlandish, as inspection of the figures in this paper will show. However, Phillips [3] has recently shown, beyond reasonable doubt, that these structures have high stability when the energetic principles of recent quark physics are applied [3]. Quark physics was, of - course, unknown in 1895 when the ESP observations were first published [1], which is a significant point in favour of serious consideration of these observations: stable structures could not have been imagined or fabricated.

The explanation given by Phillips [3] for the doubling of the observed structures is a "nuclear fusion" of two of the atoms or molecules under observation, caused by the ESP process. This seems an extremely ad-hoc hypothesis and it is very unlikely that on unstable molecule like acetylene (say) could undergo the energetics of a nuclear fusion and yet remain un-decomposed.

The purpose of this paper is to give an alternative simple explanation of why all structures were observed as doubled.

Oscillation Model This paper proposes on oscillation model for the atoms (OCA) and molecules (OCM) described by Besant and Leadbeater [1] in their ESP investigations. This model shows that the OCA and OCM can be apparently doubled due to a zero point energy vibration. The model removes the long standing problem of half atoms being present in OCM.

Many authors [3-10] have shown that present quark theories can be modified to predict 9 quarks in the proton. Besant and Leadbeater [1] show 18 Arnoo (quarks) in hydrogen, but ambiguity exists in that they may have been examining the H atom or H_2 molecule.

All molecules contain bound atoms and all atoms contain bound groupings of Arnoo. All bound particles, even if brought to translational rest (0°K), must contain zero point energy, according to the well established quantum theory. This takes the form of a high frequency vibration or harmonic motion, the restoring force being the coulombic and strong nuclear binding forces (but the quark-to-quark "string" bonds are excluded as they are not of harmonic oscillator type). This is comparable to a guitar string or a vibrating metal rod (Fig. 1) which appears to be double due to the string being repeatedly stationary at each end of its vibration trajectory. In between these extreme positions (antinodes), it is moving too fast to see. Similarly, the number of Arnoo seen in atomic and molecular structures would be double the actual number present. The guitar string analogy is not quite appropriate because no blurring is reported in the OC structures observed [1]. The Arnoo structures can remain at the antinodes for a relatively long time compared with the time spent in transit. Thus the oscillation is more like a square wave than a sine and the transits are relatively infrequent. They are much faster than a guitar string. There may thus be no visible blurring effect. A mechanical analogy to this is the escapement mechanism in a clock or watch. Also, during ESP the observer's sense of the apprehension of time may be enlarged or contracted to a stupendous extent and this may be relevant here. (See note at end of this paper).

Hydrogen
It is desirable that an odd number of Arnoo be present in the hydrogen nucleus, to obtain the correct spin of 1/2, although alternative but less satisfactory explanations have been suggested [11] to give spin 1/2.

Phillips [3] has suggested that the hydrogen structure [1] having 18 Arnoo is two protons, so that the proton then contains 9 Arnoo. To preserve consistency (9 Arnoo = 1 A.M.U.) it is then of course essential to assume that all structures in [1] are fused-together forms of pairs of nuclei concerned and Phillips [3] gives a theory of atom nucleus fusion to give the necessary pair fusions. However, using the vibration model outlined above, the hydrogen structure in [1] becomes the proton and it contains 9 Arnoo which are seen as 18 by analogy with Fig. 1 due to an inevitable "zero point energy". This zero point energy is exactly analogous to the well-known zero point energy of atoms in a crystal lattice, and must occur in any harmonic oscillator type of system because of the universal applicability of the Uncertainty Principle.

The oscillations (shown in Fig. 2) involve motion of the entire triangle shape and not individual quarks singly or separately, so the oscillation of the non-harmonic oscillator quarks string bonds is not proposed. This also applies in all other cases. Fig. 2 gives the vibration of the one triangle into the other, preserving the separate p' and n' quark types [3] as required by conservation rules. All the quarks or Arnoo change their chirality but not their 'colour' at each end of their vibration trajectory (+ becoming -), i.e. they reappear each time as their mirror image (+ and - here are used to indicate right or left handedness, not charge). Possible reasons for this are necessarily speculative; it suggests movement into a fourth dimension (the fourth spatial dimension mentioned in ref. 1) where it readily converts to its mirror image before reappearing in the other triangle. Eagles suggests that the oscillation may be the projection in 3-D space of a circular motion, with the circle perpendicular to this space; the potential energy of this harmonic oscillator could represent the kinetic energy in the fourth direction [12].

To give an analogy in 2 and 3 dimensions, consider the letters d b. Imagine the d is moving on a circle perpendicular to the plane of this paper; it becomes b where the circle cuts the paper again and a 2-D observer would see this d "atom" appearing as d b with their opposite chiralities (left and right handedness). Evidence for such effects exists in out-of-the-body experiences literature; in this experience rooms are seen as the mirror images of what they actually are [13]. Also, the authenticated removal of part of the single crystal of vanadium carbide from a sealed tube in laboratory conditions [14] suggests a fourth spatial dimension. To remove a solid through another solid requires either dematerialisation and miraculous reconstitution, or the use of a fourth dimension.

Recent physical theories [31] propose a multi-dimensional universe but restrict normal access to the higher dimensions to high energies - i.e. to very short distances. This could make a 4th dimension normally accessible only to fundamental particles.

The spins of the Arnoo in the triangles take the values required by the Pauli Exclusion Principle and the two triangles in Fig. 2 are equivalent to those given by Phillips [3] and follow those drawn by Besant and Leadbeater [1]. Simultaneous motion of all apices of one triangle along the arrow lines, moves it to the other position, and vice-versa, like the escapement mechanism in a watch. Alternatively, the 4-D circular motion model described above could apply.

The triangle of 9 quarks (which oscillates to give the appearance of two triangles shown in Fig. 2) is held together by very strong quark string bonds. These bonds, which Phillips [15] points out would not allow harmonic oscillation, are not stretched or broken during the oscillation of the triangle.

The same applies to the proposed oscillations of discrete quark groupings (e.g., Ad 6) in all the other OCA and OCM structures. Thus harmonic oscillations will be possible since such discrete quark groupings are bonded to other groupings by nuclear and coulombic forces and not by quark string bonds.

In building up the H_2 molecule from the two level E4 spheres [1] each containing 9 Arnoo, each being a proton or an H atom, one sphere (proton) would immediately fly off to the equilibrium inter-proton distance in the H_2 molecule, leaving the other to form the oscillating triangle picture shown in Fig. 2. These oscillations would represent the irreducible zero point energy of the proton being held "still" for observation; the proton, being bound in the molecule, must possess zero point energy (it cannot be static). Its triangle oscillations would occur in the direction of a line between the two protons, if the proton was not being externally 'held' for study by the ESP observer. But these oscillations would not represent the zero point energy of the molecule, which must all be contained in an oscillation of the other proton over a relatively large distance (atomic sized not nuclear sized), since the first proton is being held "still" for observation.

Although the term "zero point energy" has been used here in general, it is not rigorously appropriate for a proton being held still for observation, as the Uncertainty Principle requires the product of momentum change and position change to be h/4π. This means if the momentum change is near zero (proton being held still), the position change is very large (far greater than the twin triangle separation). This difficulty can be overcome by describing the triangle oscillation (to give the appearance of a twin triangle) as a form of zitterbewegung [15, 16] which is also a consequence of the Uncertainty Principle. It can be shown [16] that an electron, for example, cannot be static but has eigenvalues of velocity of + c and -c, which is interpreted as a rapid oscillatory motion (zitterbewegung).

The separate E4 spheres mentioned above would also have an internal zero point energy, internal to the proton, as it is a bound state of quarks.

The quarks are joined by the very strong quark string bonds which are not of the harmonic oscillator type and so negligibly small quark displacements would represent the zero point energy because of the very strong effective strength of the quark string bonds. The same oscillating triangle of the proton bound in the molecule could not occur in the free atom because it involves a centre of gravity oscillation of the proton which could not occur in a free atom (unless the proton were held by an external force due to the ESP observer).

If the hydrogen structure in ref. 1 is an atom, then the ovoid around it cannot be taken as its electron shell as the 1 s atomic orbital is well known to be spherical. But the 1s molecular orbital for H_2 is ovoid and the distance between the two protons in H_2 is known to be 100,000 times the proton radius.

This means that if Besant and Leadbeater were observing one of the protons as the H double-triangle structure, at the size shown in Fig. 2, then the other H double-triangle of the other proton in the H_2 molecule would be 5 miles (or 8 km) away! Thus it is easy to explain why, from H_2 gas, they selected the structure shown in Fig. 2 as the H atom and why they "did not observe H atoms to move in pairs". Two copies of Fig. 2 five miles apart would not be seen to move as a pair. If a proton in a molecule were being observed, then the ovoid could be the 1s molecular orbital, assuming that it is vastly larger than the proton. The diagrams are not to scale [1]. While observing a proton, the ovoid wall would be several miles away, on the scale mentioned above, and vice-versa. Thus the observers could hardly take in both ovoid and proton at one magnification. This means that although a proton observed in H_2 would be towards one end of the ovoid, it is very unlikely that they would be aware of the position in the ovoid. In this respect the diagram given [1] for hydrogen centered in its ovoid could be misleading. See Appendix 5A.

Ortho H_2 can be explained as a molecule with hydrogen variety 1 (spin ½) at both ends, which would have a higher entropy according to statistical thermodynamic calculations than the para form which is the stable form at 0°K (100% p-H2). The para form (least disordered) would be molecules with variety 2 at each end. This also accords with variety 1 being the more commonly observed [1] at room temperature (a 3:1 ratio of o:p is experimentally found in physical chemistry). A change in temperature changes the ortho:para ratio, which shows that variety 1 can easily change to 2 and vice-versa. If varieties 1 and 2 are identified with ortho and para H_2, then the chirality change which is evident in the hydrogen diagrams in ref. [1] for the Arnoo at the 10 o' clock position must occur readily, merely on changing the temperature. This would support the ease of changes of chirality, which is required for the oscillation model (Fig. 2).

Besant and Leadbeater [1] give a structure of two rarely observed crossed ovoids produced only during the electrolysis of water, which they thought might be deuterium. Electrolysis produces H atoms [17] which quickly combine to give H_2 molecules. If the scale of the drawing is considerably coarser than that used to draw the proton in Fig. 2, then this structure could show the distorted electron orbitals of H atoms in a 'collision complex', bound temporarily together by the London dispersion forces. It is well known [18] that atomic H gas has a life of about ½ sec., during which time it undergoes many billions of collisions. It only forms H_2, when an energetic enough collision occurs. This effect is the well-known activation energy factor of chemical kinetics. The spherical electron shell around each H atom would be squeezed out oval by coulomb repulsion, towards the 4 corners of a tetrahedron to give the charge distribution shown in Fig. 3. This is a temporary dipole attraction effect and the two would soon separate. H atoms would be very rarely found in H_2 gas from electrolysis and a more likely explanation could be that the crossed ovoids were a van der Waals coupling of two hydrogen molecules, but if only one pair of H triangles was observed in each, it must be assumed again that the other proton had escaped notice due to the vast size difference between the ovoid electron molecular orbital and the proton, mentioned above.

Adyarium (36 Arnoo observed = 18 actual Arnoo vibrating) can be identified with deuterium. It would be one end of an D_2 or a HD molecule, the other deuterium or H nucleus being very far away, as explained for H_2. The circular shell shown around Adyarium could be a nuclear binding force structure around one deuteron, not an electron shell, or it could be an ovoid seen from a different viewpoint.

Nuclei as harmonic oscillators

For the proposed oscillation model it is appropriate to note that nuclear structure models based on the nucleus as a harmonic oscillator (Fig. 8) have been fairly successful, some modifications being required for the heavier elements. Fig. 8 shows the Woods-Saxon potential energy curve, which is a widely accepted representation of the nuclear potential energy well, intermediate between a harmonic oscillator and a square well [19, 20]. Fig. 8 shows only small deviations of the potential energy curve of a proton or neutron, from the ideal harmonic oscillator curve (dotted line). The energy well is analogous to a frictionless wine glass in which a steel ball (nuclear particle) is released against one wall and then perpetually oscillates up and down opposite walls. This is an oscillation of potential and kinetic energies, their sum remaining constant at all times. There is considerable evidence that the mean free path of particles in a nucleus is large, and in many cases larger than the nucleus [19].

OCA structures on oscillation theory

Figs. 4-7 give some examples of the generation of double the 'expected' numbers of Arnoo in atoms, on the vibration theory. The drawings of the asymmetrical "stopped" structures clearly show unstable situations and the vibration to give the symmetrical stable structures is essential, which could be why zero point energy vibration cannot be removed without destroying the atom:

the Uncertainty Principle for a linear oscillator is $Px' = h/4\pi$ where P is the momentum (mv) change and x' is the position change. If v = 0, then P = 0 and x becomes infinite, i.e. the molecule disintegrates. A guitar string at its antinode is also an asymmetric unstable state and is stabilized by its transits to the opposite antinode.

(Note: the word "stopped" above and in the figure captions is only meant as a short way of indicating that the structure is seen or shown as <u>instantaneously</u> stopped at one of its antinodes. If a plucked guitar string is examined with a lens, scratches and details on its surface can clearly be seen, at these "stopped" antinodes.) The application of the Uncertainty Principle and the value of the constant $h/4\pi$ given above must be regarded as an assumption, in relation to ESP observations.

Pfleegor and Mandel [27] have shown that single photons exhibit interference even when a photon is not being emitted from one of the sources: the fact that a photon was recently emitted by that source is enough to hove set up fields which interfere with another photon from the other source; if one source is covered, all interference ceases. Similarly, the oscillation of the Arnoo structure can set up a field which stabilizes the observed double structure as if it were a permanent static double structure. The stabilities of such symmetrical structures as are shown in ref. [1] have been theoretically verified by Phillips [3] who assumed that they are double fused nuclei. The success of Phillips' model could be due to these being the stable vibration patterns or 'standing wave' antinodes of a single atom. The Uncertainty Principle forbids the holding still of a structure which contains groupings held together by harmonic oscillator type forces.

Some examples of the proposed oscillation theory giving the apparent doubling in the observed structures are given below. Fig. 4 shows the oxygen atom. It is interesting that such planes of symmetry (PS) as that through the central globe of oxygen exist for ail element structures or parts of structures in ref. [1].

Lithium (Spike, Groups and 7): There is a PS horizontally through the Li 63 spike and 0 vertical PS through the Ad6 petal cluster. The 4 Li4 globe is symmetrical about its centre.
Sodium (Dumb bell, Groups 1 and 7): See Na part of Fig. 14. There is a horizontal PS halfway down the rod of this atom. The

others in the group are exactly similar. The members of other groups are also exactly similar to the example chosen from their group (unless otherwise stated).

Calcium (Tetrahedron, Groups 2 and 6): See Fig. 5. There is a PS halfway between any two funnels so that with two PS the oscillation of only two funnels through these PS will give the appearance of four. The central globe has a centre point of symmetry so that its groups are similarly halved in their actual number present. If funnels are equated with valency, as Lester Smith and Slater [21] suggest, then Ca is divalent with two (actual) funnels.

Boron (Cube, Groups 3 and 5): There is a point of symmetry in the cube centre, so that the six funnels radiating out from the face centres can be generated by only 3 funnels oscillating about this centre.

Carbon (Octahedron, Group 4): There are three apparent PS which reduce the eight observed funnels to 4 actual, but as half the observed funnels have 27 Arnoo and the other half have 26, there will be only one true PS. See Fig. 10; but for clarity the 4 funnels in each antinode are not shown and must be imagined in the octahedron faces beneath the H or OH. The carbon centre (4 Arnoo) is a square (2 actual Arnoo).

Iron (and all bars group): The centre of the atom is a point of symmetry about which all the 14 bars oscillate (7 actual bars, numbered 1 to 7 in Fig. 6).

Argon (and all stars group): This is a flat 6-armed star of Arnoo, all arms being identical. The centre is a point of symmetry about which all the arms can oscillate to the ones opposite. So there are only 3 actual arms and half the observed number of Arnoo, Fig 7.

Compounds: Finally some compound examples will be given. The oscillation hypothesis removes the difficulty of apparent atom splitting in OC compounds, which the author was unable to explain satisfactorily earlier [11]. All the compounds given in ref. 1 can be explained as an oscillation of whole-atom structures, except that five on p. 302-310 of [1] have not yet been fully considered as insufficient details of their 3-dimensional structuring are given. These can be explained if assumptions are made about this. Even the NaCl structure [1] can be separated out on the oscillation model, but is complicated and is not shown here.

The molecules, although separated out into whole atoms, appear compressed, perhaps partly due to not drawing them to scale, and perhaps also by the actual clairvoyant holding process during observation. Further study is needed of the theoretical chemical calculations which can be mode using the structures given, e.g. for benzene the entropy can be calculated and compared with the measured value. Predictions of the numbers of isomers may also be made for comparison with the actual numbers, to determine whether the correct oscillating end structures hove been selected in this case. The question of OCM being perhaps of atomic size while OCA are perhaps only of nuclear size can be answered either by invoking the 4-D rotation model (mentioned earlier) which only requires nuclear sizes for OCM and OCA, or by using Phillips' suggestion that expansion of structures can occur during ESP observations. (In a simple oscillation model, OCM need to be of the order of atomic sizes or the Arnoo of different atoms would not get "mixed up" as observed, in compounds).

The trend in all the compounds seems to be that Nature avoids asymmetry - e.g. HCl would be very asymmetric if consisting of a H atom always on one side of a Cl atom in the conventional way. This is avoided if on oscillation similar to that of NaOH (Fig 11) occurs. Lack of space precludes inclusion of many compounds here.

The four dimensional circular motion model (mentioned earlier) can be advantageously applied to molecules - e.g. in Fig. 11 a rotation of either of the upper drawings about point a in a circle perpendicular to the plane of the paper, will produce the other upper drawing. Hence the lower, observed, OCM can be obtained by fast rotation. Similarly, Figs. 9 and 10 (half-structures) if rotated about the centre point of the pyramid bases, will give the ESP observed "double" structures (OCMs).

Appendix 5A

A clairvoyant observer, Mrs R. M. Watson of Blagdon, near Bristol, who has kindly made observations for the author before [11], was asked the following questions and has given the answers below.
Q. The H_2 molecule is something like H------------------H, where the distance between the H atoms is about 100,000 times more than the size of the H atom. Is it possible that the diagram in "Occult Chemistry" (ref 1) could be one end of this 'dumb-bell' structure, without CWL realising that there was another H atom some distance off?

A. Yes, it is possible that one end of a pair of the 'dumbell' structure is being observed. As I see it, if the hydrogen atoms were represented as pin-heads they would then be 1½ feet apart.

Q. Is it possible that the two interlacing triangles of the O.C. H atom are actually due to only one triangle, vibrating-something like a harp string which when plucked looks like two strings due to vibration. (We are looking for some reason why 18 Arnoo were observed while we think there should be only 9 in an H atom)

A. As regards the two interlacing triangles of O.C. I think your suggestion of one vibrating triangle appearing as two triangles is a valid one. The rate of vibration is remarkable.

Further enquiries, especially concerning hydrogen and adyarium are in progress.

The author gratefully acknowledges the time spent by Mrs. Watson in making these observations.

NOTE: The figures below are not to scale! Besant & Leadbeater [1] say that this would require "an absurdly small dot on a paper many yards square".

APPENDIX 5 Remote Viewing of Atoms & Molecules   317

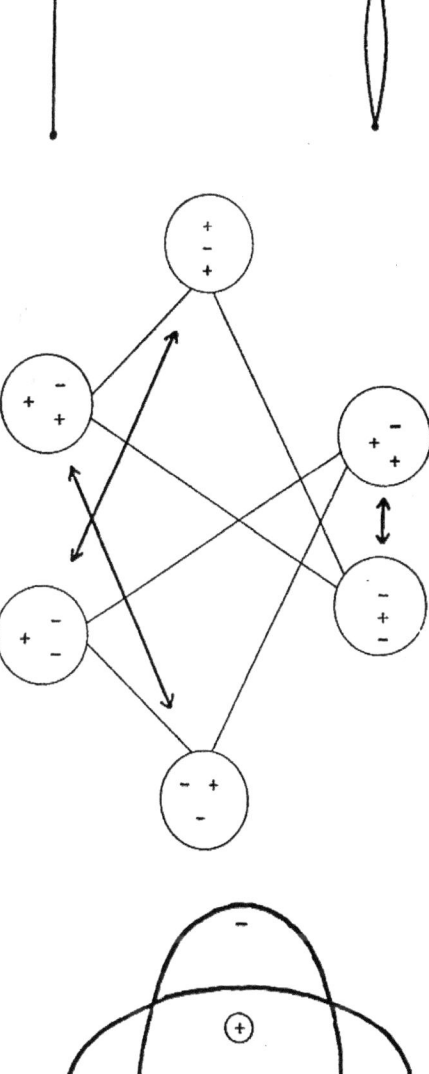

Fig. 1.
Left: guitar string, still.
Right: same string, plucked.

Fig. 2. One proton in H_2 molecule. 18 arnoo <u>apparently</u>, but 9 arnoo <u>actual</u>.

Either triangle, its apices moving simultaneously along arrow lines, converts to the other, changing all chiralities (+ & - indicate chirality here, following reference [1]).

The triangles are not in one plane [1], e.g. the lines can all be parallel in 3-D.

Actual movement will be oscillation or rotation in <u>4-D</u>, because the chiralities change.

Oscillation of + - + linear to triangular may be by a twist or change of viewpoint.

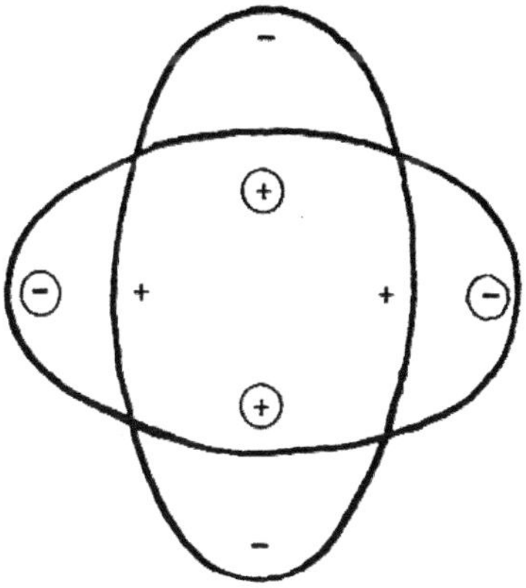

Fig. 3. Possible electrostatic London Dispersion Force Bonding of two H atoms.

The circled charges are for the Top ovoid.

+ and ⊕ indicate parts of the electron shells which are less negative than the average.

Fig. 4. Oxygen atom, O. Oxygen is a double intertwined spiral but for clarity <u>only one</u> of the spirals is shown here. Spirals might oscillate out-of-phase.

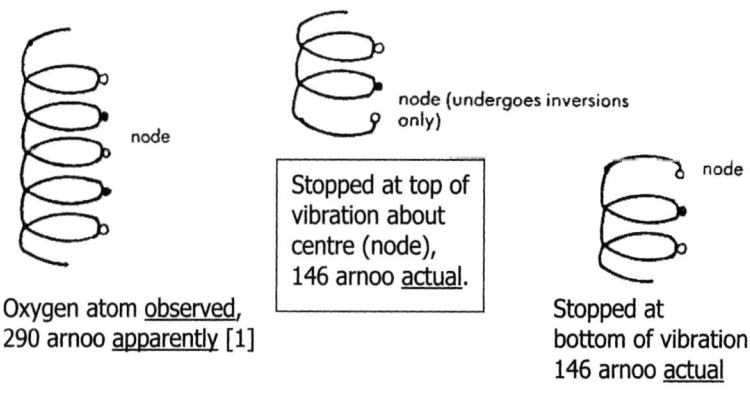

Oxygen atom <u>observed</u>, 290 arnoo <u>apparently</u> [1]

Stopped at top of vibration about centre (node), 146 arnoo <u>actual</u>.

Stopped at bottom of vibration 146 arnoo <u>actual</u>

Fig. 5. Calcium, Ca:

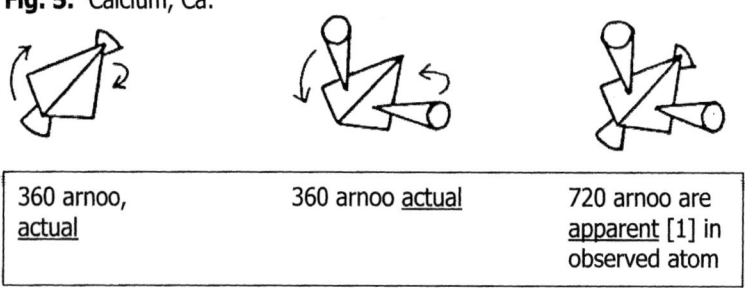

| 360 arnoo, <u>actual</u> | 360 arnoo <u>actual</u> | 720 arnoo are <u>apparent</u> [1] in observed atom |

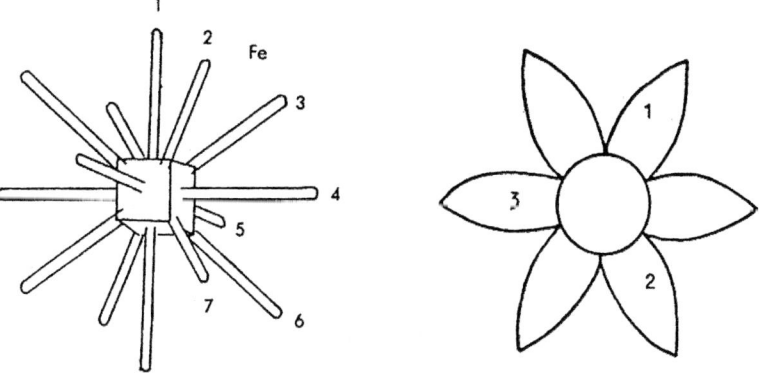

Fig. 6. Fe

Fig. 7. Ar

APPENDIX 5   Remote Viewing of Atoms & Molecules   319

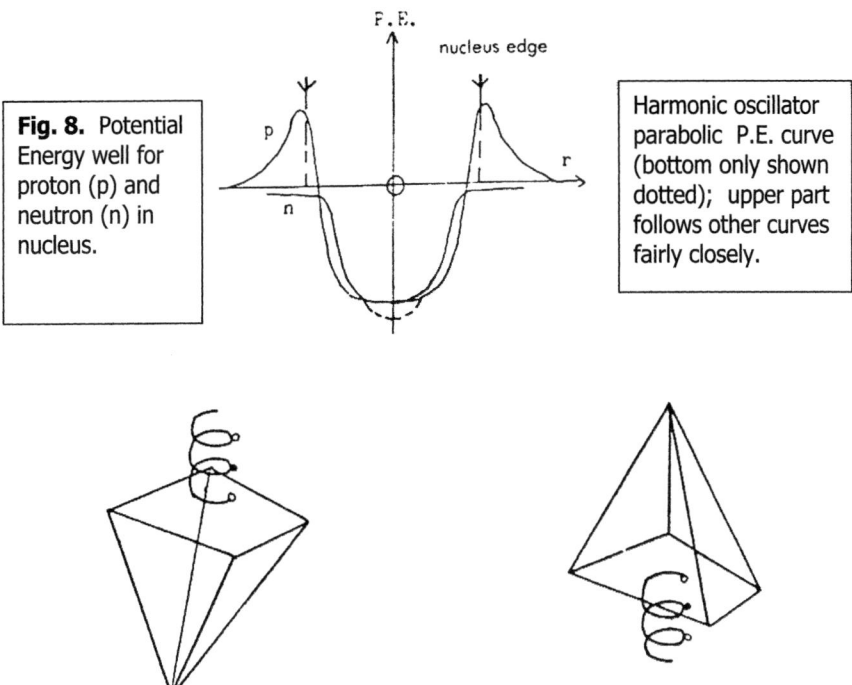

Fig. 8. Potential Energy well for proton (p) and neutron (n) in nucleus.

Harmonic oscillator parabolic P.E. curve (bottom only shown dotted); upper part follows other curves fairly closely.

Fig. 9. SnO. The O atom stands upright from the base of the Sn pyramid of funnels, spikes &c (not shown). Oscillation between these two antinodes then <u>appears</u> to contain the O atom within the tin! Cf. ref [1], page 290. (Surrounding SnO lattice atoms not shown.) This structure is compatible with the layer structure given by Wells [28].

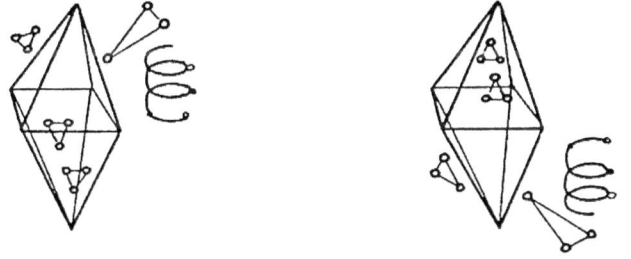

Fig. 10. CH$_3$OH. Cf. ref. [1], Fig. 196. The 3 H atoms and one OH group hover tetrahedrally over C funnels (not shown) in the octahedron, in the above antinodes.
Observed structure is the superimposition of these two antinodes. (Not to scale).

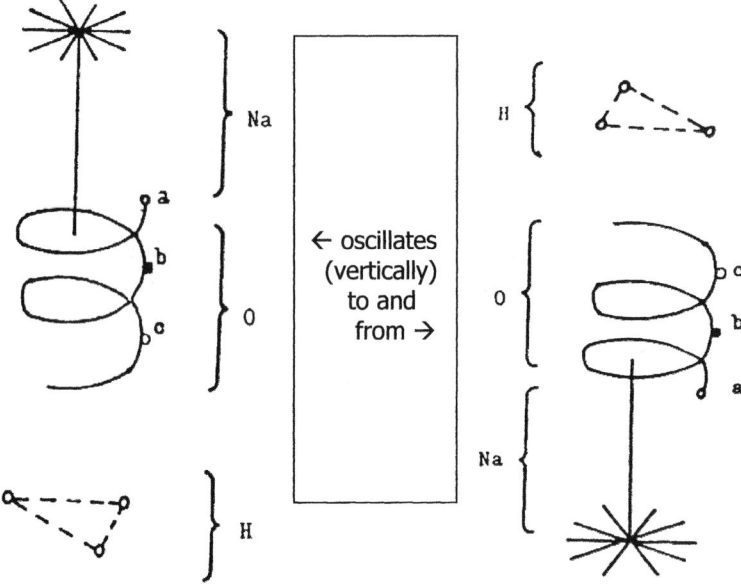

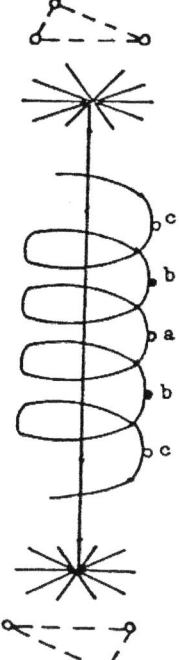

Fig. 11. NaOH

OBSERVED OCA

This double structure is apparent and is the result of vertical oscillation of the two actual above antinodes.

Cf. Fig. 1 (rotate Fig. 1 90 degrees).

A model for ESP Observation of Atoms and Molecules - Part II

Paper published in Theo. Sci., Vol 21, (5 & 6), 61 (1983)

M.G. Hocking

PART 1 described an oscillation model for atoms (OCA) and molecules (OCM) which were described by Besant and Leadbeater [1] in their ESP investigations.

The following part is not directly concerned with the oscillation model as such, but seeks to show that the identification of the OCA with the chemical atom (required by the oscillation model) is compatible with the reported (but currently undiscovered) elements named Kalon, X, Y, and Z [1].

Kalon: The Crookes Periodic Table [1] is a figure 8 spiral with 8 places around each loop, thus allowing for the filling of s and p electron shells (total 8 electrons). Transition metals and inert gases appear at crossover points. The first **d** shell to occur (3d) reaches 8 (along with 4s) at iron, but since 10 electrons fill a **d** shell, two other transition metals (Co & Ni) are placed with Fe in the Crookes table. Then 4s and 4p fill to reach 8 at Kr. The 4d and 5s shells fill to give 8 at Ru (exactly similar to Fe case above). Then 5s and 5p fill to reach 8 at Xe. The Crookes scheme then requires 5d and 6s to fill, reaching 8 at element X (there is no provision for any f shell filling). After this, 6s and 6p should fill to reach 8 at kalon. This would give kalon a structure having all s, p and d shells filled out to 6p; all known inert gases end with a p orbital filled. It is possible that a few atoms of kalon exist with this structure, as an undiscoverable inert gas (see below). Its electron shell filling order is unlikely but this is offset by its being an otherwise very stable inert gas shell. But from its mass number of 170 it would only have 98 neutrons (72+98=170) which puts it below the neutrons <u>vs</u> protons graph for stable nuclei (22). This is consistent with it being reported [1] that only one or two kalon atoms are present in a room. For comparison, the estimated abundance of tritium (relative to H) is less than 1 part in 10^{17}; this means that both kalon and tritium are undetectable against background by any scientific technique (the tritium abundance is estimated from its known half-life value). 68Er164 has an anomalously high abundance relative to 68Er162 [23], possibly due to kalon decay. On whether physics could accommodate any new "elements", the example of muonium could be cited.

Elements X, Y & Z [I]: Fig. 12 shows atomic volume vs atomic number and major volume contractions occur at the transition element groups. There are 18 places between the Fe (26) group and the Ru (44) group; there are 32 places between the Ru (44) group and the Os (76) group, because the lanthanons intervene (filling the inner 4f shell). Adding 18 on to the Ru group gives 62, 63 and 64 for the atomic numbers of X, Y and Z transition metals (bars group). These are predicted by the Crookes Table [1] which is based on a s, p, d electron shell "mainstream" evolution of the elements and takes no account of the lanthanons 4f shell filling. Fig. 12 shows a gap at the lanthanides, where X, Y and Z might have existed (dotted line). They are a possible "thermodynamically unlikely" (see below) alternative to the lanthanons and could exist in trace quantities below scientific detection limits. The situation is analogous to structural isomerism in organic chemistry, e.g. $C_4H_{10}O$ can take different structures with totally different properties. There is no indication [1] of where or in what compound the elements X, Y and Z were seen, but the investigators claim remarkable powers of collecting very rare atoms such as polonium [1].

There is a large anomalous fall in the cosmic abundances [24] of elements from mass number 145 to 175, which spans the range where the lanthanons replaced the mainstream evolution of elements which might otherwise have occurred. X, Y and Z could be the results of a suppressed mainstream evolution in parallel with the stabler lanthanons. Only the stablest alternative elements would be expected to exist in competition with the lanthanons, so only X, Y, Z and kalon are to be found in ref [1].

APPENDIX 5   Remote Viewing of Atoms & Molecules

Since nuclei are evolved (built up from lighter nuclei) in stars having very high interior temperatures which strip off most of the electrons, there is a possibility of following the Crookes mainstream periodic table across where the lanthanons are: the outer electrons are added later in the outer cooler parts of the star and the lanthanon 4f electron shell filling thus occurs only <u>after</u> the nuclei have been made. This supposes that the Crookes table is applicable for the building up of <u>nuclear</u> shells, which could then explain why the atoms (nuclei, really) seen by Besant and Leadbeater [1] have structures of types of nuclei which follow the Crookes periodic table. This table is perhaps more appropriate for predicting nuclear structures than chemical (electron shells) properties of atoms. It leaves room for X, Y, Z and kalon and explains why the observed lanthanon nuclei are spread across all the OCA chemical groups shapes (dumbell, tetrahedron, etc). The OCA groups are dependent on the atomic number (Z) of protons in the nucleus. Major closings of proton shells occur at the magic numbers such as 50 and 82, a semimajor shell closing at 64 and minor nuclear shell closings at 58 and 68 [19]. Table 1 shows relevant Crookes groups vertically; the values of (Z-n) also shown make this a "nucleus periodic table", as will be explained shortly; n is the magic or semimajor proton shell closing number next below each (Z). Exact agreement is found between successive periods of the table for 13 of the vertical groups (underlined), if subtraction of 1 unit is allowed in 4 consecutive groups (asterisked). This explains why the OCAs of the lanthanons fall into the (nuclear) groups predicted by the Crookes table. However, there are 4 groups for which no agreement is obtained: Pd-Z-Pt (not lanthanons), Ag-Sm-Au, Cd-Eu-Hg and In-Gd-TI. The last 3 groups involve the lanthanons immediately following X, Y and Z and the non-agreement could be due to the disturbance to the scheme due to X, Y, and Z. It is curious that if Sm, Eu and Gd had (Z) = 65, 66 and 67, as predicted by Crookes, exact agreement would occur for the Ag-Sm-Au and Cd-Eu-Hg groups.

X, Y and Z have such low abundances that they are undetectable by scientific techniques; they are unstable. Y and Z have higher atomic number to mass ratios than is stable (unlike Rh and Pd). It is significant that X, Y and Z immediately follow 61-Pm (Illinium) which is the only (conventionally known) element (from H to U) which is not detectable in the lithosphere by scientific techniques.

Pm follows Nd which has a magic neutron number of 82 in its most abundant mass 142 isotope; but this is similar to Tc following Mo (magic neutron number 50 in abundant 92 mass isotope) and yet Ru, Rh and Pd bars group follow Tc normally.

It is thus suggested that the peculiar lanthanon 4f electron shell filling is incompatible with a bars structure nucleus and that this structure rearranges to give 62Sm147 (from 62X147), 62Sm148 (from 63Yl48 by positron emission, since 63Y148 is on the high side of the stable proton: neutron ratio), and 62Sml50 (from 64Z150, similarly to Y), these Sm isotopes being the stablest available (Eu and Pm have none suitable). It may be significant that the abundances of Sm148 and Sml50 are well above those of other elements formed by neutron capture processes (23), suggesting another path for their formation. Present theory on the effect of nucleus shape on nucleus-electron interaction energy predicts an effect too small to affect the stability of 6s or 4f states, but the extremely non-spherical bars type of nucleus may generate additional terms.

The relation of nucleus shape to chemical properties, which is predicted by the characteristic OCA group shapes and by the Crookes table as shown above, suggests that the addition of one proton to change (say) chlorine into an argon isotope, involves more than just adding 1 proton to 17 others all alike, and calls for some drastic rearrangement such as is shown between the dumbell (CI) group and star (Ar) group. This effect masks any recognition of the proton added to the 17-CI when the 18-Ar star is examined. This opens up the tricky question of whether or not protons and neutrons exist as such in nuclei. Beta emission of electrons from nuclei does not mean that nuclei contain electrons, and one must beware of simplistic views, but the present position is fairly summed up at a recent nuclear physics conference as follows:
"The nucleus is composed of neutrons and protons. This statement would be accepted by most people and yet it conceals a great deal of ignorance. . . .
To what degree is it proper to picture the nucleus as a collection of neutrons and protons? To what degree are essential changes introduced by the background field in which the nucleons are immersed? These are questions of immense importance that we

shall certainly not answer without going inside the nucleus (using. probe beams) and looking at nucleons as individuals - if such they are - ... ." [25].

TABLE 1

Relation of Crookes Table to **nuclear** shell filling.

(Agreements underlined; brackets mean n is a minor shell closing no.)
(Z) is first no. under each element; (Z-n) is second no. under each element.

```
==================================================
Zr Nb Mo Tc    Ru Rh Pd    Ag Cd    In Sn Sb Te  I      Xe      Cs Ba La
40 41 42 43    44 45 46    47 48    49 50 51 52 53      54      55 56 57
0  1  2  3     4  5  6     7  8     9  0  1  2  3       4       5  6  7
--------------------------------------------------
Ce Pr Nd Pm    X  Y  Z     Sm Eu    Gd Tb Dy Ho Er      Ka      Tm Yb Lu
58 59 60 61    62 63 64    62 63    64 65 66 67 68      72      69 70 71
8(0) 9(1) 10(2) 11(3)  12(4) 13(5) 0   12(4) 13(5)    O  1*  2*  3*  4*   8(4)    5  6  7
--------------------------------------------------
Hf Ta W Re     Os Ir Pt    Au Hg    Tl Pb Bi Po At      Rn      Fr Ra Ac
72 73 74 75    76 77 78    79 80    81 82 83 84 85      86      87 88 89
8  9  10 11    12 13 14    15 16    17  0  1  2  3      4       5  6  7
==================================================
```

Acknowledgements

Thanks are due to Dr. S.M. Phillips and Dr. D.M. Eagles for very helpful correspondence and discussions, and to Dr. V. Vasantasree for help with some of the structures given here.

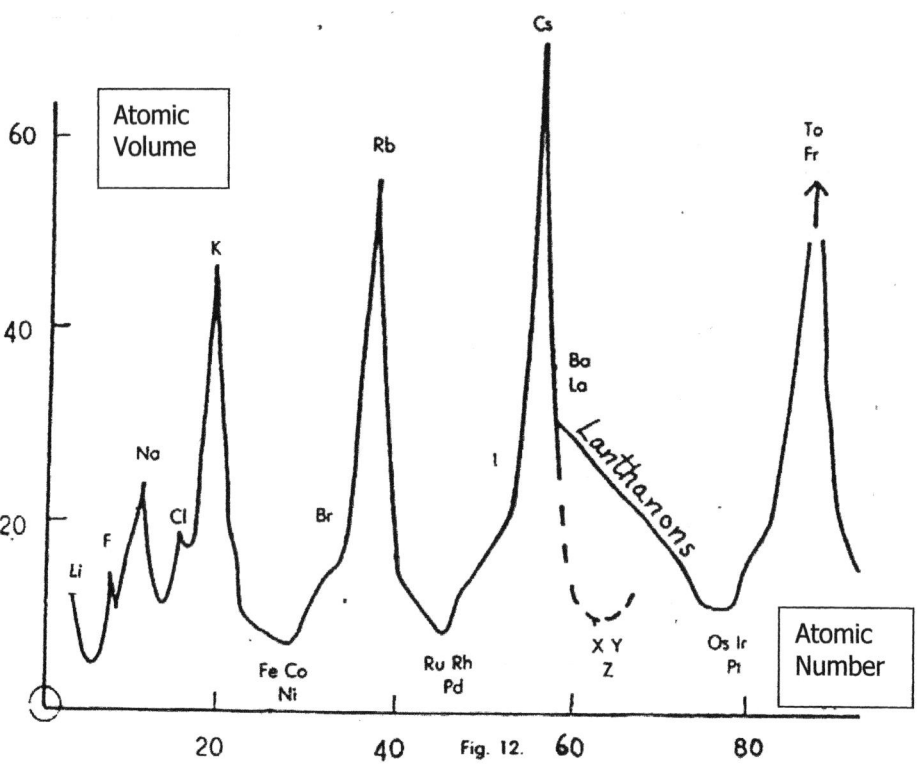

Fig. 12. Graph of Atomic volume (y-axis) vs Atomic Number (x-axis)

Appendix 5B

Some difficulties of the double nucleus fusion hypothesis of Phillips [3] are given in the questions which follow:

(1) How could a single free Arnoo be observable:
 (i) if all structures should only be observed as fused doubles,
 (ii) if it is bound to other quarks by a strong quark string bond?
(2) On dissociation to atoms, how can a single H atom, or a proton, be seen on level E-4?
(3) A single oxygen spiral (rather than a fused double) was seen to go off when ozone decomposed [1].
(4) If the clairvoyant's will causes fusion of two atoms, there should be many occasions when fusions of 3 or more atoms could have been observed. The ternary collision frequency in a gas at 25°C is 10^{25} per cc per sec; the binary frequency is 10^{28} [26]. Even if the clairvoyant does not have to rely on such collisions but simply "collects" a few atoms or molecules himself, there is still this problem of why triple and higher fusions are not obtained.
(5) If the Uncertainty Principle perhaps operates to cause fusion by spreading out the wave functions of atoms, why should its application cease after only 2 atoms fuse, i.e. it could be applied over again to the fused di-nucleus, prohibiting it from being observed at rest and requiring *ex hypothesi* a third atom to fuse to it, etc.
(6) If occultum were to be identified as tritium, it could not be observed as a fused di-nucleus since hardly one HT or T atom is present in a large room. This means that tritium cannot be observed clairvoyantly, using Phillips' fusion mechanism.
(7) Why are protons and neutrons not observable in the structures of O.C? This question applies to both hypotheses, if protons and neutrons are indeed present as such in nuclei, as is the orthodox view.
(8) A doubling or fusion process would have been noticed as the atom was slowed down for observation. But no mass or size change has been reported [1]. Single atoms or protons were seen on level E-4.
(9) It seems unlikely that the valency funnel structure (and their orientations, giving 2 funnels = 1 valency) could survive a nuclear fusion in which all the protons and neutrons are disintegrated.

(10) It is worth noting that if the triangles of 9 Arnoo in Fig. 2 are each a separate proton (or fermion of any kind), as originally suggested by Phillips [3], there would then be two protons overlapping each other and with the same spin ($\frac{1}{2}$). This is forbidden by the Exclusion Principle. If the spins are opposite (as originally in Phillips' hydrogen variety 2 double nucleus) the pair of protons would have insufficient binding energy. These are reasons why a pair of protons is not a known bound state [29]. The hypothesis of Phillips [3] has since been changed, avoiding this problem.

(11) The very strong quark string bonds which bind protons (and neutrons) would have to break during the fusion process, since protons and neutrons are not evident in OCA (except hydrogen). In the 'compound nucleus' of high energy nuclear physics, the neutrons and protons are still intact, suggesting that the energy required to break quark string bonds is very high.

(12) It is unnecessary to postulate a fusion of two atoms satisfy the Uncertainty Principle. The vibratory motion of the internal parts (quarks or Arnoo) of an atom will satisfy that principle, the atom being at translational rest. This is true for atoms in solids, liquids and gases (i.e. equally true for iron and for argon). Phillips' static OCAs violate the Uncertainty Principle.

(13) If nuclear fusion of two different atoms explains the unidentified elements X, Y, Z and kalon, then this type of fusion should have generated more than just these few examples.

(14) It is inconsistent (or, at least, just too 'convenient') to say that the observer inhibits the Meissner Effect between the quarks which make up the nucleons, but not later that between the quarks of the OCA! How could the supposedly released quarks join again to form the OCA if the Meissner Effect is inhibited by the observer?

(15) If chemical bonds are broken during OCM formation [2], there would be irreversible decomposition to products like CO_2 and steam, or pyrolysis, with unstable organic compounds. There would be no chance of preserving organic compound structures during the high energy processes of OCM formation required by Phillips' model. 'Equipartition of Energy' would feed enough energy to chemical bonds to break them. It would be incredibly unlikely that molecules like acetylene and ether would survive with their (doubled) valency structures. The chemistry involved has been overlooked.

(16) Eagles [30] points out an important difficulty in Phillips' theory concerning the valence structure of the tetrahedral, cubic and octahedral funnel structures observed in di-, tri- and quadrivalent OCMs. If the postulated fused compound nucleus is seen by the valence electrons as a single charged entity, then we would not expect the n valence electrons for a single atom to become 2n for a fused compound nucleus. Instead, we would expect the same number of valence electrons as there would be for an atom of twice the atomic number of the original atom, which (except in special cases) will not have 2n valence electrons.

References

I. A. Besant and C.W. Leadbeater, Lucifer, **Nov 1895**.
Idem "Oc. Chemistry" T.P.H., London (1908), Second edn. (1919), Third enlarged edn. TPH Adyar (1951), reprinted (1994).
2. Patanjali, Yoga Aphorisms.
3. S.M. Phillips. Theo. Science $\underline{22}$, 35 (1978); $\underline{23}$, 80 (1979). "ESP of Quarks" publ by T.P.H., Wheaton, IL (1980).
4. See Sec 2. of H. Terazawa, Phys. Rev. D 22, 184 (1980) for a review of some subquark theories. Some other theories with nine subparticles in a proton which are not mentioned in this review include those shown in Refs. 5 to 10.
5. D.M. Eagles, Review: "Nine subparticles in protons or neutrons?" (Sept. 1980) (available from Dr Eagles).
6. M.H. MacGregor, Phys Rev D 9, 1259 (1974); D 10, 850 (1974); "The Nature of the Elementary Particle", Springer, New York, (1978).
7. E.J. Sternglass, Int J Theor Phys 17, 347 (1978)
8. S.M. Phillips, Phys Lett 84B, 133 (1979).
9. H. Harari, Phys Lett 86B, 83 (1979).
10. M.A. Shupe, Phys Lett 86B, 87 (1979); J.G. Taylor, Phys. Lett 88B, 291 (1979).
11. M.G. Hocking, SGJ, 12, 99 and 143 (1968), publ by TS, available at Library, 50 Gloucester Pl., London, W1H 3HJ.
12. D.M. Eagles, private communication, (1981).
13. C. Greene, "Out of the Body Experiences", Inst. of Psychophys Res, Oxford (1968); R.A. Monroe " journeys out of the Body", p. 172, Anchor Press/Doubleday, NY (1977). Note: people have also reported that text seen while dreaming is often a mirror image of normal.
14. J.B. Hasted, D.J. Bohm, E.W.Bastin, P. O'Regan and J.G. Taylor, Nature 255, 470 (1975).
15. S.M. Phillips, Private Communication (1978).

16. P. Marmier & E. Sheldon, "Physics of Nuclei and Particles", vol. 1, p. 633, Academic Press (1969).
17 J.O'M. Bockris and A.K.N. Reddy: "Modern Electrochemistry'", Plenum Press (1970).
18 N. V. Sidgwick: "Chem Elements and Compounds", Oxford (1962)
19 A. Bohr and B.R. Mottleson, "Nuclear Structure" Vol 1, Benjamin (1969).
20. J.M. Irvine. "Nuclear Structure Theory", Chapter 13, Pergamon (1972).
21. E. Lester Smith and V. W. Slater. "The Field of Oc. Chemistry", 2nd Ed, TPH, London (1954).
22 W.J. Moore. "Physical Chemistry", 3rd Ed, Longmans (1962).
23. E.M. and G.R. Burbidge, W.A. Fowler and F. Hoyle, Rev Modern Phys., 29, 547 (1957)
24. C.S.G. Phillips and R.J.P. Williams, "Inorganic Chemistry", Oxford (1966).
25. D.H. Wilkinson: Panel Proceedings, Future of Nuclear Structure Studies, IAEA, Vienna (1969), p.27.
26. S. Glasstone. "Textbook of Physical Chemistry", p. 277, 2nd Edn. Van Nostrand (1951).
27. R.L. Pfieegor and L. Mandel: J. Opt. Soc. Am, 58, 946 (1968); Phys. Rev., 159, 1084 (1967).
28. A.F. Wells. "Structural Inorganic Chemistry", p 368 and 636, 2nd Edn. , Oxford (1950).
29. R.P. Feynman, R.B. Leighton and M. Sands: "The Feynman Lectures on Physics, vol. 3", page 4-15. Addison-Wesley (1966).
30. D.M. Eagles: Paper given at Mosman, N.S.W., Australia, (Dec. 1980).
31. S Weinberg; Lecture at Royal Society, London, May 1983 (in press).

NOTE ADDED in 2004:

One comment received on this paper was that the proposed oscillation model (for zero point energy to explain the apparent doubling of ESP-observed atoms) requires the groups of quarks (Arnoo) to pass through each other and that this seems very unlikely. But this comment misunderstands the oscillation - it is postulated to occur via the 4th spatial dimension (a "quantum mechanical tunnelling"), in which state there would be no such problem.

Eagles has suggested, as an alternative, that there could be a continuous rotation into and out of 4-D. A rotation in 4-D would appear as a vibration in 3-D but only the antinodes would be seen.

No "blurring" is reported during the ESP observations (unlike for a plucked guitar string's stationary antinodes), which is consistent with transit between the nodes via a 4th spatial dimension. It would be more like the escapement mechanism in a watch than vibration of a guitar string.

NOTE ADDED in 2010:

Phillips has published the following (below) after the above paper was written in 1983, but the same negative comments as above still apply and are not replied to in his later publications below. Phillips makes no mention of any of the author's (MGH's) publications above.

S.M. Phillips, "Anima", publ by TPH, Adyar, India & Wheaton, IL, USA (1996).
S.M. Phillips, "ESP of Quarks & Superstrings", publ by New Age Intl Publishers, Bangalore, India (1977).

Chapter 3 above, in the present book, gives a further discussion.

-=-=-=-=-=-=-=-=-=-=-=-=-=-=-=-

APPENDIX 6

M.G. Hocking, Journal of Scientific Exploration 21 (1), 13-26 (2007)

Linking String and Membrane theory to Quantum Mechanics and Special Relativity equations, avoiding any Special Relativity assumptions

M.G. Hocking
Materials Dept, Imperial College, London SW7 2BP
Email: m.hocking@imperial.ac.uk

Abstract

M-brane quark string theory and the Supergravity theory require 10 spatial dimensions. But if dimensions greater than 3 do exist, this must have important effects in other branches of physics, as quark theory cannot be compartmentalised-off. This paper shows how the concept of multi-dimensional space, essential to explain particle physics phenomena, removes conflicts between quantum theory and relativity. This leads, extremely simply, to both Schroedinger's Equation <u>and</u> to the Special Relativity equations in terms of absolute motion instead of assuming the two Principles of Relativity. The origin of the Big Bang provides an absolute spatial reference frame. A theory of rest mass is also given.

Keywords: Schroedinger, relativity, M-brane, supergravity, particle physics, multiple dimensions.

Introduction

String theory and M-brane theory predict 10 or 11 dimensions but suggest that 7 spatial dimensions are coiled up to a very small diameter so that we only perceive the remaining 3. If it is assumed that the extreme temperature of the Big Bang prevented any complexity of structure, then it is likely that at the beginning there was no coiling and so matter initially was in 10 dimensional space. As temperatures dropped, structuring became possible and on this model matter then evolved into the lower dimensions. In this case, ordinary 3 dimensional matter is formed by an energy entering from the next higher 4th dimensional space. This model leads to the equations of quantum mechanics and Special Relativity in 2 lines but without requiring either of the two Relativity Principles.

About 96% of the matter in the universe is described as "missing", meaning missing from 3-D space, but this could be because it is distributed among the higher 7 dimensions, which gravity can access but not photons, electrons etc, so it would be apparent only from gravity measurements and be missing from all other observations.

Schroedinger's Equation: $d\psi/dt = (hi/4\pi m)[d^2\psi/dx^2]$
is functionally exactly like Fick's 2nd Law of Diffusion***:
$dC/dt = (const.) [d^2C/dx^2]$, if C (concentration of a diffusate) is replaced by ψ.
This comparison suggests the following simple model:

A hypothesis of Dirac (1962) is that an electron resembles a bubble, rather than a point of matter, and this idea also accords with current membrane theories of space. "An atom is a hole with a tenuous envelope around it" - Schroedinger. This is supported by other indications that space is not "empty" but is filled with a continuous all-pervading background medium (Besant & Leadbeater, 1994), in which bubble-like particles move. Their movement through such a "space" (even in a vacuum) must then be by a diffusion process and hence Fick's Laws of Diffusion would

---

*** Readers not familiar with diffusion please see page 352-353.

be expected to apply, and in fact Fick's Law does appear, in the form of Schroedinger's Equation which is Fick's Second Law of Diffusion but with an imaginary diffusion coefficient. 3-D matter in space would then be bubbles in the continuous medium of space, inflated by containing an energy of creation (rest mass) welling up from 4-D space as mentioned above.

Schroedinger's Equation

Individual pollen grains in air diffuse jerkily due to molecular kinetic motion. Their diffusion follows Fick's Second Law of diffusion, $dC/dt = D(d^2C/dx^2)$. But there is no wavelike effect at all in the microscopically-observed diffusive jumps. Fick's Equation (1855) is exactly similar to Schroedinger's Equation (written 70 years later) which describes the motion of an elementary particle through free space, except that the "diffusion constant, D" becomes imaginary. Nature may be trying to tell us something here.

Fig.1 shows a solution of Schoedinger's Equation for the motion of an elementary particle in free space and ψ is imaginary in between the points on the trajectory where $\psi = 1$. This is like diffusive jumps but where the particle is imaginary during its jump. An obvious interpretation of this imaginary feature is that the particle may perform its diffusion jumps via a hidden dimension, the 4th dimension, in which it is momentarily absent from 3-D space and hence is "imaginary" during its jumps. This is an interpretation of Schroedinger's Equation. In earlier years, before the advent of M-brane quark string theory, which requires multi-dimensional space, many standard textbooks avoided this problem by an (unjustified) assertion that "the particle must be somewhere" at all times and they then (in effect) square ψ to prevent it from being imaginary (nowhere in 3-D space).

Pursuing the analogy between Fick's Law of Diffusion and Schrodinger's Equation, assume that elementary particles move in a similar way as diffusive jumps, but at their size, comparable to a 4th dimension's coiling-up size, there is some accessibility to a 4th spatial dimension (thus appearing to us as a "quantum mechanical tunnelling"). An ad-hoc assumption of diffusive jumps into 4-D is not required if the Zero-Point energy oscillations routinely involve very frequent regular excursions into 4-D where motion is not restricted by the 3-D background medium - a continuous medium would trap bubbles (see Dirac's hypothesis mentioned above) static in 3-D, like tiny air bubbles are trapped static in a block of ice (discussed later).

It would be strange if the existence of higher spatial dimensions required by string and membrane theories had no effect at all on fundamental physical processes such as atomic-scale motion.

So here now is a 3-line derivation of the Schroedinger Equation for motion in "free space" of an atomic size particle, which does not require any kind of wave:

A remarkable equation in pure mathematics (Euler's Equation) is:

$\exp[-2i\pi] = 1$ ....(i),

Write: $\exp[-2i\pi\{x/\lambda - t\nu\}] = \psi$ ....(ii),

so that whenever the item in { } brackets is an integer, then $\psi = 1$, but ψ otherwise contains an imaginary component.
<u>Ψ is not a wave</u> (in 3-D space): see Fig. 1.

The choice of $x/\lambda - t\nu$ for the term in {} brackets is explained as follows:

x is the distance of a moving elementary particle along a free-space trajectory and t is its time along that trajectory.
λ is the jump distance of the particle along its trajectory and ν is its jumping frequency - a diffusion-type model.
So x/λ is an integer if the distance x is a whole multiple of λ.
$t\nu$ is the number of jumps in time t.

Whenever x/λ and $t\nu$ are both integers, the particle is at a jump halt and is considered to be "present" (here in 3-D), but otherwise it is in transit and is considered to be in a higher (4th) spatial dimension and thus not present (imaginary) in our 3-D world.

The difference of two integers is also an integer, so they can be conveniently combined as in Equation (ii) above, to represent travel through both space and time. Fig. 1 plots this equation, showing ψ is unity where and when x and t both correspond to an integer number of jumps, but ψ contains an imaginary component elsewhere as required on the above model.

Finally, apply De Broglie's Equation to the x/λ term,
and Planck's Equation to the tν term (explained below), to get:

$\exp[-2i\pi\{xmv/2h\} - tE/h\}] = \psi$ ……(iii), which is a well-known solution of Schroedinger's Equation, where E is kinetic energy only:

Schroedinger's time equation is $d\psi/dt = (hi/4\pi m) [d^2\psi/dx^2]$.

Cf Fick's Second Law: $dC/dt = D[d^2C/dx^2]$, derived 70 years earlier.

As mentioned above, diffusion of pollen grains, or of ions jumping through a lattice, have no wavelike character, so Schroedinger's Equation need not have, either.

Schroedinger's Equation gives correct results for atomic-scale phenomena and so must form a part of any valid theory of Nature.

Planck's Equation and its counterpart, De Broglie's Equation, are easily explained. A "longwave" radio signal of 1 km "wavelength" would be a "photon" whose jump intersects ordinary 3-D space at 1 km intervals, whereas a light of "wavelength" 1 micron would be a photon which intersects 3-D space at 1 micron intervals. Clearly the perceived energy of the light photon would thus be 10^9 times greater than that of the radio photon, as it appears 10^9 times more frequently in the ordinary 3-D world. So E is proportional to ν (Planck's Equation). There is no wave, only jumps, rather analogous to the diffusion jumps of (say) copper ions through a solid cuprous oxide lattice.

APX 6   3-Line derivation of Schroedinger's Equation, &
1-Line derivation of "Special Relativity" Equations

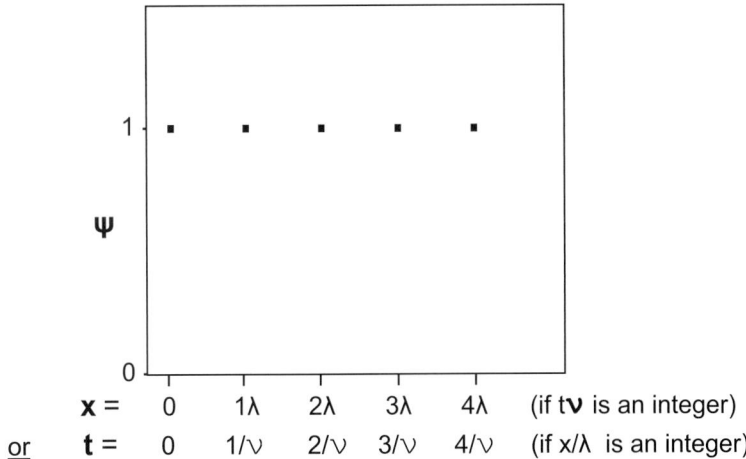

Fig. 1. Plot of equation (ii),   ψ = exp[-2iπ{x/λ - tν}]
or equation (iii),   ψ = exp[-2iπ{(xmv/2h) - tE/h}]

ψ = 1 whenever the particle has made an integral number of jumps, n (of length λ),
which is:
when its distance travelled = x = nλ, or,
when its time of travel = t = n / ν.

No Wave Function:

Thus the "diffusivity", D, of a moving elementary particle is imaginary, meaning simply that it does not continuously exist in 3-D space. Prior to the introduction of 10-D space by quark string theory, the imaginary values of ψ embarrassed physicists, who only considered 3 dimensions and thus decided in effect to square ψ to force it to be real and thereby artificially created "matter waves". They called this process "normalising" ψ and it compelled ψ to conform with the then "world view" of what Nature was felt to be. This understandable attitude at that time (that there are no higher dimensions) is very well illustrated by many standard textbooks which assert that "the particle must be somewhere", to "justify" effectively squaring ψ to prevent it from being imaginary (nowhere in 3-D space)! This procedure discounts the possibility that it actually could sometimes be nowhere in our 3-D space, if it oscillates or spins in and out of 4-D space. This "normalising" approach artificially creates a fractional probability (i.e. an uncertainty) that a particle is present at any given location, which creates the notion that particles can somehow exist as waves and leads to interpreting ψ as a "wave function". But De Broglie intuitively said that "matter waves" are "*ondes fictives*".

The following assertions are cited from classic texts which pre-date quark string theory and are based on the then "world view" of Nature. In considering these, a remark by Haldane should be recalled:

"Nature is not only stranger than we have thought,
It is stranger than we can think!"

Moelwyn-Hughes (1961) asserts, "the particle must be somewhere".
Margenau & Murphy (1961) remark, "if initially there was a certainty of finding a particle somewhere in space, there might later be uncertainty, this is a situation which would clearly be physically untenable".
Cottrell (1960) asserts, "… the chance of finding the free electron somewhere in the metal must be unity".
Moore (1962) remarks, "ψ must be finite and continuous for all physically possible values of x. The requirement of continuity is helpful in the selection of physical reasonable solutions for the wave equation".

They then all effectively use $ψ^2$ to ensure that this view prevails and discard ψ.

Margenau & Murphy (1961) grumble that a function like equation (ii) above, which, when plotted, is a series of horizontal points separated by imaginary gaps, "is a monstrosity"! It is shown in Fig.1. But it comes directly from Euler's Equation, whose equations (18th century) are also used in modern quark string theory.

Feynman (1966) avoids being so blunt, but instead asserts, "*We want* a function to be *zero* everywhere but at a point". But he admits, "there is *no* mathematical function which will do this!" (his ! mark). But instead of accepting this strong hint from Nature not to do it (you can lead a horse to water but you cannot make it drink!), the unnatural step is then customarily taken of artificially *defining* such a made-up function, called the Dirac delta function.

Schroedinger (1926) with some insight said, "One may be tempted to associate ψ with a vibrational process in the atom, a process possibly more real than electronic ..." and "The ψ function itself cannot and may not in general be interpreted directly in terms of 3-D space - because it is in general a function in configuration space and not in real space". Before the advent of 10-D quark string theory, "configurational space" was the only term that could be used.

Quantum-Mechanical Tunnelling

"Quantum mechanical tunnelling" (well named) then becomes the ability of an elementary particle to pass in a non-3D-material form, from one 3D location to another without moving through any of the three 3D dimensions - i.e. via a micro-"worm-hole" in 3-D (but without any Relativity connotations - its motion is absolute – see below). There is no need for a "wave-nature" explanation for quantum mechanical tunnelling.

In diffusion through oxide layers for example, quantum mechanical tunnelling allows electrons to reach the outer surface of the thin highly insulating oxide film on aluminium and thus creates a billion volts/metre field, which then drives further oxidation unless prevented (Moussa & Hocking, 2001). These electrons cannot have reached the outer surface of the alumina layer by moving through the alumina, as there is no electronic conductivity.

Special Relativity Equations derived assuming absolute motion: Rest Mass; Length and Time Dilation; $E = mc^2$

On the basis of the "Big Bang" theory with its residual microwave radiation, it is concluded that there is an absolute reference point of origin (the "Big Bang" site) in space. This negates the First Principle of Special Relativity, which denies an absolute reference point in space. A Big Bang point of origin in 3-D space would also be accessible in higher dimensional spaces.

APX 6 3-Line derivation of Schroedinger's Equation, & 1-Line derivation of "Special Relativity" Equations

Although Special Relativity is an idealisation for gravity-free space, and so strictly does not apply to the Big Bang universe, it is nevertheless widely used in practice in physics and should not thus be "sheltered" from the existence of a point of origin of the Big Bang. To ingeniously avoid the problem (for relativity) of having a central reference point, the analogy is sometimes given of the universe being like a balloon being inflated, starting at a point (Big Bang origin), but later when large (when the universe had expanded) anyone anywhere on the surface of the balloon would think his location was the original centre. But if space pre-existed the Big Bang, this balloon model would be wrong. Who can say?

A two-line derivation is given below, of the mass, time and length dilation formulae of Special Relativity but <u>without assuming any relativity</u>.

2-D space is not viable for the existence of life forms because the complexity required for brain interconnections, digestive tracts, etc requires 3-D. Simple calculations show that electron orbitals in atoms would not be stable for dimensions higher than 3, which makes only 3-D space uniquely suitable for life**:**

The electrostatic force falls off as the inverse square of distance in 3-D but it would fall off as the inverse cube in 4-D space (it would then be too weak to bind electrons to their atoms). The inverse square arises simply because a given flux through unit element of area on the surface of a 3-D sphere is spread out in proportion to the square of the radius, as the area of a 3-D sphere is $4\pi r^2$, but the volume of the 4-D analogue of a sphere is proportional to r^3. (Cf the perimeter of a circle is proportional to r, for a 2-D case.)

A 3-D elementary particle and derived particles like atoms and molecules cannot make up a 4-D object, because they have no extension in the direction of a 4th spatial dimension. So they (and any larger body they constitute) are thus confined to 3-D space only and so cannot enter 4-D space, with the one very localised exception described in the section on Rest Mass below, as part of a very small amplitude oscillation. For a larger scale excursion into a 4th or higher dimensions, the 7 orders of coiling-up of the 7 higher dimensions in 3-D particles must be reduced by 1 order, each time the next higher dimension is reached.

Rest Mass

In 3-D space, elementary particles which constitute molecules, etc, are proposed in the Introduction above as being like gas bubbles in a continuous medium (Dirac, 1962; Besant & Leadbeater, 1994). However, a continuous medium cannot be described as a "fluid" because a fluid is able to flow and to thus permit particles to move through it due to mobile atomic-size "holes" in it (in the conventional well-known "hole theory" of fluid flow). E.g. a solid metal does not flow - its viscosity is extremely large, but in the liquid state metals contain a large proportion (about 10%) of "holes", which confers a very low viscosity to them and they then flow very easily.

An analogy is the common observation of a solid block of ice which has a few tiny bubbles of air trapped in it - these bubbles are "locked up solid" and cannot move at all.

Thus it is proposed that 3-D elementary particles (bubbles) in the continuous background medium can only have a zero velocity in it. Actual physical movement which is of course commonly observed in 3-D space can then be postulated as occurring by the following mechanism, which is necessarily similar to diffusion (being movement through a medium). This accords with the identical functional forms of Fick's Second Law of Diffusion and Schroedinger's Equation:

If 3-D space consists of a 3-dimensional continuous background medium (Besant & Leadbeater, 1994) as explained above (Cf. air bubbles in block of ice model) an elementary particle (bubble) would be unable to move in any of the 3-dimensional directions. But if it were able to jump out as part of an oscillation into a higher spatial dimension where there is no such continuous medium, it could then move and then land back in the 3-D space medium in a different place.

APX 6 3-Line derivation of Schroedinger's Equation, &
1-Line derivation of "Special Relativity" Equations

An elementary particle might be rotating and vibrating continuously (even if at rest in 3-dimensional space) in a path which takes it continuously in and out of the fourth dimension (an effect similar to zitterbewegung). "Zero-Point Energy" means that even at zero degrees Kelvin "rest", a particle is still oscillating incessantly (called "zitterbewegung", *Ger.* "trembling"). If the energy (welling up from a 4th spatial dimension) creating the 3-D bubble, has a characteristic velocity of **c**, then an observed average velocity v through the 3-dimensional medium would consist of periods at zero velocity in 3-D (due to its very large viscosity) alternating with jumps at velocity **c** in 4-D. A characteristic velocity of **c** is not extraordinary - e.g. a photon in free space has only got this one velocity, **c**, the velocity of light.

Jumps into the next higher dimension would only be possible for elementary particles as the amplitude of an excursion into 4-D space would be limited to the very small diameter of the coiled-up 4th dimension for 3-D particles, and not available to large bodies, and it is called "quantum mechanical tunnelling" in physics but not yet interpreted as involving jumps into 4-D space. If there are higher dimensions, it would be very odd if they were not involved at all in atomic-size processes. They cannot just apply to quark physics and nothing else.

Such a model leads immediately to Schroedinger's time and distance equations (for a case with zero potential energy), as shown above. It also provides a theory of rest mass, and leads to the same experimentally verified <u>equations</u> of Special Relativity but for <u>absolute</u> motion. The derivation is far simpler than that from Special Relativity. This absolute motion derivation uses the assumption of quark string physics that there are more than 3 dimensions in space:

The mass, length and time dilation equations are easily obtained immediately by solving a Pythagorean triangle with sides m_0c, mv and the resultant mc (Fig. 2):

APX 6 3-Line derivation of Schroedinger's Equation, & 1-Line derivation of "Special Relativity" Equations

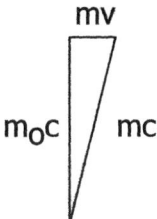

Fig. 2. $(m_0c)^2 + (mv)^2 = (mc)^2$.   Rearranging:

$$m_0 = m\sqrt{(1 - v^2/c^2)}.$$

m_0c can be regarded as the momentum of creation of a particle at rest in 3-dimensional space, due to an energy welling-up from the direction of a fourth spatial dimension (which is at right angles to any 3-D direction); m_0 is the rest mass of the resulting stationary particle in 3-dimensional space, which this force creates. If the particle is then made to move in 3-D space, by giving it momentum in a direction in 3-dimensional space, it will then have an <u>extra</u> momentum mv (see Fig. 2), at right-angles to its rest-mass (4-D) momentum-of-creation vector, where m is its mass and v is its observed velocity in 3-dimensional space. The resultant <u>total momentum content of the particle</u> due to these two momenta is mc (see Fig. 2), m being the dilated (increased) mass of the particle due to incorporation of its extra energy of motion in one of the 3-D directions (this is additional to its rest-mass energy welling up from the fourth dimension).   The momentum of creation must be at 90º to any momentum of motion in 3-D, because the 4th dimension direction by definition is at 90º to all 3-D directions - hence the Pythagoras triangle in Fig. 2.

So, from Fig. 2: $(m_0c)^2 + (mv)^2 = (mc)^2$, which rearranges

to: $m_0 = m\sqrt{(1 - v^2/c^2)}.$

This is the well-known and experimentally verified "relativistic" mass dilation formula but has been derived above for absolute motion in only 2 lines and without assuming the two principles of Special Relativity.

Time Dilation

Time dilation will also occur, because when a particle (e.g. a meson) is jumping in the fourth dimension, its internal decay processes will be frozen for the duration of that jump and so its lifetime will be extended. The well-known time dilation formula can then also be obtained immediately, as above, from a Pythagorean triangle (Fig. 3) with sides t_o, t_{OUT} and t, as explained below.

$$t_{IN} = t_o \quad\quad t_{OUT} \quad\quad t$$

Fig. 3. $\quad t^2 = t_o^2 + t_{OUT}^2$. Rearranging:

$$t_o^2 = t^2 - v^2 t^2 / c^2.$$

To explain this, pursuing the analogy with diffusion, assume that the motion of an elementary particle occurs by very short jumps alternating with longer stationary periods, thus allowing any observed overall velocity to be made up, modelled on the conventional mechanism of diffusive jumps of atoms or ions through a lattice, from site to site. The identical functional forms of Fick's Second Law of Diffusion and Schroedinger's Equation was discussed earlier. In diffusion of an ion through a lattice, a jump down a concentration gradient occurs when the chemical energy gradient (Gibbs Free Energy) gives sufficient activation energy for a jump to the next lattice position. This model is used below for motion of a particle, where a mechanical energy gradient drives it.

There are only two velocities possible, zero for the periods at rest in the 3-D world, and **c** for the periods when the energy constituting the particle is moving in 4-D space. Any actually observed overall velocity, v (0<v<c), is then made up of rapidly alternating combinations of these two values. The moving particle

travels in a series of very small jumps each of which is at velocity **c**, separated by a series of short pauses at velocity zero (analogous to the movement of the frames of a cine film - a film strip), so that the overall actually observed velocity is apparently v. This 4-D jumping model is consistent with the explanation given of the imaginary values of ψ given above.

A moving atomic-size particle is thus a "particle" when stationary and may appear to be an apparent "wave" (a non-particle) when jumping. Light photons alternately jump a distance λ in λ/c seconds followed by a stationary instantaneous wait or appearance. It is thought that photons (unlike gravitons) cannot move appreciably away into 4-D and so are bound to continually intersect our 3-D world.

Let the total stationary time (spent residing at successive positions) be t_{IN} and the total transition time (spent in jumping between these positions) be t_{OUT}.
t_{IN} is the inactive stationary time elapsing between jumps, and can have any value ($0 < t_{IN} < \infty$). t_{OUT} is the time taken for a transition or jump between residences, and represents a non-material (non-particle, apparently wavelike) condition in between the physical sites at which the moving particle successively resides. It means that there is no physical movement at all and that all actual movement occurs during the time when the particle is in 4-D, by a series of non-material (non-3-D) jumps. It is somewhat analogous to the conventional diffusion mechanism for an atom or ion diffusing between fixed lattice sites. If Zero-Point energy involves continuous vibration, or rotation, into and out of 4-D, then this process is facilitated by that and does not need a separate ad-hoc mechanism for it.

Consider now the motion of a mechanical clock which contains a balance wheel. On the proposed theory, the balance wheel (=**B**) jumps have their specific discrete **B** activations (see activation energy, above), but when the clock (clock = **C**) as whole is also set in motion, specific discrete **C** activations will occur additionally. Any jump activation which becomes due to cause an imminent balance wheel (**B**) jump during the course of a clock (**C**) jump, would be inoperative, as the clock is "frozen" - already engaged in a jump and so its balance wheel cannot also simultaneously move

then. Consider the clock to be moving much faster than the balance wheel rotations. Then the balance wheel (**B**) jump frequency is comparatively very low and those **B** jumps which arise during a regular **C** jump will be lost. The consequent loss of some **B** jumps will (in effect) slow down the balance wheel. Consider now the motion of a clock whose tick-tick period is t_o <u>at rest</u>, which corresponds to t_{IN} as defined above. Let this clock travel with a constant overall velocity v and record the passage of one tick-tick time period t_o during its travel through a certain distance s. The total **C** jumping time (at velocity c) which is non-material (being in 4-D), is not sensed or recorded by the clock (by **B** jumps, as explained above) is t_{OUT}
where t_{OUT} = distance/velocity = s/c = vt/c ....(iv).

A stationary observer would have a total time t in (iv) above, elapsed on his watch, as being the time taken for the moving clock to travel the distance s. Now, $t > t_o$ or t_{IN} due to the additional time t_{OUT} taken for the journey, noticed only by the stationary observer, which must be added to t_o. This addition must be vectorial, because as the moving clock does not sense or record t_{OUT} there is no break (in its sensation of time) at which t_{OUT} can be added in a scalar manner. t_{OUT} and t_{IN} have no component in common and must thus be added as vectors at right angles (Fig. 3).

This gives:

$t^2 = t_{IN}^2 + t_{OUT}^2$ ....(v), by Pythagoras' Theorem,

or $t^2 = t_o^2 + t_{OUT}^2$.

Substituting t_{OUT} from (iv) above, **$t_o^2 = t^2 - v^2t^2/c^2$,**

which is the well-known Time Dilation formula of Special Relativity, but all the assumptions of Special Relativity are avoided. This equation has been well-verified experimentally, e.g. by the increased lifetimes of decaying mesons which are moving very fast, compared with slow-moving mesons.

The time dilation formula can also lead to an alternative derivation of the mass dilation formula, already derived otherwise, above.

Fitzgerald-Lorentz Length Contraction Equation

Similarly, the length of a moving body will contract (only in the direction of travel) due to the interatomic cohesive forces pulling in its length across planes of jumps when it is in 4-D space (where it is not affected by 3-D electrostatic cohesive physical forces; only gravity can enter 4-D space and gravity is not involved in cohesive forces).

A similar Pythagorean triangle gives the well-known length contraction equation. The length of a moving object is proportional to the number of moving elements materially present ("IN") in it along any given line in the direction of motion. The term "moving element" merely refers to an elementary particle of the moving object. Along any such line through the object, some of its moving elements will be jumping ("OUT") and thus materially absent from the object. At a steady velocity there will be a steady proportion of moving elements thus missing, and a consequent shrinkage of the length of the object in the direction of its motion (due to the attractive forces of cohesion acting across the OUT gaps). Planes of OUT gaps (analogous to vacancies) would be expected to sweep through the object (which is not imagined to jump all at once, but as individual particles or moving elements) in the direction opposite to that of the motion; the planes of moving elements would be set perpendicular to the direction of motion; thus there is no reason for shrinkage of the object in other directions than that of the motion. Consider now a moving object, of rest length L_o measured in the direction of its motion.
At rest, $L = L_o$ and $t_{IN} = t$.

The number of planes (perpendicular to the direction of motion), of moving elements which are materially present (IN), at velocity v, is $n = n_o(t_{IN}/t)$ where n_o is the number of such planes present at <u>rest</u> (for which state $t_{IN} = t$).
$n_o \propto L_o$ and $n \propto L$, where n and L are number and length respectively, at a steady velocity v).

Thus, from $n = n_o(t_{IN}/t)$ above, we have:

$L = L_o(t_{IN}/t) = L_o[\sqrt{(t^2 - t_{OUT}^2)}]/t$, using (v) above,

so $L = L_o\sqrt{[1 - t_{OUT}^2/t^2]}$, and then using (iv) above we obtain:

$L = L_o\sqrt{(1 - v^2/c^2)}$, which is the Fitzgerald-Lorentz length contraction equation.

An alternative approach also follows from the assumption that when an object is travelling, some of the elementary particles constituting it are engaged in a jump in 4-D and are thus "missing" from the 3-D object, as suggested by the interpretation of Schroedinger's Equation given earlier. Consider the number of elementary particles in a line in its direction of travel to be n_o at rest and n at velocity v, where $n < n_o$ as some of them are jumping. n and n_o are their numbers in 3-D space.

As mass in conserved, $n_o m_o = nm$, [where m is the enhanced mass at velocity v given in $m_o = m\sqrt{(1 - v^2/c^2)}$].
Then, as $n_o \propto L_o$ and $n \propto L$, for a line in the direction of motion of the object, $L_o m_o = Lm$ and so **$L = L_o\sqrt{(1 - v^2/c^2)}$**.

This is the Fitzgerald-Lorentz length contraction equation.

$E = mc^2$ derivation

The well known Relativity equation $E = mc^2$ can also easily be obtained (for absolute motion), by elementary algebra:

From the Pythagoras triangle of the Rest Mass section above,

$(m_o c)^2 = (mc)^2 - (mv)^2$ (See Fig. 1)

Take differentials: $0 = 2c^2 m\,dm - 2mv^2\,dm - 2vm^2\,dv$

Divide both sides by 2m: $c^2 dm = v^2\,dm + mv\,dv$ ....(vi)

By definition, force is rate of change of momentum, so

$F = d(mv) / dt = m(dv/dt) + v(dm/dt)$

By definition, a force is also an energy field or gradient, dE/ds

and velocity $v = ds/dt$ where s is distance.

So $dE = Fds = m(dv/dt)ds + v(dm/dt)ds = mvdv + v^2dm$

Compare this with equation (vi) above:

$dE = c^2dm$, so, integrating, **$E = mc^2$** (Einstein's Equation).

The integration constant is zero, as $E = 0$ when $m = 0$.

Heisenberg's Uncertainty Principle

Heisenberg's Uncertainty Principle takes on a new meaning: a moving particle will actually spend most of its time at rest (punctuated by very short times at c), but its experimentally observed velocity is measured as v and so a measure of the uncertainty in its velocity at any instant will be $v - 0 = v$. (This uncertainty depends on exactly when an observation is made and so is in the mind or control of the observer and is not a property of the particle.) From de Broglie's Equation, mv is proportional to h/λ, and so the Uncertainty Principle becomes an expression of de Broglie's Equation if λ is interpreted as the moving particle's smallest jump length on the above diffusion model for motion.

Spin

An object in 3-D requires a rotation of 360° to return it to its original position, but a bizarre 720° of rotation (not just 360°!) is required to bring fermions ("spin-½" particles, such as protons) back to their original state. This is easily explained as follows, on the above model:

For clarity, a 2-D / 3-D analogue will be used, instead of 3-D / 4-D. If a lower- case letter "**d**" is lifted out of its 2-D paper sheet and turned over in 3-D space and then put back as a "**b**", then this would appear to a 2-D inhabitant to be a **d** ←→ **b** vibration with only its antinodes (**d** & **b**) being visible. If this **d** ←→ **b** vibration is analogous to Zero-Point Energy vibration, then if the "**d**" is also spinning in 2-D (**d** ←→ **p** ←→ **d**), then after 360° of spin in 2-D it could have simultaneously rotated to a "**b**" by the 3-D rotation, which means that the 360° spin in 2-D did not return the "**d**" back to its initial state and that a further 360° of 2-D spin is needed by which time the "**b**" would have rotated back to a "**d**" in its simultaneous 3-D rotation. Thus a bizarre (to a 2-D observer) 720° of spin is required for a "**d**" spinning in 2-D space to return to its original "**d**" state.

With this preamble, for our case in 3-D space, an observed (in 3-D) rotation of 720° is needed to return a proton to its original state, which can easily be explained analogously to the example above.

In 3-D to 4-D terms, this means that (to give an analogy) a tennis ball spins in 3-D and 4-D simultaneously but after 360° of observed (in 3-D) rotation the ball would be everted (i.e. having its fluffy side inside and smooth side outside, without loss of the gas pressure which it contains) by the simultaneous 4-D rotation and so clearly a further 360° of observed (in 3-D) rotation would be needed for it to return (by further 4-D rotation) to its original state with the fluffy side outside, making a total of 720°!

This can only be understood in terms of the existence of 4-D space and it happens routinely for elementary particles, which have access to 4-D space.

<u>Note:</u> A <u>rotation</u> in 3-D could only be perceived by a (hypothetical) 2-D observer as a <u>vibration</u> (like Zero-Point Energy). And a rotation in 4-D could only be perceived by us (in our 3-D world) as a vibration (Zero Point Energy).

Access of large objects to 4-D space is problematical. Eversion of tennis balls has been reported anecdotally which is, of course, not scientifically acceptable, but there is a report in Nature by Hasted et al (1975) of a refractory crystal of vanadium carbide being removed from a sealed tube in laboratory conditions, without any contact being made with the tube, which could only be feasible by transfer out via 4-D space.

APPENDIX 6A The nature of velocity:

Consider a moving particle. The substitution of <u>wavelength</u> and <u>frequency</u> of a supposed "wave", by a <u>jump length</u> and <u>jump frequency</u> of a particle, is consistent with Planck's and de Broglie's Equations and with the kinetic energy of a particle being $\frac{1}{2}mv^2$:

(1) If a particle moves by jumping distances of length λ at a jump frequency of ν, then its velocity is v = (distance) / (time)
= $(\lambda)/(1/\nu)$. = $\lambda\nu$ ....(a).

(2) On a common sense approach, if the energy of a particle is doubled, this would be manifested by it appearing twice as frequently in 3-D space.
So the kinetic energy E of a particle $\propto$ the number of times per second (ν) that it appears in 3-D space, giving Planck's Equation, $E = h\nu$, where ν is the jump frequency.
From equation (a) above, $\nu = v/\lambda$, so $E = hv/\lambda$.
$E = \frac{1}{2}mv^2$, where v = velocity, so $\frac{1}{2}mv^2 = hv/\lambda$, so $\frac{1}{2}mv = h/\lambda$, which is de Broglie's Equation.

(3) Now suppose the velocity v is doubled:

$E = \frac{1}{2}mv^2$, where v = velocity. If v is doubled, then because the jump length (λ) is halved, each particle must jump 4 times more often in order to move twice as fast.
If v is doubled, the energy is 4 times higher (since $E = \frac{1}{2}mv^2$).

References:

Besant A. and Leadbeater C.W. (1994), http://www.4-D.org.uk
Cottrell A.H., "*Theoretical Structural Metallurgy*", Arnold (1960).
Dirac P.A.M., "*The Conditions for a Quantum Field Theory to be Relativistic*",
    Proc Royal Soc (London) Series A $\underline{268}$, 57 (1962); see
        also Stedile E., "*Quantum Aspects of the fundamental Dirac Membrane Model*", Int J Theor Phys $\underline{43}$, 385 (2004).
Euler L. (18th Century).
Feynman R.P., Leighton R.B. & Sands M.: "*The Feynman
    Lectures on Physics*", Volume 3,   Addison-Wesley (1966).
Hasted J.B., Bohm D.J., Bastin E.W., O'Regan P. and Taylor J.G.,
    "*Recent research at Birkbeck College, University of London*",
    Nature $\underline{254}$, 470 (1975).
Margenau H. & G.M. Murphy G.M.: "*The Mathematics of Physics & Chemistry*", van Nostrand (1961).
Moelwyn-Hughes E.A., "*Physical Chemistry*" (chapter: "Mathematical
    Formulation of the Quantum Theory"), Pergamon Press (1961).
Moore W.J., "*Physical Chemistry*", Longmans (1962).
Moussa S. & Hocking M.G., "*Photo-inhibition of localised corrosion of stainless steel in NaCl solution*", Corrosion Science $\underline{43}$, 2037 (2001).
Schroedinger E.: "*Quantisation as a problem of Eigenvalues*", Annalen der Physik $\underline{79}$, 372 (1926);
    and Schroedinger E.: "*Quantisation as a problem of Eigenvalues*", Annalen der Physik $\underline{81}$, 135 (1926).
    [Quoted by M. Jammer in "Conceptual Development of Quantum Mechanics", p. 372, 267, McGraw-Hill (1966).].

-=-=-=-=-=-=-=-=-=-=-=-=-=-

Expanatory Note on the laws of diffusion:
This is written (on the next page) for those who are not familiar with the laws of diffusion.
It is a simple derivation of these two laws.

APX 6 3-Line derivation of Schroedinger's Equation, & 1-Line derivation of "Special Relativity" Equations

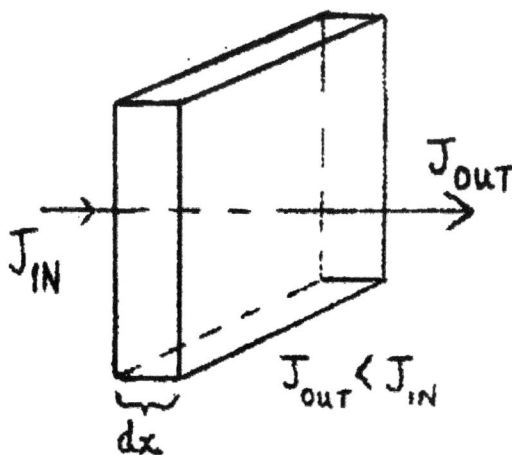

J = flux of diffusing material through the 1 m² area.
Volume of box = $(1) \times (1) \times (dx) = dx$ m³.
$J_{IN} - J_{OUT} = dJ$ = no of atoms accumulating per second in the volume of dx m³ (the above box).

So dJ/dx = no of atoms accumulating per second in a 1 m³ volume, which we write as $-dC/dt$,
and so $dC/dt = -dJ/dx$, ...(i)
where the negative sign is because the direction of increasing concentration C with time t is that of decreasing J with distance x.

Fick's First Law is: Flux is proportional to (concentration gradient),
i.e. $J = -D(dC/dx)$.

Apply this to (i) above, to get $dC/dt = d(DdC/dx)/dx = D(d^2C/dx^2)$, if D is constant.

Fig. 4. Fick's Second Law of Diffusion: $dC/dt = D(d^2C/dx^2)$.

APPENDIX 7

Inventions of Dr Nikola Tesla for transmitting energy by radio waves

M.G. Hocking (G8IFR)

Email: m.hocking@imperial.ac.uk
Imperial College, London SW7 2BP, UK

Introduction

Tesla was a prolific inventor and wrote about 1200 patents over many years around a century ago and some surprising results have been recorded, such as the remote lighting of some incandescent lamps at about 26 miles distance and driving an electric car (without using an electric power source located within the car) for 50 miles in 1931.
It has been suggested that he had invented a new source of energy, but on researching these reports they can be explained in a more conventional way. His method gives a way to avoid global warming as no CO_2 is evolved.

The Tesla electric car

The Tesla electric car is anecdotal because Tesla was very secretive. Tesla is reliably reported to have driven 50 miles at up to 90 mph in his (silent running) electric car in New York State in 1931. He had invited the press to record the event. The small auxiliary battery in the car could not possibly have powered the car for more than a few yards, and the motive power source is unknown. Tesla was a genius in electro-magnetism. He had a bad experience some years earlier, when someone stole one of his inventions, and after that he became very secretive. E.g. Marconi had claimed to have invented and used equipment to transmit signals across the Atlantic in 1901, but the USA Supreme Court in 1943 dismissed this claim and ruled that the equipment had actually been previously invented by Tesla. Tesla invented the laser in 1894 and a radio controlled boat in 1898! Radio transmission was discovered by Henry in 1842 followed by Morse in 1847 and then Loomis in 1872 who sent messages 14 miles.

Tesla invented the multi-phase AC generator and induction (brushless) motor but others claimed its invention. Tesla's AC generator allowed long distance distribution and displaced Edison's DC system which was only local due to i^2R losses. Tesla's aim was to produce a wire-less system for electric power distribution.

Tesla's electric car power unit is an unknown (or, at least, presently un-applied) means for transmitting power remotely by radio waves. The latter is suggested by his writings [see references appended] and is explained below.

Before 1900, Tesla had lit 200 incandescent bulbs remotely, without any wires, at a distance of 26 miles! After his death in 1943 Tesla's papers were confiscated by the USA Government but some went to Serbia and thence to the then USSR.

The "woodpecker", known to amateur radio as a very powerful short wave transmission around 16 MHz from NW USSR in 1977 to 1980s, was a series of modulated carrier wave dashes with a repetition frequency of 9.615 Hz (measured April 1977) and other frequencies from 5 to 15 Hz, around the alpha brainwave rhythm of about 12 Hz which has led to suggestions that it was an attempt to lock people's minds into a non-thinking alpha state. More recently, mobile phone designers have (foolishly) chosen brainwave frequency pulsing for their phones. In 2007, the Sleep Laboratory at Loughborough University found that proximity to mobile phones can lead to insomnia; this could be due to the phone pulse frequency "driving" brainwaves at around the beta "wide-awake" state or beta frequency. But more likely than for brainwave interference, the "woodpecker" was probably an attempt by the USSR to resonate at the Earth's frequency (the Schumann Resonance) of 7.8 Hz (fundamental) which Tesla had earlier discovered and pioneered for his wireless power transmission method. Tesla believed that a high power high-frequency carrier wave pulsing at about 8 Hz would resonate such that its power could be captured almost unattenuated at any point on the Earth's surface, but the highest power would probably be available near the source of the transmitter. A frequency of 7.8 Hz has a wavelength of 38,333 km, which relates to the Earth's circumference of about 40,000 km or its diameter of about 13,000 km. This would resonate and a global size standing wave would be set up between these two points. His antipodal point was in the

Southern Ocean near New Zealand; the Northern point would actually be a somewhat diffuse zone of many miles radius around Tesla's transmitter near Niagara Falls hydroelectric power station and a car near there could be driven from this power by rectifying the signal to get the modulated signal on the 7.8 Hz carrier.

Tesla pulsed his 30MV carrier wave which was in the 2 metre band region, as Tesla's electric car is reported to have had a box with a 6 foot long metal rod protruding from it, presumably an aerial. Due to the enormous amplitude of a 30 million volt carrier wave, there would be plenty of power available in the rectified signal, enough to drive a car. Tesla had a tower with a copper globe atop, and using his special anti-flashover Tesla coil design he was able to drive a 30 million volt carrier wave relative to a good ground-plate, at about 8 Hz. He deduced this Schumann frequency value from his own measurements on how natural lightning strikes (listened to on a radio) resonated tellurically. Tesla remarked that it only needs a few tens of kW to maintain such a telluric oscillation. There are large natural concentrations of ions washing about in the upper atmospheric regions which would respond to a Schuman resonance excitation. Tesla used the corresponding earth conduction wave as the means of power propagation, rather than the conventional above-ground waves propagation used for radio communication. This seems the most likely way he powered his electric car in 1931. As mentioned above, it was repeated much later in Russia, but it is not known to have been repeated in the West.

It seems likely that the "woodpecker" oscillations emanating in the 1980s from near Riga in Latvia, may have been USSR attempts to develop a Tesla power transmitter. Many complaints were made to the USSR about the "woodpecker" interference to shortwave communications and I believe they have now ceased (?). The antipodal point for Riga is in the southern Pacific Ocean.

Tesla's earlier and much larger 200 foot mast at Wardenclyffe (NY), designed for about 7 GW pulsed output, (probably) at about 12 Hz, had a 120 foot shaft below it to give a total of 300 ft, half the 600 ft that he had hoped for, due to funding problems. Morgan (the New York financier) mistreated Tesla financially and a few days after that tower was finished and being tested, bailiffs arrived and removed the equipment -- a tragedy for radio engineering. Thus that particular

superb tower was never used. The tower had 16 iron grounding pipes radiating underground from the bottom of the shaft and 4 stone-lined tunnels, possibly for adding salt water to improve earth contact conduction, which was the main mode of propagation (ground wave).
See also the References section below.

Possible Application: At present, much research is done on batteries and fuel cells for electric cars. Suppose there are a million vehicles moving in London and each one needs about 5 kW, then 5 GW is needed. This could be possible with a Tesla woodpecker system, and if a GPS system (as proposed for road charging) were used, billing could be possible. This would greatly reduce global CO_2 emissions. Possible non-ionising radiation safety problems near the transmitter need investigating but these could easily be avoided by siting the transmitter in a remote place some distance from the hydro-electric plant.
Ideal locations for transmitters would be large hydroelectric power stations, e.g. at Niagara Falls, Victoria Falls, and many other falls worldwide. Power distribution (being world-wide) would then supply cars, factories and homes located in different time zones around the world, which thus advantageously spreads the load demand around the clock.

There are very large powers already washing around naturally in the ionosphere, such as the best-known one of March 1989 which caused power outages over a wide area of North America, including blacking out Quebec and NYC. If these ion flows were marshalled and controlled by a Tesla woodpecker resonant driving oscillation, they could be put to use.

Due to the very great importance of global warming prevention, any device that has even a moderate chance of success should by investigated, and should not be summarily dismissed.

A pilot plant should be constructed to test the method.
If successful, contacts in the Zambia government have said they are seeking investors in hydroelectric power generation and have very good investment incentives to welcome proposals. The Kariba Dam (down river from the Victoria Falls in Zambia) generates 1.32 GW and annually supplies 6400 GWh typically.

For comparison the total UK electricity grid at full power runs at somewhat over 10 GW.

But it is expected that the Tesla woodpecker resonance method will extract energy from the ionosphere, so 1 GW put in would extract much more than 1 GW of power.

Also, as mentioned, different time zones would spread the load demand around the clock because demand from Europe occurs when there is little demand from USA and China.

Additional Notes: The power generation system described above is predicted by Besant & Leadbeater (in "Man, Whence, How and Whither", published in 1913), and although it is still in our future, they say it was abandoned when a limitless energy source was discovered (also further into our future!). Some clues about the latter are that it is available as a hand-held device, as well as in larger versions.

Tesla had visionary experiences and super-acute hearing (see these in Appendix 8).

REFERENCES for Appendix 7:

The following 5 books are available in the Science Museum Library, London:

M.J. Seifer, "Wizard, the life & times of Nikola Tesla", Carol Publ Gp, NJ (1996).

J.T. Ratzlaff & F.A. Jost, "Dr Nikola Tesla" Tesla Book Co, Millbrae, CA (1979).

M.J. Seifer, "Nikola Tesla, psychohistory etc", Microfiche D178 (University Microfilms, Ann Arbor, MI, USA) (a detailed account).

I. Hunt & W.W. Draper, "Lightning in his Hand", Sage Books, Denver (1964).

D. Peat, "In search of Nikola Tesla", Ashgrove Press, Bath, UK (1983).

Other references:

P.S. Sidky & M.G. Hocking, "Contactless heating of moving wires", High Temperatures High Pressures, $\underline{33}$, 627-630 (2001).

Schumann Resonance:
http://en.wikipedia.org/wiki/Schumann_resonance

Appendix 7 Inventions of Dr N. Tesla

Fig. 1. Dr Nikola Tesla.

Fig. 2. Tesla's Wardenclyffe power transmitter. The circuit schematic is shown in Fig. 3.

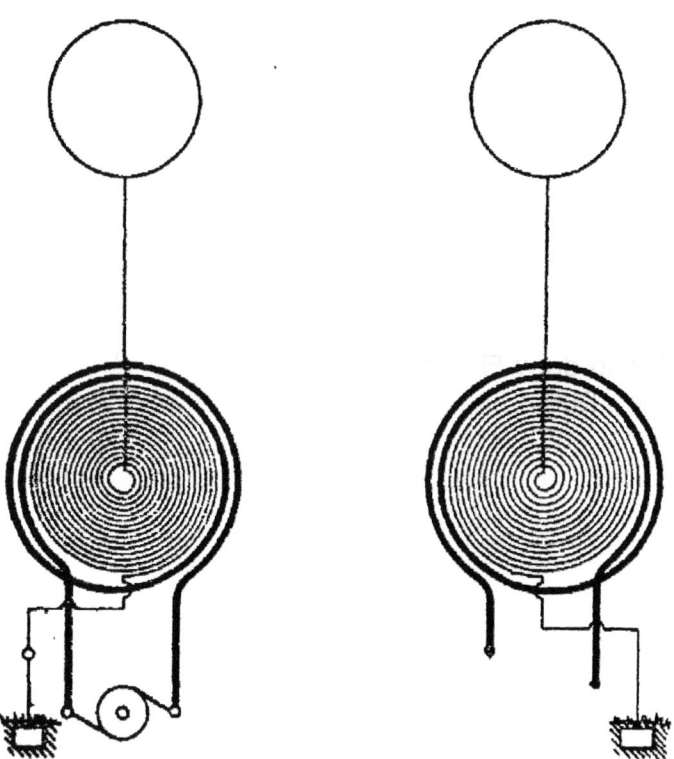

Fig. 3. Tesla's power transmitter (left) and receiver (right) schematics.

Transmitter: The spiral secondary flat coil design prevents flashover of the multi-million volts used.
The transmitter secondary connects between Earth and a dome very high above the coil (see Fig. 2).   The flat spiral coil diameters used were typically from 2.4 m to 7.6 m !
The primary is shown as a thick line.   A frequency of 925 Hz (supplied from a commutator), with wavelength of 200 miles, and a secondary coil wire length of 50 miles, were typical values used.
Receiver:
The primary is a flat spiral coil and the secondary (thick line) is connected to a bank of light bulbs or motors, in parallel (not shown). US Patent No 649,621 (1900).
Reference: J.T. Ratzlaff & F.A. Jost (see in references list).
Note: Tesla pointed out that it is the ground wave that carries the power, not the aerial wave.   See text above for more details.

APPENDIX 8

Remote Viewing Mechanism

"Remote Viewing" is a modern re-name of "Clairvoyance", as the latter word has unfortunately become associated with fairground trickery. There are many non-genuine "clairvoyants".

The author can personally vouch absolutely for the genuineness of the artist who drew Figures 0.1 & 0.2 in this book (on pages 8 & 9): Heather DuPré, in London. The subjects shown in these Figures were stable and persisted long enough to be drawn, and they disappeared if she closed her eyes and re-appeared when opened again. A scientific explanation is given below under "Etheric vision".

The author has also met others (who also were not taking any drugs) and has also asked them the same question on whether their visions are still visible if they close their eyes, and they have also reported that their visions disappear if the eyes are closed and re-appear when opened. This means that these visions were not hallucinations, in the author's opinion, and were of the 'etheric' type (see below). 'Astral' (dreamstate) visions do not disappear.

Mention of fairies brings a disbelieving smile to most people's faces!! But one of these observers reports the ability to see "nature spirits" (fairies) and also said that these are still seen if viewed via a normal glass mirror, instead of directly, so the author's conclusion is that they are a reflectable near-optical electromagnetic energy (to which their eyes are sensitive). In some country districts nature spirits are more commonly mentioned, worldwide, and the many types have been classified into very many specific names (see page 24 and Fig. 1.3). Reports were much commoner before the Industrial Revolution when rural people were probably more sensitive. This is discussed later, in a separate section below. The evidential basis for Remote Viewing is given in Chapters 1 & 3, and Besant & Leadbeater report that Nature Spirits, including fairies, do exist.

Leadbeater [1], one of the Remote Viewers whose results are reported in Chapters 3 & 4, gives the following information, *in italics*, brought up to date (in non-italic passages) by the author (MGH):

"Etheric" vision: The range of visible wavelengths for most people is from red to violet but (genuine) clairvoyants have an extended range outside this, which confers an ability to penetrate solids (like the ground-penetrating radar now used on building sites and by archaeologists; this is the microwave region, below infra-red). *Using etheric vision, veins of metal below ground can be perceived, small burrowing animals are seen below ground, etc.* Water divining (used officially by many local government councils) is a non-visual aspect of this (sense of touch instead of vision).

If a (mechanical) *clock is viewed from the back, the back face is penetrated by etheric vision and the working parts can be seen, and beyond these the front face of the clock is seen, with the figures seen written backwards. This differs from "astral" (4-D) vision (see below) which does not use the physical eyes and so which does not disappear if the eyes are closed, unlike etheric vision. Etheric vision uses only the retina (which contains light-sensitive cells* and also brain-type cells*). Etheric vision is entirely physical* and can be imitated by physical devices like ground-penetrating radar.

Clairaudience uses the sound receptors in the ears and requires an extension (rare) of a person's audio response into the ultrasonic. The artist of Fig. 0.1 & 0.2 (pages 8 & 9) has super-acute hearing, extending into the ultrasonic range when tested, & is clairaudient.

With this introduction, an explanation can now be given of the visions shown in Fig. 0.1 & 0.2. An entity is being observed which is composed of a state of matter not visible to most people whose vision range is from red to violet of the visible spectrum. If an entity is composed of neither solid, liquid nor gas, but of (cold) plasma (the so-called 4th state of physical matter), it will not be seen by normal eyesight but only by eyesight with a range beyond the normal red to violet. It disappears from perception if the eyes are closed, and so it is not a hallucination, and is seen by the retina, and should be visible to another (similarly gifted) observer.

Appendix 8   Remote Viewing Mechanism

Rarely, photographs have been taken of nature spirits but are controversial due to suggestions them being fake photographs. Probably the best-known are the Cottingley Fairies [2], discussed in a separate section below.

Folk stories mention use of eye ointments which enable ordinary people to see nature spirits, which can be well-understood today as due to minute concentrations of psychoactive drugs which may extend the perception of the retina (author's speculation).

<u>Astral vision:</u>   This is non-physical and does not use physical receptors like the eye retina, but uses the "chakras" (described in Chapter 1, page 24).   *If the clock example is used as above, with astral vision, which is 4-dimensional, the clock is seen from <u>all sides</u> simultaneously and the figures on the front face are seen correctly and not seen "backwards" as with etheric vision.*

A 3-D cube is seen from all its 6 faces simultaneously. The 4-D analogue of a 3-D cube is shown on the front cover of this book as a perspective drawing in 2-D of this 4-D object, which is called a tesseract, and is explained as follows:

If the outer cube is in ordinary 3-D space, its 6 faces will each be one face of another 6 cubes extending "within 3-D space" into a 4th spatial dimension, and the 6 far faces of these 6 cubes will form another cube which is totally remote in 4-D space.   Total number of cubes = 8.   The angles are all right angles in the tesseract! This is further discussed below.

2, 3, & 4-D Space: The square, cube, & tesseract.

To explain a 4-D tesseract, consider the analogous explanation for telling a hypothetical 2-D space inhabitant what a 3-D cube is:

To a hypothetical 2-D being, the 3rd dimension is that direction in which <u>he</u> cannot point, and for us the 4th dimension is that direction in which <u>we</u> cannot point. The eyes of a hypothetical 2-D being cannot record a 3-D object, simply because the only light rays which can be focussed by a 2-D camera are those moving in the plane of the paper. Imagine a ray passing through the paper from the 3rd dimension -- it would intersect the paper only at a point and so could not be focussed. See Fig. 1.2 on page 23.   By analogy, we cannot see or photograph any 4-D object.

To a hypothetical 2-D being, living in the plane of this paper, Fig. A8.1 below is how we could explain to him what a cube is: the outer square is in his 2-D world but the inner square is in a parallel 2-D world lying above (or below) his 2-D world, remote from it in the 3rd dimension.

The six enclosures are all squares and all the same size, but are of course distorted by the perspective of the 2-D drawing. <u>We can visualise Fig. A8.1 in 3-D, as a cube - just stare at it for 10 seconds !</u> But <u>he</u> cannot visualise it.

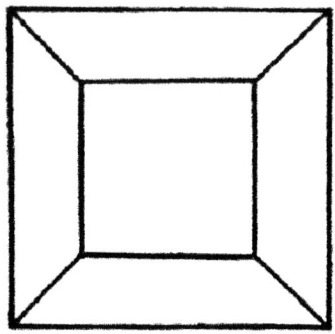

Fig. A8.1. A cube drawn on paper.

A further explanation to a hypothetical 2-D person is that a 2-D figure Fig. A8.2 below must be imagined to have its 4 outer squares bent up (or down) at right angles out of the 2-D plane of the paper and their 4 furthest edges will then form another square but entirely remote in 3-D space above (or below) the central square:

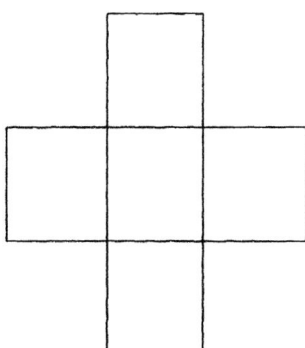

Fig. A8.2. Construction of a cube.

This can be drawn on his 2-D paper as shown in Fig. A8.1 above, or, as our normal perspective drawing of a cube:

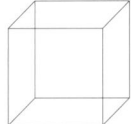 But <u>he</u> would not be able to understand that all its angles are right angles!

Similarly, by analogy, 3-D beings (ourselves) can try to visualise a tesseract, which is the 4-D analogue of a cube, as follows. See the tesseract figure below and imagine a cube is forming inwards into 4-D space from each of the faces of the large outer 3-D cube, and these six new cubes will then form another smaller cube (having 6 faces) totally remote in the 4th dimension. But in a tesseract, all angles are right angles and all 8 cubes it contains are the same size! The perspective drawing below distorts the angles and sizes. This is a drawing on 2-D paper of a 4-D tesseract:

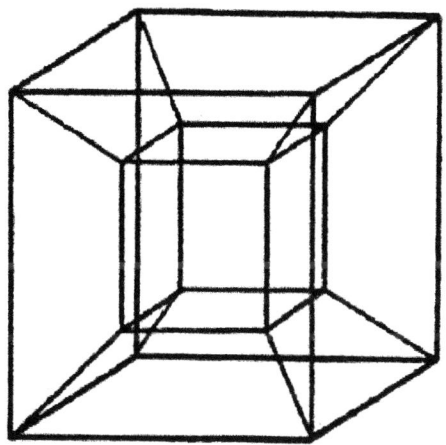

Fig. A8.3. Tesseract, the 4-D analogue of a (3-D) cube.

Similarly, a 5D analogue of a tesseract can be obtained by imagining a tesseract going into 5D space from each cube of the 4D tesseract, and the far cubes of such a structure then form another tesseract totally remote in 5D space. But this is too complex to draw on paper and is not shown here!

The Cottingley Fairies: Pictures: See Fig. 1.3, near page 63.

In 1917 and again in 1920, a total of five photographs were taken by two girls, aged 9 and 15 in 1917, apparently showing fairies. The last of these (Fig. 1.3, near p. 63) is "unfakeable"; the fairy wings are transparent and grass stalks are seen behind and in front of them. But the surprising story of the first four is summarised below. All 5 were taken at Cottingley, near Bradford, where Elsie lived. Frances lived with her cousin Elsie at Cottingley from 1917 to 1919. In 1919 her father returned from World War I and Frances then went to live in Scarborough. Frances visited Elsie in Cottingley for the last 2 weeks of August 1920.

Two photographs taken in 1917 apparently showing fairies were shown in 1920 to E.L. Gardner [2]. He sought a photographer who was knowledgeable about fake photogaphy, H. Snelling, who was professionally able to examine the photographs fully. Snelling's considered judgement, in his letter to Gardner of 31 July 1920, was: "These two negatives are entirely genuine unfaked photographs of single exposure, open-air work, show movement in all the fairy figures, and there is no trace whatever of studio work involving card or paper models, dark backgrounds, painted figures, etc. In my opinion, they are both straight untouched pictures." [2].

The last three of the 5 negative plates printed in Gardner's book [2] were (secretly) pre-marked plates which were given (in 1920) to the girls by Gardner and 3 photographs (C, D & E) were taken in the glen at Cottingley. Elsie had prepared paper cut-outs for photographs C & D only. After exposure, the 3 plates were developed by Elsie's father, who was an amateur photographer, and returned to Gardner in London. Gardner then had the marked plates confirmed as being three which had been secretly marked by the manufacturer (Illingworth). Photographs C & D had been taken during the week and E (Fairy Bower) was taken by Frances on the following Saturday when she saw misty objects beginning to appear in harebell-filled grass and she instinctively reached for her camera and took the photograph from about 3 feet [4]. It is significant that this differs from the other photographs by having no girls in the picture. Frances returned home to Scarborough the next day, Sunday and the plates were posted back to London on

Appendix 8   Remote Viewing Mechanism

Monday or later in the following week; Frances did not see her uncle develop the plate but received a copy on her return home.

The earlier (1917) two (unmarked) plates (before Gardner's involvement) were loaded by Frances' uncle Arthur Wright) and developed by him. (He did not want any publicity and always indignantly declined to receive any money throughout the whole series.)

Gardner instructed Snelling to enhance photos A & B (the two 1917 photos). These were then re-photographed with a camera different from Arthur Wright's Midg camera. The Midg was only used to take the two 1917 photographs. From these two new negatives, Snelling made lantern slides; he also made photographs for sale at Gardner's lecture at Mortimer Hall, London. This lecture brought them to the notice of Sir Arthur Conan Doyle.

Following Snelling's opinion, it was proposed that if the negatives survived a second expert's judgment, preferably Kodak's, then Edward Gardner and Conan Doyle should join forces and make the photographs a leading feature in a Strand Magazine article. An appointment was made with Kodak's manager, Mr West. His studio chief and two other expert photographers were also present. The negatives were examined by all at some length, and the results of the inspection were as follows, all agreeing [2]:
(1) The negatives are single exposure.
(2) The plates show no sign of being faked work, but that cannot be taken as conclusive evidence of genuineness.
(3) Kodak's were not willing to give any certificate concerning them because photography lent itself to a multitude of processes, and some clever operator might have made them artificially.
(4) The studio chief added that he thought the photographs might have been made by using the glen features and the girl as a background; then enlarging prints from these and painting in the figures; then taking half-plate and finally quarter-plate snaps, suitably lighted. All this, he agreed, would be clever work and take time.
(5) A remark made by one was, "after all, as fairies couldn't be true, the photographs must have been faked somehow."

This last remark shows obvious negative prejudice.

Some historical background is in an extract from "Chronicles & Stories of Bingley and District" by Harry Speight (1904):

"Around Bingley (Yorkshire) there used to be, and possibly there still is, a strong belief in the existence of fairies. In Gilstead Crags there was an opening in the rocks known as "Fairies Hole", and it was said that the tiny creatures used to trip and dance and play their merry antics in the bright moonlight. Anyone who intruded at such a time, it was said would lose their sight. At Harden, in a secluded part of Deep Cliff, it is said that the fairies could sometimes be heard clanging musical tongs and what looked like tiny white garments hung out on the trees could be seen on bright nights." (taken from reference [5])

The author's (MGH's) comments on this are that, "the opening in the rocks" is reminiscent of the experience of Yajna (see Chapter 4, page 133-135 and see especially p. 136, in the present book). The fairies could be in a parallel 3-D world (see Chapter 1). The comment about "loss of sight" above is also mentioned by others and by Leadbeater [1] who said that <u>etheric</u> vision (see above) uses the physical eye and its retina, and if these are damaged (only temporarily, hopefully) then sight will be lost.

In <u>1983</u>, sixty-six years later than when she first saw the fairies, Frances prepared notes for a book which was subsequently published in 2009 by her daughter [4], from which the following details are given:

She said that she (aged 9) often visited the stream alone because Elsie was normally working. She was daydreaming one day and then saw a miniature man about 18 inches tall walking down the bank. He held a leaf in his hand and crossed the stream by just walking over it, walking on the water, and quite unaffected by the current which she expected to carry him downstream. This first time she saw him, he stared at her and was aware that she could see him. But on later occasions he gave no indication that he knew she was seeing him, but she sensed that he knew.

His face was that of a typical working man and was neither ugly nor friendly. Later she said she saw him leading other miniature men, dressed like him in green clothing. She watched them until they went behind some plants out of sight. Frances said she did not tell Elsie for a long time and it was her own secret, which she did not want to share. She said Elsie never mentioned seeing the miniature men or "the pretty fairies" when she was there with her when they appeared.

It turned out later that Elsie never saw any fairies and only Frances had that gift of "second sight".

A website [5] reports that in the 1980's, a forester (Mr Ronnie Bennett) in Cottingley Estate Woods, reported having seen fairies in those woods. He claimed that he saw elf-like figures while working there: "When they showed themselves about nine years ago there was a slight drizzle around. I saw three fairies in the woods and I have never seen them since. They were just about ten inches tall and just stared at me. There is no way the Cottingley Fairies is a hoax." [5].

The author (MGH) comments that drizzle was present during this sighting. There is a waterfall in the beck (see it on left side of picture 10436087 in Fig. A8.4 between p. 63 & 64), and it was raining for the week before the day (in August 1920) when the last, "unfakeable" photograph was taken (Fig. 1.3 between p. 63 & 64, and enlarged in Fig. A8.5 & A8.6 loc cit). This means there would have been high humidity in the air, rising from the rain-saturated ground.

The waterfall (out of focus) is seen on the far left of picture 10436087 (Fig. A8.4 between p. 63 & 64), about 20 feet behind Frances, age 9, and rocks are on the right. It is known that breaking water into tiny drops creates a negative charge cold plasma ion cloud. So the waterfall produces ideal conditions for a normally-invisible cold plasma (if nature spirits have bodies composed of that) to become normally-visible and photographable. Millikan's cloud chamber is used in modern physics to see plasma particle tracks and if the nature spirits (composed of cold plasma) were in a high humidity atmosphere or something else within the wavelength range recordable by a camera, then they could become photographable.

In the summer of 1917, Frances was 9 and Elsie was 15. Fairies are traditionally said to be seen by some pre-puberty children but rarely by adults. There were no more photographs after the summer of 1920. Four years later, in 1921, Frances returned to Cottingley for a week. During this week G. Hodson (clairvoyant) was with them at all times at the beck but no photographs were taken.

It emerged later [4] that Elsie had never seen any fairies at any time but she felt sorry for her younger cousin who was always in trouble for slipping and getting her clothes wet in the beck. Following a particularly difficult incident, Elsie had the idea of making paper cut-out fairies and taking a photograph of them with her father's camera, thus showing the grown-ups that this was why Frances was always going near the beck.

Elsie, liked jokes, and set up fake pictures with paper cut-outs, as she could not get fairies to appear. See Fig. A8.4. She asked Frances to agree a children's pact never to disclose this, but many decades later Elsie disclosed it and the press latched onto it, claiming fraud. Accounts of the controversy that this caused are found on the internet (search for: Cottingley Fairies), but for an accurate account see the book written by Frances and her daughter [4]. The first 4 photographs were of paper cut-outs which Elsie had produced. Elsie and Frances had remained silent until 17th February <u>1983</u> when Elsie said in a letter that all the photographs were a hoax, claiming that they had drawn the fairies, cut them out and fastened them to the ground with hatpins. This was not true for the 5th photograph. Elsie was not telling the truth in 1983 when she said that <u>all</u> 5 photographs were fakes. Frances, who was the first to see the fairies (when she was alone), said in her last television appearance in <u>1986</u>: "there were <u>fairies</u> at Cottingley".

The expert who had supplied the last 3 (1920) marked plates confirmed they were the ones he supplied; he would not commit on plates C & D but said the last one (E) was, "an impossibility to fake" [2] (see Figures 1.3 & A8.5 & A8.6 & captions, near p.63).

Critics have suggested that the fairies' hairstyles, being those of the time (1917), means that the fairies are cardboard fakes made by the girls, but it is reported by Remote Viewers that fairies like to copy humans (whom they can see), so the hairstyles do not prove anything. Mediaeval reports of fairies describe them in mediaeval attire.

In the years, after Conan Doyle's 1920 & 1921 articles in Strand Magazine, and Conan Doyle's book "The coming of the Fairies" in 1922, the girls were hounded by the press, to the extent that Elsie emigrated to America to escape it. A video [3] was made about 60 years later by Warner Bros, press hounding re-started but the girls were then much older women in their 70s and Elsie began to say that it was all a hoax. This turn of events was of course a great dis-service to the subject, but as always, it is not "allowed" for an incontrovertible proof of such phenomena as fairies to be given; an element of doubt is always inserted!

But, as a suggestion by the author (MGH) for another attempt, if hot plasma (excited atoms & ions) is available in a visible form as a flame (unlike an invisible-to-us cold plasma), then "salamanders" (specialised nature spirits) could possibly manifest and become visible in flames to any observer who has only normal eyesight. It is said that if a fire is lit, then salamanders from miles around rush into it. But nature spirits reportedly keep well away from humans, except young children (example above), as they don't like their disturbing thoughts nor their iron objects (being a sign of the presence of humans), so a view of them is a very rare occurrence. This suggests use of a method like a "hide", as used for photographing birds. For photographing salamanders, a camera could be set up some distance from a fire out in the wilderness and left running with no humans nearby. This experiment has probably never been done. (Note: Caution is required with an unattended fire -- CCTV is a way of supervising it remotely).

The above suggestion is facilitated by the specific (reported) attraction of flames, for salamanders, which is not available for fairies, who can be located anywhere (so one would not know where to place a camera and so many attempts would be required). But there is also the probability for fairies to become visible to normal vision or photographable in a mist and perhaps in a plasma generated by the aura of a "sensitive" person (like Frances); maybe both are required. It is hoped that someone may try these experiments.

It is reported by Besant & Leadbeater [6] that many millennia ago most people were able to see nature spirits. In recent times this is quite rare [7]. But over many decades and even centuries, very many accounts exist of fairies still being seen by a significant number of people, e.g. see examples in reference [2], but they are all anecdotal and so do not pass the criterion for science that any observation must be reproducible by anyone. But this does not mean that they are untrue, of course.

Sceptics studiously ignore all the evidence in these persistent accounts down the centuries, which, although anecdotal and thus "not accepted by science", are nevertheless "witness statements" which are a class of observation routinely considered valid in a court of law.

Acknowledgement: The author is grateful for additional unpublished information from Mrs Christine Lynch, the daughter of Frances Griffiths. See especially reference [4] below.

References for Appendix 8:

1. C.W. Leadbeater, "Clairvoyance", publ by TPS, Adyar, India (1899 & later), available as a free download from: www.AnandGholap.net
Note: This website has many free downloadable books, but see important comment in General References below.
2. E.L. Gardner, "Fairies", publ. by Theo. Publ. House, London (1945). Contains photographs. See: http://www.cottingleyconnect.org.uk/fairies.htm
3. "Fairy Tale, a True Story", Video S015879, by Warner Bros (1997).
This video is about Conan-Doyle's (actual) investigation of the fairies.
4. "Reflections on the Cottingley Fairies" by Frances Griffiths (ISBN 978-1-8992 28-06-5) in her own words with additional material by her daughter Christine - £8.99 (postage/packing + £1.50 UK. + £2.50 Europe, + £3.99 Rest of World - More than 1 copy please request p&p rates) from Christine Lynch at JMJ Publications, 1 Thornhill, Malone, Belfast, BT9 6SS or email: orders@cottingleyreflections.com   Published 2009.
Website: www.cottingleyreflections.com
Also available from Saltaire Bookshop and Media Museum, Bradford.
5. Website: http://www.cottingleyconnect.org.uk/fairies.htm
6. A. Besant & C.W. Leadbeater, "The Lives of Alcyone", publ 1924, downloadable free from www.AnandGholap.net   but see comment in ref. 1 above.
7. See espec. D. Van Gelder, "The Real World of Fairies", Quest Books (1999).

Appendix 8   Remote Viewing Mechanism   373

-=-

Copyright Disclaimer

The author was unable to find the copyright owner of Figures 4.2 to 4.8 & Fig. 4.10 in the present book, which were taken from "Through the Eyes of the Masters" by David Anrias, published by Routledge & Kegan Paul in 1932. Routledge & Kegan Paul were taken over by Taylor and Francis who have no record of that copyright. The Figures are given for general information purposes only and are provided by the publisher of the present book and we make no representations or warranties of any kind, express or implied, for any purpose. Any reliance placed on who owns the copyright, if anyone, is therefore strictly at the reader's risk if copies are made.   In no event will we be liable for any loss or damage including without limitation, indirect or consequential loss or damage, or any loss or damage whatsoever in connection with any conclusions about the copyright of Figures 4.2 to 4.8 & 4.10 in the present book.   No copyright of these figures is claimed by the present publisher.

Note: No copyright is claimed by the author or publisher on any figures in this book, except where stated in a figure caption.

-=-

GENERAL REFERENCES

NOTE: References are listed at the end of each chapter, not here.
The references listed here below are in addition to those at each chapter end. They are not listed in any particular order but are all considered valuable.

Note: Many books by Besant & Leadbeater are available as free downloads from: www.AnandGholap.net and other websites:

E.g. "The Lives of Alcyone" and "Man, Whence, How & Whither".
CAUTION !!! : The books were scanned and an optical character reader programme was used to convert to editable text (ASCII), but there are some errors on the www.AnandGholap.net website. **Careful checking against the printed source book is needed, where scanning has been used.**
But the author (MGH) **strongly congratulates** Anand Gholap on his major efforts to make many books available as free downloads.

The original source (1924) by A. Besant & C.W. Leadbeater is long out of print, but has been scanned and is available in 3 parts as "The Lives of Alcyone", from: www.AnandGholap.net   but this free download contains some un-corrected errors due to the optical character reading process (e.g. "kind" may appear as "king", "our" may appear as "out", etc, and occasionally lines are repeated or omitted.

Note: The texts used in the present book were not taken from the AnandGholap website but were scanned from the original printed books and then carefully checked by the author (MGH), to avoid such errors.

-=-=-=-=-=-

Some readers may like suggestions for music – a list is given below the books.

-=-=-=-=-=-

GENERAL BIBLOGRAPHY

"The Inner Guide Meditation" by E.C. Steinbrecher, ISBN 0 85030 780 5, publ by Aquarian Press (1988) **Especially recommended.**
"The Secret Gospels of Jesus" by M. Meyer, publ by Darton, Longman &Todd Ltd, London (2007)
"Exploring the Sub-Conscious using New Technology" by M.G. Hocking, published by CMC Ltd., London, 2nd edn (1993).
C.W. Leadbeater, "The Masters and the Path", TPH, Adyar: **Especially recommended. Available as a free download from:** www.AnandGholap.net
C.W. Leadbeater, "The Chakras", and other books, free downloadable, loc.cit.
C.W. Leadbeater, "Invisible Helpers", free downloadable, loc.cit.
C. W. Leadbeater, "The Astral Plane", free downloadable, loc.cit.
C.W. Leadbeater, "Clairvoyance", free downloadable, loc.cit.
A. Huxley, "The Doors of Perception" , Penguin, (1960). .
M. Hutchison, "Megabrain", Morrow, NY (1986).
E & A Green, "Beyond Biofeedback", Delacorte Press, USA, (1977). Also see Research Centre J. (T.S.), <u>16</u>, no 4, p 87 (1972).
C. Maxwell Cade & N. Coxhead, "The Awakened Mind", Element Books, UK, (1989). & G.G. Blundell, C. Maxwell Cade, "EEG Measurement", Published by Audio Ltd, London.
J. C. Lilley, "In the centre of the Cyclone", Paladin (1973).
R. A. Monroe, "Journeys out of the Body", Anchor (1977).
O. Fox, "Astral Projection", University Books, NY and Citadel Press, NJ.
Yram, "Practical Astral Projection", Weiser, NY (1979).
J.H. Brennan, "Discover Astral Projection", Aquarian/Thorsons (1991).
J.H. Brennan, "Astral Doorways", Aquarian Press (1971).
R.Bruce, "Astral Dynamics", publ by Hampton Roads Publ Co Inc, Charlottesville (2009) (on **OBEs**).
M. Sadhu, "Concentration", Unwin Paperbacks (1977) **Especially recommended.**
Mouni Sadhu, "Samadhi", Allen & Unwin (1962)
Swami Vivekananda, "Raja Yoga", publ Bharatiya Kala Prakashan, or others (same text) (first publ about 1880). Note**:** does not warn against excessive use of pranayama (breath control). And:
Monks of the RamaKrishna Order, "Meditation", publ RamaKrishna Vedanta Centre, London (1972). **Especially Recommended.**

H. Hewitt, "Meditation", Teach Yourself Books, Hodder & Stoughton (1978).
A. E., "The Candle of Vision", University Books Inc., NY.
R. M. Bucke, "Cosmic Consciousness", Dutton, USA, (1969).
C. Castaneda, "The Teachings of Don Juan", (& other titles), Penguin (1977).
Paramhansa Yogananda, "Autobiography of a Yogi", Rider, London (1961).
Maharishi Mahesh Yogi, "Commentary on the Bhagavad Gita", Penguin (1969).
"Highways of the Mind", "The Shining Paths", "Inner Landscapes" and other books by D Ashcroft-Nowicki, Aquarian Press (1980s).
K. Harary & P. Weintraub, "Lucid Dreams in 30 Days", Aquarian Press (1989)
M. Hutchison, "The Book of Floating", Quill - Morrow, New York (1984)
Michael S Gazzaniga "The Split Brain in Man", Scientific American, 24, Oct. 1967.
M. Zdenek, "The Right-brain Experience", Corgi (1983)
N. Drury, "Music for Inner Space", Prism Press, UK (1985)
Lama Anagarika Govida, "The Way of the White Clouds" Rider (1966)
I.Tweedie "The Chasm of Fire", Element Books, UK (1979)
Gopi Krishna, "Kundalini", Robinson & Watkins (1971)
Gopi Krishna, "The Secret of Yoga", Turnstone Books, London (1972).
Dr R.A. Moody, "Life After Life", Bantam Books, (1977) (a best seller)
Dr R.A. Moody, "Reflections on Life After Life", Bantam Books.
H.Wambach, "Life Before Life", Bantam (1981)
Dr M.B.Sabom, "Recollections of Death", Corgi, UK (1982)
D.Scott Rogo, "The Return from Silence", Aquarian Press, (1981)
   and: "Life after Death", idem, Harper-Collins, Glasgow, (1992)
Suzuki, "Zen mind & beginner's mind", (& other titles), Weatherhill, NY, (1970)
E. Herrigel, "Zen in the Art of Archery", Routlege & Kegan Paul (1959)
Z'ev ben Shimon Halevi (Warren Kenton), "Kabbalah", Thames & Hudson, London (1979).
S. Court, "The meditator's manual", Aquarian Press, UK (1984).
P. Russell, "Meditation", published by BBC, London (1979).
P. Brunton, "In search of secret Egypt" (& other books).
R.O. Becker & G. Selden, "The body electric" and R.O.Becker, "Cross currents".
Patanjali, "Yoga Aphorisms", published about 400 BC.
M.G. Hocking, "ESP observation of atoms & molecules", Bull Th Sci Stdy Gp $\underline{21}$, 53 (1983) & $\underline{22}$, 5 (1984). Free downloads available from: www.4-D.org.uk
M. Meyer, "The Secret Gospels of Jesus", publ by Darton, Longman &Todd Ltd, London (2007)
K. Taylor, "The Breathwork Experience", publ by Hanford Mead, Santa Cruz (1994) (Holotropic breathing, overbreathing)
G. Minett, "Exhale", publ by Floris Books, Edinburgh (2004) (Holotropic breathing)
I. Donnelly, "Atlantis", publ by Sidgwick & Jackson Ltd, London (1950): **especially recommended.**
J. Grant, "The Winged Pharoah", a classic from about 1930 giving memories of previous lives (claimed), from Remote Viewing, a best seller. Whole series in print from Ariel Press, USA: **especially recommended.**
A. Avalon, "The Serpent Power", a yoga classic, several publishers.
G.Hodson, "The Kingdom of the Gods", publ by Theos Publ House, Adyar, India.
G. Hodson, "Clairvoyant Investigations", publ by Quest Books, Wheaton, IL (1984).
Swami Vivekananda, "Raja Yoga", a classic, any edition.
P. Devereux, "Stone Age Soundtracks", publ by Vega, London (2001).
Lao Tsu, "Tao Te Ching", a classic, various publishers.

"The Breathwork Experience", by K. Taylor, publ by Hanford Mead, Santa Cruz (1994)   (Holotropic breathing, overbreathing).
"Exhale", by G. Minett, publ by Floris Books, Edinburgh (2004) (Holotropic breathing, overbreathing).

<u>NOTE</u>: The above list is very incomplete! There are many other very good books, not listed here.

MUSIC – CD list

Some readers may like a list of CDs which have an uplifting effect. The list below is of course very incomplete, but a **** recommendation is given:

********* "Dorje Ling", D. Parsons, Fortuna 17076-2; Tracks 1 & 4 (<u>must</u> be heard with eyes closed). Note: Deep bass & infrasound produce devotional feelings (some mediaeval cathedral organ pipes are in the infrasound acoustic frequency range!).
******"Vespers ... 1610", Monteverdi, Apex 2564 61429-2; track 10.
******"Parsifal", Wagner, Conducted by Barenboim, Berlin Phil, very good slow version, Teldec 4509-97910-2.
******"Gloria", Vivaldi, <u>very</u> best version is film music LP record of "Runaway Train", available on internet, Enigma Records, Capitol Records – EMI Inc, SJ 73200.
*****"Pilgrim's Progress", Vaughan Williams, EMI CMS 7 642122; Act 4 Scene 3.
*****"Spem in Allium", Thomas Tallis, (best version) Gimell CDGIM 006.
****"The Encircled Sea", Boyle, Silva Screen Film CD 076; tracks 8, 12.
****"Faust Cantata", Schnittke, Malmo S.O., BIS-CD-437; tracks 9, 10.
****"Miserere", Allegri (very special version) Ambionay Naïve E 8846 www.naïve.fr
****"Responsories for Tenebrae", Victoria, Decca CD 425-078-2.
****"Gathering of Spirits", R. Fox, FX Music FXCD4; track 3.
***"Gotterdammerung", Wagner, Cond. Solti, Decca 455 569-2; Act 3, trk 11.
***"Koyaanisqatsi", P. Glass, Island Masters IMCD 98 (814042-2)
*** Organum – Compostella (chanting), Ambroisie 3 760020 170660; tracks 1, 8, 11.
***"La Poeme Harmonique, Boesset, Alpha 057, track 20 (Nos Spirites Libres).
**"Atlantic Realm", Clannad, BBC CD 727.   And other Clannad CDs.
**"Akenaten", P. Glass,
**"Turn of the Tides", Tangerine Dream, Miramar CD 09006-23088-2.
**Angelic music (various CDs) by Iasos (reported by NDE people as nearest to what they heard during an NDE!)

<u>See also the book</u>: N. Drury, "Music for Inner Space", Prism Press, UK (1985).

For those who want to compose music but have difficulty in writing it down, a programme called Akoff Music Composer, which converts humming or whistling a tune into music (e.g. violin, + notation), is available at: www.akoff.com

INDEX

Note: Chapter 4 has extensive quotations from previously published books and so is not comprehensively indexed below. Those books* are available as free downloads and are computer-searchable: www.AnandGholap.net but see important note at end of page 373.

Below, **F-7** means Foreword **page 7**; **2-57** means Chapter 2 **page 57**, etc.

4-D**	1-20, 4-136, Appendix 8
Abortion	2-58, 2-62, 6-258
Algeria	4-171
Ancient of Days	4-99, 4-102, 4-104, 4-110, 4-195
Apportation	5-249, 5-250
Arabia	4-90, 4-101, 4-107, 4-108, 4-115
Aramaic	5-254
Arnoo	3-67, 3-70 to 3-73, et passim
Astral	1-28, 1-34, 1-42, 1-73, 4-92, 4-97, et passim
Astral vision	Appendix 8-362, 363
Astrology	4-114, 4-142, 4-155, 4-164, 5-243
Athanasius	F10, F12, F14
Atheism	2-53, 2-61, 6-259
Atlantis	4-88, 4-90, 4-91, et passim
Atomic Weight	3-70
Acts	F13
Alcyone	4-89, 4-90, & everywhere in Chapter 4, Fig. 4.9 (p. 207)
Alcyone's lives list	4-90
Alpha waves	1-32, 1-33
Anoxia	1-29, 1-30
Arnold, Edwin	2-56, 2-57, 5-252
Art, psychedelic	F8, F9
Aura	Colour Figures 4.11 & 4.12 near p. 63
Avesta	4-89
Banana	4-100
Bees	4-100
Beta waves	1-32, 1-33
Bhagavad Gita	2-56
Blake, William	2-56, 6-259
Black magic	4-92
Bodhisattva	4-89, 4-103, 4-107, 4-109, 4-116, 4-225, 5-250
Breathing	1-29, 1-30, 1-35, 1-43
Brihat	4-89, 4-90, 4-101, 4-127, 4-142, 4-144, 4-155, 4-173 to 176, 4-229, 5-246, &c, Fig. 4.10 (p.228)
Buddha	2-56, 4-89, 4-221, 4-227, 5-250
Buddhism	F11, 1-34, 2-55, 2-56, 2-60, 4-225, 5-244, 5-246
Brainwaves	1-32, 1-33, 7-261
Canon	F10, F12, et passim
Carthage	F16

Cellular memory	2-49
CERN	3-67
Chakra	F7, 1-24 to 29 &c, 3-68, 5-246, 7-266, Fig. 1.4, 1.7, 1.8 near p. 63
Chaldea	4-100
Chemistry	Chapter 3, Appendix 5-306
Cheops	4-100
Chernobl	5-255, 5-256
Chinese restaurant	1-35
Chirality	3-72, 3-77, 3-82
Chittagong	4-155
Christ	F6, F10, 2-54, 4-89, 4-109 &c, Chapters 4 & 5, 6-259, Fig. 5.3 near p.63
Christian Canon	F10, F12, et passim
Christianity	Chapter 5, 6-259, et passim
CIA	F6, F7 to 10, 1-24, 3-66 to 69, Appendix 2-292, Fig. A2.1 near p.63
City of Golden Gates	4-91, et passim
Clairaudience	Appendix 8-362
Coil	3-73
CO_2	1-29, 1-30, Appendix 7-354
Codex	F14
Constantine	F15
Constantinople	F11, 2-54
Corona	4-89 to 91, 4-93, 4-98, 4-101, 4-194, 4-196
Cottingley	Appendix 8-363 & 366
Crop Circles	4-137
Crucifixion	5-253
Cygnus	4-90, 4-93 to 4-98, 4-169 to 188, 4-218
Daita	4-91
Damasus	F16
Dangers	1-28, 1-30
Dark matter	3-72
Delerium Tremens	1-28
Delta waves	1-33
Dhruva	4-89, 4-90, 4-117, 4-128, 4-137
Diffusion	3-81, 6-352 & 353, Appendix 6-333
Dimensions	F6, F10, 1-20 to 22, 3-72, 3-73, 4-136, 5-242, Apx 8-362
Doubling	3-70, 3-71, 3-77 to 79
Dream membrane	1-28
Dreams, lucid	1-32, 1-34, 1-35, 1-36, 7-260, 7-273
Drugs	F7
Earthquake	6-258, 4-99, 4-206
Edison	1-32
EEG**	1-33
Einstein	F10, 1-19, 1-32, also see Relativity & General Relativity
Egypt	12, 16, 20, 24, 42, 4-89, 100, 107, 115, 153,173, 219-221, 243, 264
Elementary particles	F6, F7, 1-20 to 23, 1-29, 2-47, 3-64, 3-66, et passim
Elijah	F11, 2-54
Epistles	F13
Etheric vision	Appendix 8-362

Fairy	1-24, Apx 8-366 onwards, Figures: 1.3 & A8.4 to A8.6 near p.63
Fides	page 384-5
Flotation tank	1-31, 7-260
Galantamine	1-34 to 36, 7-273
Ganzfeld	7-270
General Relativity	Appendix 3-292
Gobi Desert	4-90, 4-101
God	1-20, 2-46, 2-48, 2-52, 2-61
Global warming	2-58
Gravity	3-73, 4-100, 4-233, 4-240, 4-242 & 247, 5-249, Apx 3-292, Apx 7-262
Guitar string	3-71, 3-78, 3-81
H atom	3-66, 3-69, 3-71, 3-77
HCl	3-71, 3-74
Heart transplants	2-49, 2-51, Appendix 1-279
Heaven	F6
Helios	page 384-5
Hell	F6
Herakles	4-89, 4-90, 4-98, 4-140, 4-203 &c in Chap 4, Fig. 4.9 (p. 207)
Herod	5-242, 5-243
Herbal	1-32, 1-35
Hinduism	F11, 2-55
Homosexuality	4-230
Hypnosis	F11, 2-48, 2-59, 5-267, 5-253, 7-265
Ice Age	4-88, 4-131, 4-141, 4-230
Imagery	7-266
Imaginary	2-47, 3-80, 3-81, Appendix 6-334
India	F7, 1-28, 2-56, 4-88, 4-90, 4-100, 4-158, 4-207, 4-220, et passim,
Infrasound	376
Ireland	4-90, 4-116, 4-234, 4-239, 6-259
Iron Age	4-88, 4-230
Isotope	3-69
John	F10 to F13, 1-20, 2-54, 4-87 to 89, 4-137, 4-222, 5-246, et passim
Judaism	2-55
Judas	F14
Jupiter	4-89, 4-90, et passim
Kabalah	2-55
Karma	2-46, 2-52 to 57, 4-88, 4-101, 2-58, 2-60, 5-244 to 247, et passim
Kennedy	2-59
Khufu	4-100
Krishna	2-56, 4-87 to 89, 4-227, 5-242, 5-247, 5-250, 5-255, 6-259
Kundalini	1-26
Leo	F12; 4-90, 4-101, et passim
Levitation	5-247, 5-248
Lincoln	2-59
Lo Han	1-35
Luke	F13, 1-21, 1-25, 2-54
Magnetic	1-41
Magdalene, Mary	F14, 4-135, 4-137

MahaGuru	4-89, 4-194, etc in Chapter 4
Mansions	F10, 1-20
Manoa	4-90, 4-103, et passim
Manu	4-89, 4-101, 4-193
Mark	F13, 1-25, 2-54
Master	4-89
Matthew	F11, F13, 1-25, 2-54
Mars	4-88 to 91, 4-98, 4-101, etc in Chapter 4, Fig. 4.2 (p. 94)
Mary Magdalene	F14, 4-135, 4-137
Memory, cellular	2-49
Mercury	4-89, 4-90, 4-101, 4-110, etc in Chapter 4, Fig. 4.4 (p. 143)
Methane	3-75
Mexico	4-90, 4-139
Miracles	5-247, 5-250
Mizar	page 384-5
Mummification	1-42
Names	4-89 & page 383-5 (full list)
Neptune	4-89, Fig. 4.8 (p. 189)
New Zealand	4-137
NDE**	1-29
North America	4-141
Nuclear winter effect	5-251, 5-256
OBE**	1-32, 1-38, 1-42
Octahedral	3-75, 5-329
Old Testament	F13, F16, 2-57, 2-61, 5-243, 5-245
Origen	F11, F15, F16
Ormuzd	4-113
Orpheus	4-89
Osiris	page 384-5
Overbreathing	1-29
Pan	4-92, 4-99
Parallel 3-D Worlds	4-136
Past lives	F11, 2-48 to 53
Paul	F13, F14, 4-89, 4-189, 5-257
Persia	4-90, 4-107, 4-110
Peru	4-100
Philippines	4-106, 4-253
Planes, dimensional	3-73
Plasma	4-240, Appendix 8-362
Poltergeist	3-83, 5-249
Poseidonis	4-91, 4-99, 4-176
Pranayama	1-29, 1-43
Prophesies	2-60, 5-243, 5-245, 5-246, 5-251, 5-253
Proton	F10, 1-22, 3-66, 3-67, 3-71, 3-77, 3-84, et passim
Psalms	2-55, 2-62
Purpose of life	2-46, 2-47, 2-54, 2-62, 5-245
Pyramids	4-100
Quark	F6, F10, 3-66 to 3-78, 7-267, Apx 3-292, Apx 5, Apx 6

Radar	Appendix 8-362
Red Spider Lilly	1-35
RAS**	7-269, 7-270
Reincarnation	F11, 2-46, 2-48, 2-53, 2-54, 2-55, 2-61, 5-245, 5-252, Apx 1
Relativity	1-20, 3-69, Appendix 3-292, Appendix 4-302, Appendix 6-332
Relaxation	1-32, 7-265, 7-267
Remote Viewing	F6 to 12, 1-24, 3-65, 4-87, 5-251, 7-272, Apx 2-291, Apx 5, Apx 8-361
Resurrection	4-137, 5-254
Revelations, Book of	F13, 2-54, 4-87, 5-245, 5-254 to 257
Ruta	4-91
Sainthood	4-89, 5-250, Fig. 4.1 near p. 63
Safety	1-29, 1-30, 1-42, 7-277
Salamander	Appendix 8-371
Salvation	2-55, 2-62, 6-258
Samaritans	F17
Sanat Kumara	4-99, 4-102, 4-104, 4-110, 4-195
Saturn	4-89, Fig. 4.6 (p. 172)
Schroedinger	3-69, Appendix 6-332
Selene	page 384-5
Self-consciousness	1-28, 1-32, 1-34
Self-hypnosis	7-267 (see also hypnosis)
Sensory deprivation	1-31, 1-32, Chapter 7
Shamballa	4-101, 4-102, 4-165, 4-193
Siddhi	F7, F10, 3-67
Sirius	4-89, 4-90, 4-101, et passim
Snowdrop	1-35, 1-36
Soul	2-47 to 50, et passim
Space	3-71
Sphinx	4-100
St Francis	5-249
St Patrick	5-255, 6-259
St Theresa	5-249
String	F7, 3-66, 3-68, 3-71 to 3-83, 7-267, Appendices 4, 5 & 6
Suffering	2-60, 2-61
Suicide	2-53
Surya	4-88, 4-89, 4-91, 4-102, 4-227, etc in Chapter 4, Fig. 4.3 (p.109)
Synod	F15, F16
Talmud	2-55
Tao Te Ching	1-28, 4-165, 4-227
Telekinesis	3-76, 3-83, 5-249
Tennyson	5-244
Tesla	Appendix 7-354
Tesseract	Appendix 8-363 to 365
Tetrahedral	3-76
Theology	Chapter 2
Thoth	4-89, 4-107, 4-110
Tibet	4-165
Torah	F17

Trent	F15
Touch, sense of	1-32
Thomas	F10, F12, F15
Tolkein	4-136
Toltec	4-91, 4-140, 4-176, 4-181, 4-196
UFOs	4-233, 4-235, 4-238, Appendix 3-300
Uncertainty Principle	3-79, 3-80, Appendix 5-308 to 328, Appendix 6
Underground World	4-136, 4-137, 4-241
Universes	1-20, 1-21, 1-23, 4-136, Fig. 1.1 at page 63
University	4-171, 4-180
Vajra	page 384-5
Vega	4-101, et passim in Chapter 4
Venus	4-89, Fig. 4.7 (p. 177)
Vesta	page 384-5
Vimana	4-99, 4-100, 4-230 to 233, et passim
Viraj	4-89, Fig. 4.5 (p. 166)
Visualisation	7-262, 7-265 to 267
Vulcan	4-89
Vyasa	4-103, et passim in Chapter 4
War, ancient	4-235
Water divining	Appendix 8-362
Waves	3-79 to 82
Werewolf	4-97
Witches	4-97
Wormwood	5-255 to 257
Yajna	4-133, Appendix 8-368
Yellowstone	5-256
Yoga	1-31, 1-43
Zero Point Energy	3-71 to 76, Appendix 5
Zarathustra	4-89, 4-108 to 116, 6-259
Zend Avesta	4-89
Zitterbewegung	3-79, Appendix 5, Appendix 6
Zoroaster	(see Zarathushtra)

** Acronyms:

- 4-D 4th spatial dimension
- EEG   Electro-encephalogaph (brainwave)
- OBE   Out of the body experience
- NDE   Near-death experience
- RAS   Reticular activating system

* Two books extensively quoted from, in Chapter 4:

A. Besant & C.W. Leadbeater, "Man: Whence, How and Whither", published by TPS, Adyar, India (1913).

A. Besant & C.W. Leadbeater, "The Lives of Alcyone", (2 volumes, 740 pp), published by TPS, Adyar, India (1924).

Available as free downloads, from: www.AnandGholap.net   but see important note at end of page 373.

Glossary of assigned names [by Besant & Leadbeater]:
(Some pictures of these individuals are in Figures 4.2 to 4.10)

The "full" list below only includes those who were alive between about 1870 and 1924, with some exceptions (e.g. Julius Caesar).

Notes:

(i) Some pictures of the individuals below are in Figures 4.2 to 4.10, in Chapter 4. The names below were (necessarily) assigned by B&L to identify individuals in successive lives.

(ii) "Master" in Table 1 below, means one who has achieved salvation/sainthood, and so is not required to be reborn again into our world. During the 70,000 year period covered by the investigation, many of those who are now Masters, were ordinary people like ourselves.

(iii) The term Bodhisattva (Sanskrit) means World Teacher and is a Master who has achieved this status.

Mahaguru = The World Teacher (Bodhisattva) of the 4th period, appearing as Vyasa, Thoth (Hermes) (40,000 BC), Zarathrushtra (Zoroaster) (29,700 BC), Orpheus (6980 BC), and finally as Buddha (623 BC).

Surya = The World Teacher (Bodhisattva) of the 5th (present) period, called Maitreya or <u>Christ</u>, who appeared (voluntarily – see John 9:18) as Krishna and later as Christ.

Manu = Vaivasvata Manu, Founder and Head of the 5th (current) period.

Viraj = The Maha-Chohan, of equivalent rank to a Manu or a Bodhisattva.

<u>Athena</u> = Now a Master known on Earth as Thomas Vaughan, pen name "Eugenius Philalthetes" (1621-1666).
<u>Brihat</u> = Now the Master Jesus (explained in Chapter 4).
<u>Dhruva</u> = The Master of whom is now the Master "KH" (see below).
<u>Jupiter</u> = Now a Master, residing in the Nilgiri Hills.
<u>Mars</u> = Now the Master known as "M".
<u>Mercury</u> = Now the Master "KH" (previously Pythagoras).
<u>Neptune</u> = Now the Master Hilarion.
<u>Osiris</u> = Now the Master Serapis.

Saturn = Now a Master, called "The Venetian" in some books.
Uranus = Now the Master D.K.
Venus = Now the Master Ragozci (Comte de S. Germain in the 18th century)
Vulcan = Now a Master (Sir Thomas More in his last life).

Alba = Ethel Whyte
Albireo = Maria-Louisa Kirby
Alcyone = J. Krishnamurti.
Aletheia = Johan van Manen.
Altair = Herbert Whyte.
Arcor = A.J. Willson.
Aurora = Count Bubna-Licics.
Capelle = S. Maud Sharpe.
Corona = Julius Caesar.
Crux = The Hon. Otway Cuffe.
Deneb = Lord Cochrane (Tenth Earl of Dundonald).
Eudox = Louisa Shaw.
Fides = G.S. Arundale.
Gem = E. Maud Green.
Hector = W.H. Kirby.
Helios = Marie Russak.
Herakles = Dr Annie Besant (photograph in Chapter 3, p. 65 of ref. [3]).
Leo = Fabrizio Ruspoli.
Lomia = J.I. Wedgwood.
Lutea = Charles Bradlaugh.
Lyra = Lao-Tze (author of **Tau Te Ching**).
Mira = Carl E. Holbrook.
Mizar = J. Nityananda.
Mona = Piet Meuleman.
Norma = Margherita Ruspoli.
Olaf = Damodar K. Mavalankar.
Pallas = Plato (Note: his "Timaeus" date of 9600 BC for the sinking
   of Atlantis was possibly obtained by a Remote Viewing ability)
Phocea = W.Q. Judge.
Phoenix = T. Pascal.
Polaris = B.P. Wadia.
Selene = C. Jinarajadasa.
Sirius = C.W. Leadbeater (photograph in Chapter 3, p. 65 of ref. [3]).
Siwa = T. Subba Rao.　　　Spica = Francesca Arundale.
Taurus = Jerome Anderson.　　Ulysees = H.S. Olcott.
Vajra = H.P. Blavatsky.　　　Vesta = Minnie C. Holbrook.